FORECAST 2025

ROBERT SHAPIRO spent four years (1997–2001) as US Under Secretary of Commerce for Economic Affairs. He is the co-founder and chairman of Sonecon and currently advises US and foreign businesses, governments and non-profit orga...

FUTURECAST 2020

A Global Vision of Tomorrow

ROBERT J. SHAPIRO

P

PROFILE BOOKS

First published in Great Britain in 2008 by
PROFILE BOOKS LTD
3A Exmouth House
Pine Street
London EC1R OJH
www.profilebooks.com

First published in the United States of America in 2008 by
St Martin's Press

1 3 5 7 9 10 8 6 4 2

Printed and bound in Great Britain by
Clays, Bungay, Suffolk

The moral right of the author has been asserted.

A CIP catalogue record for this book is available from the British Library.

ISBN 978 1 86197 968 1

FSC
Mixed Sources
Product group from well-managed
forests and other controlled sources

Cert no. SGS-COC-2061
www.fsc.org
© 1996 Forest Stewardship Council

The paper this book is printed on is certified by the © 1996 Forest Stewardship Council A.C. (FSC). It is ancient-forest friendly. The printer holds FSC chain of custody SGS-COC-2061

For my parents and teachers

CONTENTS

ACKNOWLEDGMENTS

This book began in a moment of humility and private discomfort several years ago. I was on a panel in Washington, speaking about various economic and political challenges facing the United States, when I realized that I was saying much the same things about the American and global economies that I had said a decade or more earlier. I couldn't deny or ignore that much had changed and that all the changes required serious rethinking. I listened to the others on that panel and on more panels in succeeding months, and I realized that most of the views offered by other policy analysts and policy makers also hadn't changed for more than a decade. So I set out to reconsider what I knew and believed by reconsidering the factors and forces that determine the basic economic and political conditions in my own country and in the world's other major societies.

What began as an exercise in rethinking turned into a genuine re-education. The aging of every major society in the world, which was fully in place in the early 1990s, was now actually at hand. The globalization we all saw emerging in the early 1990s had been transfigured into something vastly more powerful and momentous by the fall of communism and the unprecedented modernization of China. And the world's geopolitics had changed most of all, with the end of the Soviet Union and the communist bloc, and the rise of the world's first sole superpower with no near peer in more than a thousand years.

No one but this author is in any way responsible for what this book holds to be the significance of these developments for the world's major countries. But the analysis and the book itself would not have happened at all without the help, support, and encouragement of many generous and smart people. My friend Ivan Kronenfeld goaded me to write the book from the beginning and provided consistently insightful advice every week for three years. My agent and friend, Frank Weimann, convinced me that the answers to the questions I asked myself had the makings of a book, and he convinced

St. Martin's Press of the same. My editor at St. Martin's, Michael Flamini, has been preternaturally patient and helpful, and no one would be reading this book but for his efforts.

My many friends and colleagues helped me in ways most of them are not aware of. Kurt Almquist, Amy Armitage, Antonia Axson-Johnson, Frank Arnold, Tim Barnicle, Evan Bayh, Eric Bergloff, Alan Blinder, Andrei Cherny, Bill Clinton, Hillary Clinton, Peter Harter, Kevin Hassett, Robert Hormats, Vicky Johnson, Elaine Kamarck, Greg Mertz, David Miliband, Gary Mathias, Shelby Perkins, Justin Peterson, Jonathan Powell, Mark Penn, Nam Pham, Sigrid Rausing, and David Rothkopft all provided valuable insights in offhand conversations in meetings, over dinner, or even late at night. I want to especially acknowledge Paul Stockton, of Stanford University, who understands geopolitics as well as anyone in the world, and Al From, of the Democratic Leadership Council, and Simon Rosenberg, of NDN, who understand American politics better than anyone else in Washington. And two people very special to me, Bonnie Anthony and Annie Maccoby, deserve special mention for encouraging me when I was tempted to give it all up and for bringing me back to reality when I wanted to take another year or two to finish. My sister Betty and brother Alan always indulged me as I went on much too long about whatever subject I was dealing with at the time.

Figuring out what is actually happening in worldwide demographics, globalization, and geopolitics required enormous research, and a legion of talented young people helped me figure it out. First among equals is Giwon Jeong, now at the Institute for International Economics, who did superlative research on Asia. I also learned enormously from the research on Europe's economies by Timo Behr from the School for Advanced International Studies. My thanks also to Georgina Voss, at the University of Sussex, and Greg Tenentes, of the U.S. Bureau of Economic Analysis, for their research on technology. Meaghan Casey, now at Kinsinger McClarty Associates, and Giannis Doulimis, at George Washington University, helped me understand terrorism. I also want to thank David Datelle for his extensive and intensive research on health care.

Finally, thanks to Josie Weefus and Amber Leggett for their love throughout the process.

CHAPTER ONE

THE GLOBAL BLUEPRINT

ENORMOUS CHANGES HAPPEN to most people over any ten- or fifteen-year period. Your brain and body, wants and desires, circumstances and prospects, all change as you grow from a five-year-old to a young adult of twenty, from a new husband or wife of twenty-five to early middle age, and from a fifty-year-old at the height of your career to a grandparent of sixty-five. Societies usually change less than most individuals over fifteen-year periods. Most societies have much longer lifetimes, too—although occasionally a great event like a war or an exceptional boom can shake up a nation's conditions and prospects in short order. Broader and more powerful forces, beyond the control of any country or its leaders, can shift the course of development of the entire world. Such developments—some new force changes the way people approach family life, new technologies arise that change fundamental factors in all economies, or a global empire falls—usually happen once in many centuries. We live in a period when three such tectonic developments are all unfolding at the same time.

Three great, global forces are currently reshaping our near future. The first is an extraordinary global demographic shift. Over the last half-century, every advanced society and most developing countries have experienced a baby boom followed by a baby bust. The plague killed one-quarter of Europe's population in the thirteenth century, and the world wars of the twentieth century decimated the young male population of several countries. But what's happening today—across most of the world, the greatest aging of national

1

populations ever seen, along with the smallest relative numbers of working-age people on record—is without historical precedent. It will have large effects on both how fast different countries grow and the basic capacity of governments to meet the needs of tens of millions of newly elderly people.

The second great force is globalization, principally, the rapid advance of hugely complex, worldwide networks of money, resources, production, and consumer needs. There have been several previous periods in which world commerce and communications suddenly expanded—especially in the era of exploration in the seventeenth century, and the spread of the telegraph and electricity in the late nineteenth century. The current phase is more far-reaching, with new information technologies affecting more societies more quickly. It's also more comprehensive, with 151 countries agreeing to the World Trade Organization's general rules that open each of them to foreign investment and to much greater foreign and domestic competition. Whether most people in Europe, Japan, or America thrive or merely get by will depend on how and whether their governments and societies find a way to prosper under these rules—and to compete with China and India's unprecedented combinations of advanced technologies and huge numbers of low-wage, skilled workers.

The third great historic development is the fall of the Soviet Union, its European empire, and its political ideology. Empires and ideologies have fallen many times before, but not since the time of Rome has such an event left one global military and economic superpower with no near peer. And the rise of Rome didn't coincide with a period of world-changing economic globalization. Nor was the fall of past empires accompanied by other seismic geopolitical changes of the sort we see today—namely, the rapid makeover of the world's biggest country from socialism to a supercharged form of capitalism, and a shift of the center of global politics from the Atlantic nations to the Pacific rim. In place of the cold war and everything it demanded of most countries, the geopolitics of the next fifteen years will be driven by a combination of all of these extraordinary developments.

These developments and their combinations and interactions will have profound effects on the course of every major society and the daily lives of their peoples, and no nation or person can opt out of the consequences.

Strictly speaking, they will not determine anyone's precise future. Wars and new alliances, booms and recessions, social progress and terrible domestic conflicts will come to pass in the next fifteen years, and each one will arise from countless decisions and events no one can know today. But whatever the near future holds precisely for all of us, it will unfold in a world where these three great forces shape its outset and outcome.

The powerful effects of these forces come from their depth and breadth. The family-planning decisions of billions of people over more than two generations have gradually shifted the demographics of most nations and produced the unprecedented aging of their populations. The roots of modern globalization, and the reason it's not going away, are similarly elemental. Globalization as we know it today comes out of the responses of tens of thousands of businesses to the new availability of low-cost skilled labor abroad, the growing capacity of developing nations to attract foreign capital and technology, breakthroughs in manufacturing that have enabled producers to break up and distribute the parts of their production process across plants in different countries, and the spread of information technologies to manage and coordinate global networks. And America's position as the world's sole superpower has come from not only the epochal collapse of the Soviet Union but also sustained decisions over a half century by America to spend whatever it takes to be stronger than other nations, and by Japan and Europe's major powers to depend on the United States for security. The deep sources of these developments give them enormous momentum as well as power.

Over time, every nation has little choice but to respond in some way to the pressures created by these historic forces. They can affect the way these forces influence their societies through policies that change the behavior of large numbers of their people or companies. But such changes take time to put in place and more time to affect such deep developments—in our judgment, a good ten to fifteen years. This potential for change limits our foresight to roughly the period between now and 2020.

Over that span, for example, nothing can change the demographic dynamics producing the current aging of all major societies and its impact on a country's savings and growth rates and the sustainability of pension and health-care systems. Japan or the major countries of Europe could ease restrictions on immigration to expand their baby-bust labor forces and trim their state pension and health-care obligations. If such changes were to come about—and there's little sign of any of this today—the early economic and social effects might be felt in about a decade. Similarly, France, Germany, or Japan could today undertake the difficult economic reforms necessary to make themselves more successful players in globalization—at least France now talks about it—but it would take at least ten years to see an impact on their growth rates and the incomes and productivity of their people. China's and America's strengths in the new global economy also could unravel over time if, for example, China were to confront serious social unrest that slowed its modernization or, even less likely, if America turned inward economically. The key roles the two countries play in globalization, however, are very unlikely to change materially in a decade's time. In geopolitics, it's conceivable that China or the European Union could permanently double or triple their military investments, starting today. It's unlikely because it would slow China's modernization and redouble the pressures on Europe's social-safety-net systems. Yet if they did so, it would take a generation of such investments to build the capacity to successfully challenge American superpower on even a regional basis.

Looking further ahead over the next one or two generations, profound changes are certain to unfold in many societies. We cannot know what life will be like in America, Europe, Japan, or China in 2035 or 2050. But the aging of national populations, the opportunities and pitfalls of economic globalization, and the consequences of a geopolitics organized around a sole superpower are forces sufficiently entrenched and far-reaching to provide a unique blueprint of their likely course over the next ten to fifteen years.

THE EARTHQUAKE OF DEMOGRAPHIC CHANGE

An historic aging process is taking place in virtually every society today. Across the globe, the average age of the population in almost every nation is rising. This kind of change has occasionally happened in the past to one or a few countries at a time, when a war or epidemic has wiped out part of a younger generation. Germany and France each lost about 10 percent of their young male populations in World War I—more than 3 million between them—and the Spanish flu of 1918–1919 claimed 30 million lives, the majority ages fifteen to thirty-five, from more than ten countries. This time, for the first time in recorded history, almost every country in the world is experiencing an unusually large generation followed by an unusually small one. Both the causes and the economic results, however, differ among countries.

In Western Europe, America, and most of the world's advanced countries, the initial baby boom came out of decisions by tens of millions of couples to defer marrying or having children during the Depression and the Second World War—and once prosperity and peace returned, they choose to have many more children than their parents did. The same outcome was unfolding at roughly the same time in dozens of less-developed societies, but from different sources: There, tens of millions of young couples discovered that many more of their children were surviving than in the past, principally because of large improvements in basic health and sanitary conditions. A second factor also emerged across Asia, Europe, and North America: Medical advances were increasing life spans, especially at the other end of life. From 1960 to 2000, the life expectancy at birth of Americans, Japanese, and Europeans increased by seven to thirteen years; and reached eighty-three to eighty-five for those sixty-five or older. Over the same period, the life expectancy of the Chinese at birth nearly doubled, from thirty-six to seventy years.

Finally, a separate series of factors brought these baby booms to screeching halts in the following generation, when expanding economic opportunities—especially for women—better infant care, and inexpensive contraceptives led baby boom women around the world to choose to have many fewer children than their mothers. Put together all of these forces, and the neat order of generation succeeding generation—the stable age structure that most countries had since the early 1800s—changed radically across the world.

What would happen to your family if, following a few decades of growing paychecks earned by you and your mate, all at once one of you could no longer work full-time and a relative with no resources had to move in? To begin with, it's a family crisis, and unless you have a plan to somehow raise your income, the material conditions of your family's life would deteriorate fairly quickly. Family meals would become more basic and vacations less frequent. The cars you drive would grow older and break down more often. You would have to live with more stress and your general health might well worsen. The time and energy you would have to help out at church or your children's school could slowly disappear. Ultimately, such a crisis breaks up marriages and, almost always, children suffer both at home and at school.

The results are much the same for a country—when the number of working-age people who produce everything from food and pharmaceuticals to education and supersonic bombers, falls, just as the number of people who need and expect a lot of help from government rises. Over time, the quality of national life may deteriorate, too. With fewer people producing wealth and more older people needing income support and medical treatment, taxes rise and investment slows. Lower investment rates mean that over time, most people's paychecks will grow more slowly—or stop growing altogether—even as their taxes increase.

Demographic shifts are very powerful forces, but their ultimate outcomes are not predestined. Starting with Thomas Malthus's famous essay in 1798, people have worried that the aging of a generational bulge like the one so many countries are experiencing today will badly strain a country's resources and drive down its economy, often long before the baby boomers reach middle age. Malthus is not much read today for his predictions, but his dark scenarios retain a hold on many people's imaginations. As recently as the 1970s, the U.S. National Academy of Sciences and the United Nations both warned that sharp increases in population around the world would leave most people worse off. A presidential commission of academics and public officials, chaired by John D. Rockfeller III, pronounced that "we have concluded that no substantial benefits would result from continued growth of the nation's population," and zero population growth "would 'buy time' by slowing the pace at which growth-related problems accumulate."[1] Like Malthus, they overlooked the power of new technologies to create new resources and get more out of those we have, they underestimated the wealth that could be generted by better-educating all those additional people, and they didn't understand how well some economies and societies can adjust to changing conditions.

A family, like a nation caught between less income and greater expenses, has alternatives, none of them wholly appealing. One of the parents can go back to school and learn a new set of skills that pays more. The family could take in a working-age cousin in similar straits and pool their incomes. Then there are options no one likes to consider, such as asking an aging relative to cut back on his medications. A nation, like a family, has alternatives, too, even if their responses are shaped by other factors. In the United States, the society learned new skills: Government spending and the ambitions of entrepreneurs and scientists promoted the development of computers and software that ultimately raised U.S. productivity, so that even a smaller number of working Americans could produce more. China went further along the same kind of path. There, a new generation of leaders not bound by the 1948 revolution, or Mao Zedong's perverse ideology, opened their backward economy to advanced western technology and business methods. The United States, China, the East Asian countries, and later much of Central Europe also deregulated scores of industries and opened themselves up to foreign producers.

These changes ultimately cost America about one-third of its manufacturing jobs, as they did in Europe and Japan. One difference was that Europe and Japan generally held on to most of their regulation; while in America deregulation brought on more intense competition, producing economic pressures that created a lot more jobs of other kinds. Another difference was that as the boomer generation aged, the United States opened its borders to more immigrants that supplemented its workforce, while most other advanced countries kept their borders closed. The rapid changes in China and Southeast Asia also cost many tens of millions of people their traditional livelihoods, while other factors opened up economic opportunities in much of Asia. Once falling infant mortality rates brought down the average number of births by women across much of Asia, millions more Asian women went out to work.

In some countries, the right policies helped turn their generational bulges into what pundits called economic miracles. All the investments in education and health care across East Asia in the 1950s and 1960s helped make their bumper crop of boomers healthier and more productive; and by the 1970s and 1980s, those societies had the world's fastest-growing economies. The baby boom came a little later to Ireland than to East Asia or most of Europe. Since the early 1990s, the "Celtic Tiger" has been Europe's fastest growing economy—huge tax and regulatory concessions for foreign IT companies explains much of it, but economists figure that the bulge of better-educated Irish boomers entering the workforce in the 1980s and 1990s has accounted for two-fifths of that achievement. While small-nation powerhouses like Korea and Ireland have created economies that can help support their demographic bulges as the boomers retire, countries that have taken less advantage of the new demographic conditions find themselves in real trouble. Much of Latin America ignored their own boomer bulges or approached them as a burden. Almost two generations of military dictatorships neglected mass education and the policies that might produce sensible modernization, leaving a legacy of slow growth and volatile politics for their more democratic successors, who now face rapid expansions in their labor forces from late-coming baby booms. Japan, Germany, and France invested in educating their boomers in the 1960s and 1970s. Their resistance to the demands of globalization in the 1980s and 1990s, however, has cost them a generation of potentially very large, demographically driven dividends and left them no cushion when their boomers hit retirement age over the next fifteen years.

These changes, and the varied responses of different societies, are sending shock waves through the economies and daily life of every region and country. In Japan and across most of Europe, the baby bust is beginning to shrink their labor forces; and unless robotics produces a new cyber-labor force, that factor alone will drive down savings, investment, and overall economic growth over the next fifteen years. In most other major nations, the number of new workers will continue to expand, but more slowly than before, producing some of the same effects in milder form. Between now and 2020, for example, America's baby bust will be largely cushioned by the effects of two generations of high immigration. Across China, most of South and Southeast Asia, and perhaps India—all among the fastest-aging countries in the world—great strides in health care and burgeoning market economies will enable more people to work both longer and more productively, containing many of the national costs of their baby busts. Whether a country's labor force expands or contracts in coming years will also have huge social effects as its boomers grow older. The elderly populations in all of these countries will expand by 35 to 60 percent over the next fifteen years, forcing governments nearly everywhere to expand their public spending and raise taxes—or run huge deficits—to pay for basic medical care and retirement. Where a shrinking taxpayer base and slow growth drives down national wealth, these demands will likely polarize the public debate and ultimately produce intense and nasty political conflicts.

A NEW ECONOMIC LANDSCAPE

Large and deep changes in the way the world's major economies operate is a second historic force shaping everyone's near future. For the past thirty years, trade and investment between countries has expanded twice as fast as the total growth and investment of all individual countries, creating a new global landscape. By 2020, most of what Europeans, Americans, and Asians drive, use, operate, and consume—the vast majority of manufacturing and many business and personal services—will be made or provided in factories and office complexes of fast-developing, lower-wage countries—not just China, but also places such as Bangladesh, Malaysia, Indonesia, Mexico, Brazil, Poland, Romania, Tunisia, and Ghana. Today, Renault, Daewoo, and DaimlerChrysler all produce cars in Romania. In the same small country, Vodafone and Qualcomm manufacture electronics, Hewlett-Packard and IBM produce computer parts, and Procter & Gamble, Colgate, and Coca-Cola produce their brands there as well. Goodyear tires, Procter & Gamble soaps and toiletries, and Coca-Cola that Americans and Europeans use and consume may have come from Morocco; while Pioneer Foods' StarKist tuna, Kaiser and Alcoa aluminum, and Coca-Cola are all also produced today in Ghana.

Pundits and politicians first began talking about globalization in the 1990s; but in its current form, globalization really began in the 1970s, when it morphed out of the centuries-old dynamics of international trade. One reason for the change was the unilateral decision by the United States in 1971 to end the regime of fixed exchange rates established after World War II. For the first time, western companies could exchange the profits they earned abroad for their currencies at market exchange rates that shifted little by little, day to day. Just as important, OPEC tripled the price of oil at roughly the same time. Every large company scrambled to cut other costs and began looking more closely at the most successful developing countries, where everything but energy was a lot cheaper than in the West. On the heels of these changes, foreign investments shot up. U.S. and Japanese companies expanded their foreign operations and business relationships in the developing world, especially in Asia, where the demographic shifts and local improvements in health and education were creating a large, relatively skilled yet low-wage labor force. By the end of the 1970s, western multinationals were transferring technologies and building plants across East Asia. The United States, Europe, and Japan began losing manufacturing jobs, and the miracle of the "Asian Tiger" economies took off.

Even then, there was still no real global economy to speak of. Most economic relations between countries occurred within the two great geopolitical blocs, and almost never between them. The first, reasonably dubbed the "free world," took in non-socialist, developing countries of Asia, Latin America, and southern Africa, nations that exported commodities like oil, minerals, and food—principally to the advanced economies of Europe, Japan, and the United States, which also exported more sophisticated finished goods mainly to each other. The second bloc comprised the Soviet Union, Eastern Europe, China, India, and Cuba, and involved similar patterns of trade. The two blocs faced each other in the long cold war, and for fifty years had little to do with each other economically.

These separate political and economic blocs suddenly dissolved in the 1990s, changing the basic shape of international economy. First, the Soviet Union collapsed in 1989, ending the separate, socialist economic bloc. For the first time, companies in America, Europe, and Japan gained access to the huge labor forces and resources of China and, to a much lesser extent, India, which themselves had been busily educating a slice of their own demographic bulges. The pool of potential workers available in some way to western companies expanded by hundreds of millions of people in less than a decade—an event unprecedented in economic history—all ready to work for a fraction of the wages earned in the West, or even in the Asian Tigers or Latin America.

The hitch was that, in 1990, most of those low-wage, skilled workers worked in state-owned monopolies that couldn't produce much of anything that Westerners would buy. But with the spectacular failure of communism in its founding place, market economics became the only national or global economic strategy left standing. So, in the early 1990s, the world's leading market economies, headed by the United States, led the global negotiations to establish the new World Trade Organization (WTO). Technically, the WTO simply replaced the old General Agreement on Tariffs and Trade (GATT) that countries had used for fifty years to negotiate cuts in tariffs and quotas. From its beginning, the WTO has had more far-reaching ambitions, with an operating creed that channeled the beliefs of America and its president, Bill Clinton, in the benefits and inevitability of global investment, global technology transfers, and global markets for everything.

On January 1, 1995, seventy-four countries joined up, including every advanced economy and scores of developing nations from Brazil and Korea to Bangladesh and Kenya. With most of the world's significant economies on board, the WTO began negotiating and writing the rules that every country would have to follow to be part of global capitalism, covering most major aspects of a country's economic life. Countries have to agree to gradually roll back national laws and regulations that restrict foreign imports, sector by sector, from furniture and semiconductors to government procurement and telecommunications. They also have to agree to revise barriers that go across sectors, such as restrictions on foreign ownership or domestic competition. The bottom line is that countries that want to be part of globalization, WTO style, have to roll back subsidies and protections they've used for decades or centuries to support their own domestic industries.

Developing countries like China and India have faced the biggest shocks: They've had to junk their state monopolies and, sector by sector, open themselves up to western investment, joint ventures with western companies, and their own domestic competition. Of those two, China has carried through more extensively and deeply than India, and become a powerhouse in ways that India probably won't match for more than a generation. Still, as these two huge countries, along with scores of smaller ones, generally opened their economies to markets, the lightning spread of new information technologies intensified the pace of change by giving American, European, and Japanese businesses much more capacity to shift their money and operations across continents and cultures, and keep track of their suppliers, subcontractors, and customers across the developing world.

The combination of these factors turbocharged the action in two different ways. It increased the ability of many developing countries to attract the capital to build modern factories and assemble modern business organizations. It also increased the ability of western companies to transplant their technologies and ways of doing business. Early western investors in China, such as IBM and the global water valve maker, Watts Water Technologies, didn't make much profits from their initial joint ventures of the 1990s—labor was cheap but unreliable, transport was primitive, government officials chose their suppliers from their cronies, and the Chinese business practices of the time clashed with their modern methods. Gradually, each side learned what the other needed—and not only in China, but also across much of the rest of Asia and much of Eastern and Central Europe. By the year 2000, for the first time in history, companies with state-of-the-art technologies and business methods operated legally and efficiently in countries with virtually unlimited numbers of reasonably skilled workers ready to work for a small fraction of the average global wage.

With every major nation in the world on board, the new global economic terrain will help shape the path and fate of every one of them. By 2020, major heavy manufacturing will substantially disappear in the advanced economies, and the production of most autos and steel, appliances and electronics—regardless of their western brand names—will shift forever to the developing world. Heavy goods that are expensive to ship will be made in developing economies close to advanced-nation markets. For instance, China and Indonesia will produce heavy goods for the Japanese market, Turkey and Romania for the European market, and Latin America for the American market. Global producers of worldwide commodities will always keep a production foothold in their major Western markets, so American, European, and Japanese auto and computer plants will not disappear entirely from the their own and each other's countries. Nevertheless, by 2020, most current American, European, and Japanese jobs in industries that compete directly with China and other modernizing, low-wage countries will be gone.

If there's any doubt about the power of these changes to destroy jobs, consider that the United States lost more manufacturing jobs in the three years from 2001 to 2004—some 2.8 million of them—than during the "Rust Belt" deindustrialization years, from the late 1970s to the early 1990s. And, during the recent years of strong American growth from 2003 to 2005, electronics and electrical equipment makers, auto and transportation equipment producers, and industrial and computer equipment manufacturers closed more than six hundred U.S. plants. American or European politicians and their parties cannot turn back these forces, whatever they promise, and bring back tire or big machinery manufacturing to Toledo or Liverpool. At best, they can affect how their countries respond to those forces.

Between now and 2020, globalization will perform much the same economic alchemy on many services. The foundations already exist in the extensive networks of business and financial relationships extending today from Chicago, New York, Frankfurt, and London to Shanghai, Taipei, Bangalore, and Budapest. The important new element is continuing advances in software development that enable companies to break up a complex service, such as a research-and-development project, into its constituent parts—just what happened in manufacturing—and distribute the parts to companies anywhere that can deal with them cheaply and efficiently. In a few areas, this process is already well advanced, including service centers, and software programming itself, as well as some areas of research and development. In certain areas of business, financial, and health-care services, programs allow people with fairly basic skills to carry out pretty sophisticated tasks. Software now incorporates most of the technical operations associated with preparing tax returns, financial statements and bookkeeping, and conducting medical tests.

Before the end of this decade, these programs will cover wide areas of inventory control, medical diagnostics, engineering, and legal analysis. The only barrier to shifting these tasks across borders is language, so the outsourcing of business services from the United States has so far focused initially on India, while German, French, and Italian companies look to Central Europe.

Looking out over the next fifteen years, the globalization of services could well have even broader effects than the previous globalization of manufacturing. For consumers and businesses, globalization will drive down the price of many basic personal and business services, just as surely as it drove down the price of television sets, automobiles, DVDs, and computers. The economic implications are far-reaching. Services not only make up two-thirds or more of every advanced economy, compared to manufacturing's average share of about one-fifth. In addition, the worldwide market for services is poised to take off: Incomes will almost certainly rise sharply across China, India, and much of the developing world over the next fifteen years; and as economists have long observed, when people earn more, they spend a larger share of their incomes on services.

With software programming leading this wave, this development also could boost productivity in many advanced economies. After all, America's rapid growth and rising incomes in the 1990s were fueled by the spread of information technologies. The key to why they spread so rapidly was their falling prices—once the production of most computers, storage devices, and transmission modes had shifted to low-wage Asian operations. Similar savings haven't happened yet in software and most information services because they're still produced mainly by highly paid workers on the East and West coasts of the United States, in Ireland, Sweden, Finland, and Germany—and recently, in a few cities in India.

Just wait a few years. Over the next decade or sooner, many of those jobs will shift. The good news for Americans—and to a lesser degree, for Europeans and the Japanese—is that the price of new software will fall; and that could have powerful economic effects. The falling prices of computers spread information technologies across most Western economies, but some companies and sectors have lagged behind, especially in health care and education. Falling software prices will make the technologies more cost-effective for the laggards. Since health care and education in particular are such huge parts of every advanced nation, the productivity gains, as they catch up technologically with finance and manufacturing, will produce big payoffs for both businesses and consumers. It could also improve the quality of people's health care by allowing much greater sharing of medical information between primary physicians and specialists. John Driscoll, a senior executive at Medco, the nation's largest mail-order distributor of pharmaceuticals, describes the need for more information sharing this way: "For really sick people, their primary physician knows the most about them but does the least, while some specialist knows much less and has to do much more."[2]

If the initial spread of computerization is any guide, the race to become more productive among hospitals, colleges and training institutes, and tens of thousands of small service companies will also increase competition, forcing everybody once again to rethink the way they do business. At least, that's likely to unfold wherever regulation doesn't smother competition. There, everything from medical clinics and dry cleaners to fast-food and landscaping services will create new services and provide their old ones at lower cost. And just as in manufacturing, it will also cost millions of people their current jobs.

Based on how the globalization of manufacturing has already changed the world, the globalization of services could have pervasive effects—like the butterfly that ultimately sparks a hurricane a half-world away. In a genuinely global economy like the one unfolding today, when as huge a country as China modernizes its export industries, it produces new pressures on jobs and wages across the world. In the 1990s, for example, China relaxed restrictions on domestic competition and foreign investment in apparel and furniture manufacturing, as well as in scores of other sectors. As these changes quickly made China's apparel and furniture makers more efficient, their Korean and Thai rivals faced tougher competition. Some figured out how to survive and compete and others didn't; and on the margin, investment and expertise in Korea and Thailand shifted into other industries, such as electronics and medical equipment. With more capital and expertise, the best Korean and Thai electronics and equipment makers developed new products and became more efficient themselves, intensifying the competition for their German and American competitors. In the last decade, this process has happened over and over again, across scores of countries and manufacturing industries, leaving almost everyone, everywhere, facing more intense competition. Over the next decade, this same process will extend to scores of service industries.

The hitch is what happens to jobs and wages when competition heats up: Companies find it harder to raise their prices, even when their health-care or energy costs increase, so they turn to jobs and wages to cut their overall costs. That's precisely what's been happening in the United States and much of Europe for the last five years. Their economies have been growing reasonably well. Despite the growth, the United States and Europe have seen more than a half-decade of stagnating or falling incomes and slow or stalled job creation. As globalization spreads across services over the next decade or so, these forces will intensify. For advanced countries like America and most of Europe, tens of millions of people without specialized, nonobsolete skills will no longer be able to depend on economic growth guaranteeing them well-paying jobs.

Countries facing these pressures, like individuals, have choices. They can size up the new global environment and try to adapt to it. In taking essentially that course, the United States and England largely part company with Germany, France, Japan, Italy, and others. In many ways, much of Europe has been in an extended state of denial. The depths of that denial seemed evident in France in 2006, when young people violently rioted over a modest government proposal to let companies dismiss workers in their first two years on the job, simply when it would make economic sense to do so. In May 2007, Nicholas Sarkozy ran and won the race for French president on a platform of market-friendly economic reforms. Much like Japan's reform-minded prime minister Junichiro Koizumi, Sarkozy hasn't been able to deliver on many promises, at least not in his first six months in office. But for the first time, France has a president who publicly acknowledges that globalization requires serious changes in the way European governments approach their economies.

One reason why many European countries seem indifferent to the real terms of globalization may be that most of them still trade almost entirely among themselves or with America and Japan, while a much larger share of U.S. and British trade involves developing markets. Whether it's a vestige from their colonial periods or just force of habit, many European businesses seem to think of the rest of the world as dusty outposts barely worthy of their trading attention. The same holds true in global investment. In 2004, barely 7 percent of the foreign stocks owned by Europeans—just 3 percent for Germany—were in companies in developing countries, compared to almost 23 percent of the foreign stocks held by Americans and 20 percent of those owned by Britons. The contrast is especially stark for investing in China, although Japan rather than Britain joins America here in taking the plunge: German investment in China is barely one-fifth the U.S. or Japanese level; and France, Italy, and England lag even farther behind. This record of continental Europeans suggests a form of financial cognitive dissonance, since emerging markets outperform Europe's own markets.

Where a country trades and invests matters, because it's the main way that companies and societies learn how to be part of another country's economic life. All of that U.S. and Japanese trade and investment with China is building up the knowledge, relationships, and networks of suppliers, distributors, and managers that will make American and Japanese companies integral parts of China's economy in 2020. And by competing so actively in the world's fastest-growing and fastest-changing market, U.S. and Japanese companies are being forced to stay on their toes by developing new products, services, and ways of doing business. Ultimately, the results will shape their own domestic economies as well, because businesses fully engaged in markets like China will raise the productivity and profits of their operations not only in China, but everywhere else. Unless the large European nations change their attitudes and long-entrenched trade and investment patterns, in 2020 most of them will be left fighting over their own shrinking domestic markets.

By then, advanced economies will have to concentrate not on basic manufacturing or services, but on coming up with new goods and services for workers elsewhere to churn out; running the global operations that coordinate worldwide production networks and oversee their financing, marketing, and distribution; and producing more customized local goods and personal services, especially in areas like education and medical care. Promoting this innovation will be a problem for much of Europe and Japan. There, a long-standing distrust of the freewheeling competition that drives innovation will be heightened by the slow growth and high taxes associated with their shrinking labor forces and growing numbers of elderly. The outlook is better for a handful of niche advanced economies—principally Ireland, Sweden, and Finland—that have demonstrated singular capacities for coming up with new, world-class technologies.

England is often thought to be more disposed to deregulation and innovation than France or Germany, and Margaret Thatcher and Tony Blair both talked a good game. A lot of it is still just talk. In 1999, Blair's chief of staff asked me (I was then United States Under Secretary of Commerce for Economic Affairs) to brief the prime minister's economic advisers on how America used new technologies to rack up large productivity gains. In a sunny conference room at 10 Downing Street lined with gilded portraits, I laid out how the power of competition, especially from young companies, had both driven the development of cutting-edge IT and moved other businesses to change the way they did business to make good use of the new IT. Blair's policy wonks and political operatives asked probing questions; but when we were done, they agreed that the kind of competition that forces companies to change was still distasteful to many Britons. Two years later, the Blairites were back asking for concrete recommendations on how the prime minister could break Britain's cycle of low productivity growth. I advised them that the single most powerful thing they could do would be to relax the worst bureaucratic and regulatory barriers to starting new businesses. Once again, I listened to the most powerful people in British politics rue the distate of many of their countrymen and government for the bare-knuckled competition that globalization requires.

The outlook for the United States is more bullish. Drawing on the world's largest and most variegated network for research and development, American companies should continue to produce more cutting-edge technologies and services than companies anywhere else. Already, U.S. businesses invest nearly as much in ideas as they do in plants, equipment, and land. Equally important, Americans accept harsher competition—and swallow the greater inequality and job insecurity that comes with it—than most Europeans or Japanese have ever tolerated. These differences should continue to push American businesses to raise their productivity faster than businesses in most other advanced countries. That advantage, in turn, should also continue to make the United States a strong magnet for foreign investment. A decade from now, America will still be the world's largest and most technologically advanced economy, and the one with the greatest impact on everyone else.

But nothing will stop globalization from destroying job security for millions of Americans, along with their European and Japanese counterparts. From 1997 to 2003, more than 5 million Americans lost their jobs and were never recalled in extended, mass layoffs.[3] In 2003, large layoffs of this kind hit places like Fresno and Salinas, California, and Chicago, in industries from food processing and machinery manufacture to computer makers and administrative services. Relatively free domestic markets, more innovation and more new businesses will produce more alternatives for millions of laid-off service and manufacturing workers in the United States. But the transition is almost always brutal on the workers and their families, and their next jobs usually pay a lot less. It will be especially agonizing for millions of older workers who may also lose some of their private pensions, along with their dignity. Unlike normal bouts of unemployment in the time before globalization seized everyone's prospects, they will find it much harder to land a comparable job or "retrain" for a good, new one. Politicians who will tell them that the technology and globalization uprooting their lives can somehow be stopped by a new law or government program will earn their well-justified anger.

China is the other major economy well positioned to take good advantage of globalization. Its potentially huge domestic market and low-wage, skilled workforce will continue to attract investment and technology transfers from every advanced nation. As the only major country likely to match the United States in freewheeling market capitalism (and economic inequality), China should be able to use that investment, technology, and skilled labor to establish itself by 2020 as one of the world's two indispensable economies. As we will see, in China as well, this progress involves large burdens for hundreds of millions of families forced to give up their farms or jobs with state-owned enterprises. And under current Chinese law, there's no unemployment compensation to help cushion the blow, or health-care and pension coverage for most of them. Nor is China unique in this respect—if India hopes to modernize with anything like the breadth and speed of China, it also will have to displace hundreds of millions of Indian farm workers stuck today in one of the world's least productive agricultural sectors. The great difference is that rural Chinese have little choice but to pack up and go to urban areas where they might find new jobs (at least young Chinese can do that), while rural Indians have political outlets to slow down or even stop the whole process.

For all the sparkling modernism of Shanghai, Beijing, and Guangzhou, the country's stunning, two-decade-long record of overall growth, and the enthusiasm of many western commentators, China's economic ascendancy over the next decade is not secure. Most Chinese today are still poorer than most other Asians, Latin Americans, and some sub-Saharan Africans; and their country still lacks most of the basic institutions and arrangements that comprise the economic and social infrastructure of a well-functioning modern economy. The financial system is primitive, with no real central banking, mortgage finance, or corporate bond markets; property rights, contracts, and bankruptcy laws are all haphazard; basic infrastructure outside the major coastal cities is generally obsolete or crumbling; and energy generation and transmission in much of the country is spotty. Even the great manufacturing capacities of the Chinese economy are, in one important sense or another, still largely foreign transplants. The world witnessed the extent to which China's native manufacturers lag behind the transplants—and Chinese regulators lag their counterparts almost everywhere else—in 2006 and 2007, when American, European, and Latin American authorities found deadly contaminants or defects in Chinese exports of seafood, children's clothing and toys, dog and cat food, toothpaste, vitamins, and cough syrups. Whether in 2020 China is a global economic power in its own right, or a poor country with an uncertain future, known for hosting many of the manufacturing operations of every western global company, is a serious and open question. The answer will depend to a large extent on whether it can do in ten or fifteen years what took every other modern economy and society the better part of a century, and pull it off without sacrificing the rapid economic growth on which China's leaders believe their country's social and political stability depend.

GEOPOLITICS WITHOUT COMMUNISM

The third historic force shaping the next fifteen years is the radical rearrangement of geopolitics, ushered in by the fall of the Soviet Union. It is hard to overestimate the extent to which this one event is transforming the world.

For starters, it enabled China to fully break with socialist economics and create the conditions for both its lightning economic progress and, in time, new military prowess. The Soviet demise and China's rise also are shifting the focus of world affairs to Asia, marginalizing Europe's role in global security.

In place of the cold war competition that always threatened to turn hot, the geopolitics of the near future will be defined and ordered by two new conditions. For the first time in more than 1,500 years—since the days when Rome ruled most of the world—there is a global military and economic superpower with no near peer. With the world's largest economy and markets, and with military spending nearly equal to the rest of the world combined, the United States went to war in Iraq over the objections of close allies without affecting its military and economic preeminence. Even if that war is ultimately lost, America could do it again and not diminish its military and economic dominance.

The ascendance of a sole superpower with no peer has also changed the pattern of conflicts in world politics, or at least those that matter. For the first time in many centuries, international conflicts of consequence mainly involve differences not between powerful countries, but between those connected to each other through global networks of trade, money, and information, and those left largely outside or unconnected. One signpost of this new landscape is the record of foreign military fighting over the fifteen years or so since the Soviet Union fell. Virtually every foreign military operation has been initiated by the United States, and almost every one has involved American military involvement in countries still outside the orbit of globalization—in the Caribbean rim (Haiti), sub-Saharan Africa (Somalia and Sierra Leone), the Balkans (Kosovo and Bosnia), the Middle East (Iraq and Kuwait), Central Asia (Afghanistan), and Southwest Asia. The partial exceptions involve disputes over territory between countries outside the global economic system, but where the United States was not involved directly—principally Iraq's invasion of Kuwait and incursions on their neighbors by sub-Saharan African countries.

Another signpost of the new geopolitics is the rise of new institutions to manage the economic problems that surface among countries *inside* the orbit of globalization—especially the World Trade Organization (WTO), the World Intellectual Property Organization (WIPO), the European Union (EU), the Asia-Pacific Economic Cooperation organization (APEC), and the renewed role of the International Monetary Fund (IMF) in helping countries manage stresses triggered by global flows of goods, services, and money. Much of a growing opposition to globalization itself is directed at these organizations, perhaps simply because they have proper names, real addresses, and annual meetings at which to protest. Some of this opposition, like recent attacks on traditional intellectual property rights, is fueled by Brazil, Argentina, India, and a few other developing nations' governments that want to legalize forms of modern-day piracy and counterfeiting so they can produce (and export) generic versions of the innovations coming out of developed countries. The WTO also is deeply suspect in many places, and especially among two groups. First are many traditionalists in developing countries who see the western products and business organizations imported through globalization as an assault on their values and cultural traditions—and often on their own political power inside their countries. The WTO also attracts antagonists in Europe and the United States, including those who nearly broke up the group's 1999 annual meeting in Seattle, Washington. Many of them are ideological orphans from the sudden death of global socialism who, now bereft of a real alternative, rail passionately against the injustices of globalized capitalism.

Another indicator of what geopolitics looks like in the age of globalization and a sole superpower is the rise of Islamic terrorism. With the United States the world's only great military power, the source of international violence has shifted from nations to a borderless movement. This transnational movement is centered almost entirely outside the orbit of globalization, justifying its mission and its violence by rejecting everything globalization represents and promises. Islamist terrorists are certainly no allies of the WTO's secular antagonists or WIPO's critics in the large developing nations. But they try to claim a common cause with religious traditionalists in Muslim countries, whose tacit allegiance and sometimes concrete support Islamic terrorists need, for at least reasons of credibility.

One big question all of this raises is whether or not the networks and institutions of globalization have ended military competition and conflicts among the major nations. They probably have for a while—so long as the United States remains a superpower without a near peer and ready to preside over the system. Countries outside the orbit of globalization, most notably North Korea and Iran, could find themselves in open conflict with the United States and some of its allies. But great-power conflicts are very unlikely. The next in line after the United States, China, is at least two generations away from becoming America's military and economic near peer on a global scale, despite its fast-rising military spending. Still, by 2020, China may achieve that status in its own region. Europe and the Western hemisphere will remain beyond China's military reach for a very long time; but within a decade, China could be jousting with America for influence in North, South, and Southeast Asia, including Japan and Korea—and perhaps footholds in the Middle East as well.

Even so, the flow of hundreds of billions of dollars in goods, services, investment, and ideas between the Chinese and American economies—as well as between most other economies and China, and most other economies and the United States—will mightily discourage armed conflicts between them. But that flow won't prevent dangerous confrontations from occurring between their allies—most conspicuously and perilously, Pakistan and India.

History teaches the perils of believing that globalization can guarantee peace. During the mid-to-late eighteenth century and again in the late nineteenth and early twentieth centuries, international commerce, investment, and communications all intensified as much as in our own time, or nearly so. The first period of globalization saw vast expansions in ocean trade, the first world-spanning businesses and the spread of steam power, along with the heyday of the French and Spanish global empires. Even in young America, John Jay wrote in *The Federalist Papers* about how those advances opened up trade with the world's most distant countries: "In the trade to China and India, we interfere with more than one nation, inasmuch as it enables us to partake in advantages they had in a manner monopolized, and as we thereby supply ourselves with commodities which we used to purchase from them."[4] The second great period for globalization brought electrification, the telegraph and telephone, multinational investment combines, and the crest of the British and German global empires. In both instances, all that globalization did not stop the world's major commercial and trading powers from waging terrible wars on each other—once there was more than one country with the capacity to do so with hope of success. Every day of the next fifteen years and beyond, U.S. and Chinese military planners will contemplate the lesson of history that even deep economic relationships do not preclude wars between the parties, once they're each other's near peers in military power.

Even with no direct military conflict between China and the United States over the next decade or so—as there were none between the United States and the USSR from the Cuban missile crisis of 1962 to the downfall of the Soviet Union more than a quarter-century later—the American-Sino competition will be a powerful force shaping global politics. Long before 2020, the United States and China will have to manage a fierce competition over energy resources. They will almost certainly have to work together to contain regional tensions and terrorism in a world in which nuclear proliferation has long ago failed, as they have recently with North Korea—though as yet, they haven't found common ground over the Middle East, including Iraq and Iran. America and China also will probably work together and with other major countries to head off threats to the global Internet architecture.

To the extent it serves their purposes, the United States and China, acting together, also will provide some order to global economic affairs. For example, if the two countries can agree on the future scope of intellectual property rights, the rest of the developing world will almost certainly go along. Such an agreement will be crucial to both economies, since most of the value of the industrial equipment, laptops, and other sophisticated goods that China produces (or merely assembles) comes from ideas developed in the West, while the value of those ideas forms the core strength of the western economies. Equally important, the trading and financial relationships between America and China hold a key to a potential crisis for global growth. Since globalization took off in the late 1990s, U.S. imports, now topping $2 trillion a year, have helped keep much of the rest of the world growing—with China and a few other countries financing America's huge trade deficits by loaning or selling hundreds of billions of dollars in excess savings to the United States. Unless these huge imbalances are reduced, they will eventually produce a run on the dollar that will stall out economies around the world. America and China will have to politically manage that transition together—both countries will have to agree to realign their currencies while America raises its national savings and China increases its national consumption.

In all these matters, the geopolitical marginalization of Europe seems all but certain. Europe has steadily cut its defense capacities and commitments, and the slow economic growth that lies ahead will probably mean even deeper cuts. On top of that, both Europe and Japan are likely to be preoccupied politically with the fierce domestic conflicts certain to erupt when that slow growth collides with the tax hikes and spending cuts required to keep their pension and health-care systems going. It may be hard to imagine today, when much of Europe disdains America's power and dislikes its president, but these developments could strengthen the Atlantic alliance. However much some European intellectuals complain about it, a strong Atlantic alliance will be the only way to preserve a European voice in global politics. And, if Europe's choice is America or China, it would be a step too far for Europe to ally itself with the world's fastest-rising authoritarian nation against the world's leading democracy.

China's growing military and economic power will make the same choice more uncomfortable for Japan, but no less inevitable. Russia will have different options, given its long border with China and the absence of a shared democratic heritage. In many respects, Russia will find itself in the position China occupied at the height of the Cold War—with a foot in both camps, trying to turn that position to its advantage. But the weakness of the Russian economy and military, compounded by its mounting demographic crisis, also will make the once-mighty superpower a minor player in most conflicts. Russia is losing population at an unprecedented rate for a fairly developed country—as much as one million souls a year—and by 2020, Russia's population will be 50 million less than Nigeria or Bangladesh and barely more than Mexico.[5] Moreover, with death rates among young Russian men rising to more than three times U.S. levels, experts at the Rand Corporation believe that by 2015 or sooner, Russia will have difficulties marshalling the soldiers and police required to merely patrol its borders and maintain internal order.[6]

Throughout all of these developments, the basic line in the global sand will still divide those engaged in globalization and those outside its orbit. While some form of U.S.-Sino competition will be a basic fact of geopolitical life in 2020, no nations will have more at stake than America and China in defending globalization against attackers from the outside and its doubters inside.

THE DIMMING PROSPECTS FOR EQUALITY

These three historic forces will help shape everyday life over the next decade, both directly and through a number of social conditions which follow from how they will interact. First, demographics and globalization will intensify economic inequality almost everywhere. China and the United States, globalization's two principal drivers, already are the world's two most unequal major societies. It's no accident. Wherever globalization and its technologies take hold, the return on investment rises and makes the rich richer while more intense domestic and international competition holds down most workers' wage gains even when their productivity increases. That doesn't even count the economic shock and sorrow of millions of workers in advanced countries who will lose their jobs and, in the current technology-driven competitive environment, eventually settle for lower-paying ones.

Growing inequality is also part of another dynamic tied to the information technologies that are turbocharging globalization: Ideas are replacing physical assets as the main source of wealth and growth. The idea-based economy is no longer an approximation or metaphor, but a hard reality. American corporations now invest as much in "intangibles," principally intellectual property, as they do in plant, equipment, offices, factories, and real estate. Twenty years ago, the market value of the physical assets of the top 150 U.S. public companies—their "book value"—accounted for 75 percent of the total value of their stocks. A firm was roughly worth nearly what its factories, offices, equipment, real estate, and so on could be sold for. By 2004, the book value of the top 150 corporations accounted for just 36 percent of the total value of their shares. Today, nearly two-thirds of the value of a large company comes from intangibles—what it knows and the ideas and relationships that it owns: its patents and copyrights, databases and branding, organizational arrangements and the training or human capital to use these ideas.

The ideas themselves float in the economic ether, their value sometimes protected by hard-coated patent and copyright laws. Everyone encounters these ideas through the technologies and business methods that embody them and generate the productivity and profits that give an idea-based economy its buoyancy. The good news is that new ideas tend to generate more of the same, so the number and power of the new information and Internet technologies, new nano- and biotechnologies, and new business strategies, has no certain limits even for the near future. And once a new idea proves itself in the marketplace, anyone, anywhere in the world can use it—though sometimes they have to pay to do so. The painful downside is that with more economic wealth and incomes tied so directly to all the new (and older) ideas, it will become a lot harder for anybody who has spent most of his or her life in the old economy to earn a middle-class living and support a family. Already, average workers in the United States and the best-performing European economies have seen their real wages stagnate or decline for five years, despite healthy productivity gains in most of these countries, with only professionals, managers, and investors enjoying rising living standards in almost every advanced nation.

As sure as death and taxes, these developments will intensify inequality in every major country. Nearly everywhere around the world, the market forces unleashed by globalization and technology are bidding up the value of the skills of the top 20 percent or so who work adeptly with ideas, and driving down the value of the work that many other people do. It's already here: The last six years of economic expansion have seen the smallest job gains and wage increases for most people of any expansion in more than fifty years—and the largest gains for both highly educated workers and the wealthy people with claims on the rising return on invested capital.

On top of these changes, aging societies also tend to become more unequal, because elderly people as a group have smaller incomes than working people. So, as the numbers of elderly Americans, Europeans, and Japanese increase over the next ten to fifteen years and beyond, the share of these societies living on relatively small incomes will rise. And that will be no comfort for everyone else—not only because globalization and an idea-based economy tend to hold down the incomes of many of the nonelderly, but also because everyone's taxes will go up to support the pensions and health care for all those retirees.

THE PERFECT STORM IN HEALTH CARE

The near future holds other troubling prospects, because demographics and globalization also have set every advanced country on a course that could gradually degrade medical care for most people. Three factors are converging to produce a health-care version of a perfect storm. Not only will the number of elderly people inescapably rise by 35 to 50 percent in every advanced country between now and 2020, but in addition, medical advances—all of them expensive—are turning more life-threatening illnesses into chronic conditions and allowing people with terminal conditions to survive for additional months or years. On top of that, government programs in every advanced nation guarantee that every older person can and will demand every new treatment. These cost pressures aren't limited to the elderly. For example, overall cancer rates and the costs of treating each case have both been increasing. Dr. Alan Lotvin, who heads M|C Communications, a leading provider of continuing education for physicians, estimates that, while five years ago patients with colon cancer were typically treated with two drugs that cost about $500 and on average lived about eight months, today they receive chemotherapy and live on average for thirteen to twenty months, at a cost of $300,000 to $500,000.[7]

Globalization itself is expanding the number of costly new medical advances year after year. For the first time, there is a potential global market for any new biotech and pharmaceutical product, costing hundreds or thousands of dollars for a course of treatment, and every new piece of medical equipment selling for tens or hundreds of thousands of dollars. This enormous expansion in the market for health-care technology is changing the economics of research and development, creating a reasonable risk for firms to spend a billion dollars or more to develop a new product. And since the key to promoting innovation is intellectual property protections that grant the developers monopoly prices for their inventions, the prices for successful new medical products stay high for many years.

The fallout from this perfect storm in health-care costs, especially with rapid aging, would be easier to manage if the numbers of working people paying the taxes to finance it were not also falling across most of Europe and Japan, or slowing down in the United States. As a result, soaring medical costs will almost certainly produce political storms in most advanced countries. The ensuing political conflicts won't be limited to government programs for elderly, because the same forces will continue to drive up the cost of everyone else's care. What to do about it could be the central political issue of the next decade.

Unlike budget deficits or educational performance, the crisis of health-care costs confounds the usual approaches to political compromise. With demographics alone expanding the elderly's voting clout steadily, what political party is going to cut back on their care—especially when their lives may rest on it? It may be easier to trim access for everyone else, if only because their treatment is less routinely a matter of life or death. But that doesn't mean that working families in Europe, Japan, or the United States will tolerate politicians who want to raise their taxes to pay for the medical treatment of others, especially when they're on the losing end of growing inequality and their own health care is contracting.

Unlike the demographic shifts and economic globalization that underpin this problem, there is little certainty about how precisely it will play out. If innovation responds to market pressures—something many economists doubt—we could see the development of much-lower-cost technologies. The more likely path is higher out-of-pocket payments, tax reforms that indirectly dedicate more taxes to health care and leave less for everything else, implicit rationing through longer waits to treat anything but life-threatening conditions, weaker patent rights that slow the pace of medical advance, and a quiet shift to lower-tech treatments. The only certainty is that in every country the problem will fuel ugly political conflicts.

THE BATTLES OVER ENERGY AND THE GLOBAL CLIMATE

Much like health care, both energy markets and the global climate are heading for a long-term crunch. For nearly a generation, world demand for oil has consistently grown faster than the world's capacity to produce it. Over the next ten to fifteen years, this gap will widen. Rapid industrialization and fast-growing auto ownership in China, India, and other large, developing nations will be the main sources of this growing demand. Within a decade, these nations, none with significant oil reserves of their own, will have the most energy-intensive economies in the world.

One near certain consequence will be much higher energy prices on a permanent basis, even as various hybrid or other alternative engines bury the conventional gas-combustion automobile. Europe and Japan will be less affected than the rest of Asia or the United States, thanks to high oil and gas taxes that have forced up their energy efficiency. But with more than half of global energy needs being met by the world's most politically unstable regimes in the Middle East, Central Asia, and Africa, no policy will protect anybody from the energy shocks and accompanying recessions almost certain to occur over the next fifteen years.

The successful application of technology and investment over the next fifteen years could make these challenges much more manageable. The world may face shortages of cheap light oil, but it is not energy poor. Canada, Russia, and Venezuela have heavy-oil reserves larger than all of the world's conventional light-oil reserves. Russia and the Middle East also have enormous reserves of natural gas; and comparably huge coal reserves lie in China, India, the United States, South Africa, and Russia. Governments and corporations could develop, produce, and transfer the investment, technologies, and infrastructure required to meet any global energy demand for the next fifteen years and well beyond—if they can find the trillions of dollars required to develop ways to reduce the viscosity and contaminants in heavy oil, liquefy natural gas in large quantities and transport it across the world, tap deep-ocean oil deposits, store and manage nuclear waste, and decarbonize coal with supercritical boilers and gasification. The alternative is an 80 percent expansion in Persian Gulf oil production by 2020—and those much higher prices.

The truth is, energy prices will continue to rise under any scenario. Much as the scientific debate over the risk of climate change is effectively settled, economists agree that higher prices for carbon-based energy will be required to shift businesses and consumers to alternative fuels and technologies. As energy prices rise, other alternatives will become economically practical and popular. It's impossible to know which innovations will matter until they actually do, but many experts in this area today are betting on two new sources, given the right technological breakthroughs. One leading alternative to fossil fuels is hydrogen fuel cells, which could be adapted for automobiles, laptop batteries, and home heating and cooling. For larger energy users such as office buildings and industrial plants, the smart money is betting that advances in nanotechnology-based solar energy will be able to generate steam to run turbines and electrical current across grids.

The onset of grave global environmental problems could provide more impetus for these investments, especially by governments. The political consensus about the onset and effects of global warming will almost certainly grow and deepen in coming years, triggering serious political debates everywhere. But with the timing of global warming still uncertain, no one can say whether it will take ten years or two generations to see the climate changes that ultimately will make alternative energy a universal choice—whatever it costs.

A more certain development is that the mismatch of energy supply and demand will expose the most glaring anomaly in globalization and geopolitics, namely the growing importance of the world's most unstable and economically insular region, the Middle East, to the growth and stability of the world's major countries. China and the rest of Asia are already twice as dependent on the Persian Gulf as Europe or the United States. By 2020, Asia will import 75 percent of its oil, 80 percent of it likely from the Middle East. Yet the autocratic governments of the Persian Gulf oil states, facing treacherous challenges from Islamist fundamentalists, consistently resist foreign investment and domestic competition. The region most crucial to global growth has isolated itself from the forces transforming the rest of the world.

This anomaly will make the world a more dangerous place over the next fifteen years. Most obvious is the prospect that the House of Saud could come crashing down or that one or more of its Gulf neighbors could face their own religious-based insurgencies. Even if the world beats the odds and the Gulf's royalist autocracies remain stable, energy will be a growing sore point between America and almost everybody else. With the United States playing the heavy's role in Middle Eastern public opinion, it would be only natural for Beijing, Bonn, Delhi, and other world capitals to manage their relations with oil states as independent of America as they can. Early signs are evident already in the extensive network of long-term contracts and alliances that China is building with Iran, Russia, Nigeria, Venezuela, and other big oil exporters. Within a decade, these deals and alliances will draw China into the geopolitics of not only southern Asia, but also the Middle East, Russia, and parts of Central Asia.

Yet, even with America's ongoing misadventure and failures in Iraq, the basic arithmetic of globalization and energy will intensify U.S. involvement in the Middle East. For its own interests, the United States will have to do its best to ensure that Asia can meet its energy needs and contain energy costs for itself and other major economies. If creating and managing global, cooperative arrangements for energy proves to be beyond America's diplomatic capacity—as seems likely—the United States will still have to keep the oil pumping by somehow maintaining political stability across the Middle East. Day to day and year to year, the U.S. Navy—the world's only "blue-water fleet"—will provide maritime security for oil shipments across the Indian, Pacific, and Atlantic oceans. And as the world's foremost platform for technological development, the United States will also have to commit huge resources to developing some of those alternative energy sources.

TWO WILD CARDS

There are two other large developments out there that will affect the course of nations and most people's lives over the next fifteen years. But they're not structural forces with implications that can be reasonably predicted, and their likelihood remains fairly small. The first of these wild cards is terrorism and the growing political power of Islamic fundamentalism. It is usual, but dangerous and wrong, to conflate these two developments. It's true that many terrorists in our time are Islamic; and most Islamic fundamentalists, along with terrorists groups, reject the political, cultural, and economic values associated with globalization and its leading promoter, the United States. But most fundamentalist Muslims are not terrorists in any sense—and some of the world's most violent and well-organized terrorist groups, such as Hamas and Hezbollah, are not fundamentalist. Moreover, many of the major terrorist actions of the last decade have not involved Islamic causes at all, from Timothy McVeigh's assault in Oklahoma City and the Aum Shinrikyo cult's sarin attack on Tokyo subway riders, to the November 17 attacks in Greece and the Basque separatist attacks in Spain.

Most Islamic terror groups in the Middle East and Asia have domestic agendas which, but for the small likelihood of their destabilizing or destroying Israel, should have little direct impact on life in Europe, Japan, or the United States over the next fifteen years. These groups most resemble the terrorists who dogged the peace and development of Northern Ireland and Colombia in the recent past, with no significant effects on geopolitics or the economies of most advanced countries. Al Qaeda and its imitators, comprising a few thousand followers and fighters, are different, because their agenda is explicitly global and their tactics are aimed at the United States, its major allies, and the oil producing countries. Nevertheless, the record shows that terrorists actually change a country's course of development only when their violence is both protracted and pervasive inside that country. A singular, localized attack such as 9/11 had no effect on America's macroeconomy or international power. A nuclear or virulent biological attack is probably the only way that Al Qaeda or others like it could inflict the pervasive and protracted shock and losses that actually could change the course of a large society like the United States, Japan, Germany, or Great Britain. The likelihood of that is wholly speculative and fairly remote—but it's real, especially given how little has been done to strictly control loose nuclear material around the world, especially in the former Soviet Union. Don Clark, who led the FBI investigation of the 1993 World Trade Center attack, notes that the danger of this could rise sharply if Al Qaeda once again secured a state's support—for example, through a fundamentalist takeover of Pakistan.[8]

Islamist fundamentalism represents a different wild card for the next ten to fifteen years. It has neither the ambition nor the capacity to derail globalization or the course of geopolitics. One possible scenario that keeps geostrategists awake, however, is a fundamentalist takeover of Pakistan. That event could enable Al Qaeda terrorism to use fundamentalism to secure at least temporary state protection and access to nuclear weapons. And nuclear terrorism is a wild card unlike any other—a potentially civilization-changing event that would alter the life and liberties of every advanced society.

One other easily imagined scenario that would affect virtually everyone is a fundamentalist takeover of Saudi Arabia. The world and its major powers survived Iran's takeover by fundamentalists; but the oil price shock which followed bred an inflation that affected everyone. The developed nations are much more energy-efficient today than they were in the late 1970s. But Saudi Arabia is a much bigger player than Iran was, and globalization is vastly increasing global energy consumption. An extended interruption or large cut in Saudi oil exports anytime in the next decade would play havoc with growth and prosperity around the world. Perhaps most important, it could weaken social stability in China, based today on its leadership's bet that they can maintain their monopoly on state power by producing endless high growth. Yet even such an event would not derail globalization, as the Iranian example also suggests that a fundamentalist government in Riyadh would see its way to restoring oil supplies in fairly short order.

Technological advance is the other wild card that could affect the course of our near future, the positive yin to the frightening yang of terrorism. There is no question that information technologies in 2015 and 2020 will perform tasks that can hardly be imagined today, and countries that promote their spread will be richer and stronger for it. Medical advances will end the suffering and costs of some diseases and injuries, and nanotechnology could both change the way countless products are made and lead to thousands of new products.

The social, economic, and political changes of the next ten to fifteen years also will affect the course of this technological progress. The demographic shifts are already focusing more R & D on health care for older people and on productivity enhancements for the slow-growing numbers of working people. Globalization will transmit these advances across borders, societies, and cultures with greater speed and intensity than even seen during the computerization waves of the 1980s and 1990s. And America's status as the lone superpower with no near peer may mean that relatively less of all R & D will focus on military technologies than, for example, during the space and missile races in the 1960s and 1970s.

Technology is a frustrating wild card, because its precise trajectory cannot be more than guessed at. A number of leading nanoscientists, including Paul Alivisatos at Berkeley and Sumio Iijima, from Japan's National Institute of Advanced Science and Technology (Alivistos pioneered nanoscale semiconductor crystals and Iijima developed the first carbon nanotubes) see potential near-term breakthroughs in nanotechnology that would mean much cheaper, abundant, and cleaner energy. But ten years ago, others predicted the same results from breakthroughs in other fields that never happened. And many prominent scientists focused on deciphering the human genome, such as William Haseltine, and Susan Greenfield from Oxford University, see cures for such diseases as cystic fibrosis, and Huntington's and Parkinson's diseases, in the next decade, from genetic screening, stem cell advances, and wide use of "neurochips" (genetically altered, or selected cells on integrated circuits that are implanted in people's bodies). But twenty years ago some of their predecessors expected to find a cure for cancer within a decade. No one can chart even the next five years of technological progress with any confidence, and it is beyond anyone's foresight or imagination to figure out how scientific advances are likely to affect countries and the global economy. As a wild card, the likelihood of important and economy-shaking technological advances seems pretty certain—even as their particular character and implications remain unknown.

THE CONVERGENCE OF DEMOGRAPHICS, GLOBALIZATION, AND GEOPOLITICS

Everyone's daily life will be most affected not by each of these separate forces but by their combinations—and their combinations will make it harder for governments to manage their separate implications. Start with demographics. Every year from now to 2020, the number of elderly Europeans receiving public pensions and health care will increase by nearly 3 percent a year, as the number of working-age Europeans falls by about 1 percent a year. Japan is caught in an even tighter demographic vise. These demographic changes are overtaking many countries so quickly and sharply that by 2020, the number of elderly in both Japan and Europe will equal more than half the number of their working-age people. And the problems this poses are not going away: Across Europe and Japan, the number of children and adolescents who will become the working, tax-paying adults of 2025 and 2035 is falling even faster than the number of working-age people.

As these demographic shifts collide with the pressures of globalization, Japan and much of Europe will face a vicious circle that will gradually but steadily degrade the lives of many of their citizens. Unless their economic paths unexpectedly change, they cannot avoid years of slow growth and meager productivity gains that will preclude any incremental efforts to ease their burdens by growing their economies. Instead, the natural course is that the slice of their societies' incomes and wealth that will be claimed for public pensions and health-care services will increase, forcing up taxes and slowing investment, which in turn will dampen their growth even more. The OECD sees France's economy growing by barely 1.5 percent a year by 2020, and that assumes that the next decade of globalization will exact no greater costs on

French competitiveness and employment than today. With the same caveat, Germany's leading forecaster sees growth of just over 1 percent a year by 2020, by which time public pensions and health care will consume 18 percent of German GDP. That share of GDP is equal to 80 percent of the share claimed today by the entire U.S. federal government.

Accelerating globalization could make these challenges even more terrible for millions of Europeans and Japanese. The steady shift of manufacturing and service jobs to the fast-growing developing countries will leave advanced countries with only two real options: Raise productivity for the nontradable services they still have, and develop new industries by becoming an innovation powerhouse. Both require resources and changes that today seem beyond the normal social and political capacity of Japan and most of Europe. Leaps in productivity require a willingness to change the way businesses operate, so they can take advantage of new technologies, as well as high savings and investment. Investing in technology isn't the problem: In the 1990s, Europe and Japan invested about as much as America in IT advances, although consistently a year or two later. But they achieved little productivity gains, because that would have involved large-scale firings, job reassignments and job redefinitions that their cultures and strong unions won't accept. High saving and investment also will be a problem for Europe and Japan. Their slow growth and rising taxes will pinch most people's incomes and savings, their government spending and deficits will likely rise, and the number of workers in the age group when people save the most, those in their late thirties to early fifties, will slowly decline.

An economic and political culture that drives innovation also seems out of reach for most of Europe and Japan, because it depends on laws and attitudes that directly support entrepreneurism. Layers of subsidies, regulations, and informal networks continue to protect every established business in Japan and most of Europe today, as they have for decades, and creating a new business is still almost an anomaly. In England, new storefronts are rejected because they sully the experience of a green space a half-mile away. In France, a new worker, however skilled she is, cannot earn more than half of what an average worker makes and then can't be let go for eighteen months. Europe's capital markets have long been organized to finance government deficits rather than new business ventures. The preference for the old and established suffuses entire cultures. While Americans admire Bill Gates, Sam Walton, and other rags-to-riches entrepreneurs, Japanese and Europeans are more likely to revere the families that founded industrial combines generations ago, from the Schlumbergers of Germany and the Agnellis of Italy to the Hitachis of Japan and the Axison-Johnsons in Sweden.

Nor are there signs of real change. Less than a half-decade ago, France's answer to rising unemployment was not to deregulate or encourage new businesses that might create more jobs, but to lower the mandatory retirement age by five years—think of what that will do to long-term pension costs—and institute a thirty-five-hour work week that leaves the French working 14 percent fewer hours than the average for other advanced countries. Another French reform passed in 1997 to mandate some form of private saving met with such resistance that it was never implemented and was finally repealed in 2002. Nicolas Sarkozy's election to France's presidency in 2007 could produce real reforms, but they haven't appeared yet. Similarly, in Germany, low-skilled young people lose so many public benefits if they take a job, that their posttax incomes are higher when they stay unemployed. Germany also recently passed new laws barring German companies from laying off any worker without "personal cause," and those causes don't include plans to reorganize a workforce after upgrading their technologies.

On this course, France, Germany, and others will find themselves losing even greater ground to globalization just as their demographic shifts hit their budgets with full force. To finance the pensions and health care that elderly Europeans and Japanese expect, they will have to cut those benefits. If that's not politically possible—and where has it been so?—they will have to raise taxes. With a shrinking workforce and an average tax bite across Europe that's already 45 percent, who will pay those higher taxes? The last alternative is higher government deficits year after year, which will squeeze business investment and weaken their currencies, producing even more drag on the productivity and growth that ultimately finances everything.

The next decade will find the welfare states of democratic socialism unable to honor their basic social contracts with tens of tens of millions of Europeans, much as communism could not keep its economic promises to its subjects in the 1970s and 1980s. This is a formula for serious political conflict and polarization, and one of the worst possible social and political environments for the drastic deregulation that Europe and Japan will need to grow in an era of globalization. It's already begun in virulent backlashes against Muslim immigrants in many European countries, as well as riots and demonstrations in France and Germany protesting minimal steps to loosen regulations and trim benefits.

It will require the political talents of Winston Churchill and the persuasive skills of Ronald Reagan to convince European and Japanese voters that their interest lies in accepting fewer benefits and less economic security from their governments. And the expansion of the European Union, encompassing tens of millions of lower-wage Eastern and Central European workers who will be able to move wherever they choose within the EU, will only deepen the crisis. These kinds of social emergencies sometimes bring forth a great leader, and one could emerge in one or two countries—but not in most places. After more than a half-century of social and political stability across Europe and Japan, these pressures could produce a new era of more extreme politics on both the left and right.

If the challenge for Europe and Japan was merely demographic, they could try to protect their industries and artificially stimulate growth without worrying about how foreign investors and producers would respond. If the challenge involved just the pressures of globalization, they could cut taxes and shift resources to education and investment without worrying so much about pension and health-care costs. It's the convergence of these two powerful and largely unalterable forces that is creating consequences that democratic governments find so difficult to manage. Unless most of Europe and Japan can convince their people to accept changes that they will certainly consider radical, by 2020 they will find themselves in a period of real decline.

In both places, there will be pockets of economic success. Individual companies with good leadership can avoid the pressures that grip a society through its investment strategies, and by picking up and moving operations to places where taxes and everything else are less expensive. No advanced country has been more resistant to deregulation and change than Japan, especially when it might affect its sprawling service sector—the least productive in the advanced world. But Japan's leading export companies—Sony, Toyota, and others—long ago accepted the facts of global competition. While many large Japanese companies have foundered, they manage to come up with new products and operating strategies year after year—while also moving much of their production to China and other lower-wage Asian countries.

On a larger scale, Ireland, Sweden, and Finland have developed approaches that have created successful niche economies: Generate the growth that can keep a small society prosperous by developing a few world-class technologies in great global demand or a few world-class companies producing global commodity products. Sweden slashed corporate taxes and drew on a national dedication to engineering to produce a handful of world-class companies—Erikson, Tetra Laval, and IKEA—servicing the world with cellular technology, cheap packaging, and prefabricated furniture. In Finland, Nokia, a company that used to make rubber boots, turned itself into the world's leading cell phone producer at a time when exploding demand for the product could carry a tiny economy. And in just fifteen years, the Celtic Tiger went from the EU's slowest-growing and poorest member to the top of the European heap by combining every policy that might help attract foreign investment and entrepreneurs to make that happen—cutting taxes and overall government spending (including public pensions and the payroll taxes that support them), expanding spending on education and training, abolishing most restrictions on hiring, firing, and foreign investment, and more. Ireland had the special good fortune to do all of this just as globalization took off and U.S., British, and European tech companies were looking actively for new places to expand.

Latvia, Slovenia, and even Spain and Poland send policy experts and business leaders to Dublin to help figure out how they might import the Irish miracle to their own countries—but not the EU's much larger and much richer countries: Germany, France, and Italy all could use a miracle—especially as, what works in a small country or a single sector may not be able to move large and highly differentiated economies struggling with the combination of aging populations and globalization. In its absence, they need to manage the political will to reform their taxes, regulations, and public pensions in major ways. As it is, the social contract that these countries can no longer afford, they also seem unable to fix.

In most respects, globalization, demography, and their combination find China and the United States in stronger positions than Europe or Japan. While the numbers of elderly people will increase faster in both America and China than in Europe or Japan over the next fifteen years, the number of working-age Americans and Chinese also will grow. It will make all the difference in the world. In 2020, elderly Americans, necessarily contributing little to productivity, will still number less than 28 percent of those working, and elderly Chinese will equal less than 20 percent of China's workers. Compare that to Europe and Japan, where the elderly will equal more than 50 percent of their workers. Moreover, America's social security system provides much smaller benefits than most other advanced nations (with the notable exceptions of the United Kingdom and Ireland), and the vast majority of Chinese receive no public retirement or health-care services at all. These differences may make European societies more just and equal, and certainly more generous. But the American and Chinese approaches can sustain themselves over the next generation, while Japan and Europe's systems cannot.

Today, China and the United States are the world's most profoundly pragmatic major societies. Both largely accept the world for what it is, even if America's ambitions in the world and China's domestic strategies sometimes exceed their capacities. Japan and much of Europe cling to old commitments and hopes that globalization and their own demographics will no longer let them realize, which deeply impedes their ability to change. If they cannot manage to do so in the next decade, the world as it is will inexorably erode their economic well-being, political influence, and many of their cherished social values.

Crucially and critically, China and the United States should occupy two poles of successful accommodation to the pressures of globalization for at least the next decade. China will have hundreds of millions of relatively skilled lower-wage workers producing the goods the entire world wants, with technologies and business methods imported through the greatest flood of foreign direct investment in history. The United States will have many millions more high-wage workers and managers involved in the extensive business networks of globalization than there will be in Europe or Japan. America also will have millions more highly skilled workers, managers and scientists producing new products and services that the world will also want, propelled by new businesses and the world's biggest scientific and higher education establishment. Finally, America's more deregulated system should make its companies more open to adopting new waves of technology and business methods, whether from the next-generation Internet, nanotechnologies, or entirely new areas known today only to scientists and engineers.

China's economic ascendancy will be built not just on its hundreds of millions of low-wage workers using western technologies, but on what is arguably the most single-minded modernization program the world has ever seen. Much of it would be unimaginable but for the collapse of the Soviet Union, although it began a decade before Mikhail Gorbachev, when a succession of sclerotic Soviet leaders clung fast to the utterly failed and corrupt forms of state planning developed in the 1950s and 1960s. Most westerners didn't notice through the cold war's fog, but in the late 1970s and early 1980s, Deng Xiaoping was discarding socialist purity for the demands of economic modernization. It began with a trial run in agriculture, a sensible course for a backward developing country with huge natural resources, and the experiment taught China's next generation of leaders a great deal about how to manage the much larger transitions that awaited them. With a series of ruthless dictates, Deng relaxed decades of state controls on the prices and distribution of grains, and in the process forced hundreds of millions of farmers thousands of miles from Beijing to shift out of land-intensive crops. Farm productivity went up, and the first food surpluses China had seen in centuries drove down prices, leaving tens of millions of Chinese farmers destitute. At the same time, Deng repealed Mao's bar on private enterprises in villages and towns, and some of the millions of out-of-work farmers, along with local leaders, created tens of thousands of small businesses that began supplying China's state-owned enterprises. In less than a decade, China shifted countless millions of agricultural workers to other sectors—a move that took two generations in Europe and America—and by the late 1980s, the village and town businesses were the single largest sector of China's GDP and job growth.

An even bigger leap came with the final collapse of the USSR. Over the 1990s, China threw overboard the ideological inhibitions that had thwarted Soviet economic and social progress for two generations. The Chinese leadership turned its back on Lenin's axiom that the quality of a society is proportional to the size of its state sector, and started withholding public money and loans for state-owned industries deemed less than essential to the nation. This process was so rigorous and far reaching that China today has a smaller official public sector than any nation in Europe, and one barely larger than the United States or Japan. They even quietly jettisoned the ideological primacy of workers, writing a new party platform that made the development of production and culture the top priority, and one that would be realized in the interest not of the proletariat, but of the masses. Even as some European intellectuals still clung to socialist axioms from the 1930s, Jiang Zemin signaled that a new generation of entrepreneurs, the human heart of capitalism, would be brought into the bosom of the Party and its program for modernization. Where the Russians sold their state enterprises to their managers for largely worthless shares and its natural resources to a handful of oligarchs for short-term government loans, China's leaders gradually and relentlessly have unwound much of their state enterprises and allowed hundreds of thousands of young Chinese to build new businesses to take their place. By the late 1990s, the Communist Party of China was no longer Lenin's party or Mao's. It vowed explicitly to encourage private enterprise.

The leadership backed up its words with a rapid succession of additional reforms designed to tap into a central strength of private markets—the capacity to move resources of every kind among companies and across sectors and borders. They junked the *hukuo,* the national geographic registry that restricted internal migration from village to town or city. Just as important, they opened China's economy to foreign trade and investment, so China's new entrepreneurs could create alliances with foreign companies eager to transfer their capital and technology. In doing so, they undercut the small town and village enterprises that a decade earlier had formed the first front of modern Chinese capitalism. By the mid-1990s, their markets had deteriorated sharply, overtaken first by foreign-funded businesses and then by domestic Chinese companies. By 2005, nearly fifty thousand Chinese companies had foreign partners who had invested more than $125 billion in their joint ventures, including nearly $90 billion in manufacturing.

The main western criticism of this lightning modernization, the absence of political freedom of virtually any kind, has so far been one of its secret

strengths. Unlike most of Europe, where unions, farmers, and old-line industries hold fast to regulations and restrictions that preserve a status quo ill-
suited for globalization, China's politics provides no channels for any
resistance to the massive changes shaking its society. The real test of that approach could come in the next ten or fifteen years. The changes that are making China an economic power have happened, so far, mainly in its Pacific
coastal cities and regions. Much like America in the first half of the nineteenth century, China's huge physical expanse had kept the revolutionary
changes shaking their country hundreds or thousands of miles away from
most of the people.

Over the next decade or so, those changes will come closer to home for
hundreds of millions of Chinese. In that short time, China will have to build
or create the complex financial and social institutions that the advanced countries of the West took a half-century or longer to put in place. Today, China
has no modern private banking, insurance, or mortgage finance systems to
speak of. Its financial system consists largely of a number of huge state banks
that are conduits for channeling the small savings of hundreds of millions of
Chinese into loans to fund the state's remaining enterprises. As those enterprises disappear, their many hundreds of billions of dollars of bad loans will
have to be written off. And like manufacturing before it, Chinese finance will
import modernization. Unless there's an economic crisis—say, $150-per-
barrel oil—foreign investment will take the sting out of the inevitable insol-
vency of some of the huge state-owned banks. By 2010, under current
timetables, China will open itself to the Western and Asian banking conglomerates that run finance throughout the developed world. It's already begun with joint ventures, to be followed soon by a mix of home-grown and
foreign-owned banks and insurance companies. The real challenge will be political: Saying no once and for all to more loans for the failing state enterprises
that still employ many tens of millions of Chinese.

The same pattern is playing out in infrastructure. Over the next decade,
China will build more new highways, bridges, optical cable systems, airports,
high-speed rail, subways, and water and waste treatment plants than the
United States and Europe combined. The leadership fully expects Western companies to eagerly pick up the bill for most of the telecom and transportation
construction—and they probably will. By 2005, there already were almost six
hundred major transportation and telecommunication projects with foreign
partners who had invested over $6 billion. For its more public forms of infra-
structure, China will have to raise many hundreds of billions of dollars from
a fast-growing tax base.

China needs modern roads, transit systems, airports, and treatment plants not only to make its economy efficient in the basic ways that advanced societies take for granted, but also to create a more integrated national economy out of its tens of thousands of towns and cities. Huge economic disparities between people on the coast and the rest of the country are already apparent to most Chinese. If that gap worsens, it could promote the kind of popular social demands that China has never dealt with successfully. For a leadership committed to denying political rights, modern infrastructure is their way of extending a realistic promise of prosperity to tens of thousands of smaller cities and towns in the central and western regions, and taking some of the steam out of what could become explosive resentments.

In the same decade, China also will have to build a social welfare system that can provide some form of retirement security, health care, and unemployment compensation. These systems can also help bind hundreds of millions of ordinary Chinese men and women, dislocated and disoriented by the waves of radical changes emanating from Beijing, to the state directing those changes. In time—not much time, judging by the record—China will import the services of western consulting firms and governments to help establish those systems. In 2005, more than 4,500 foreign joint-venture projects in social services and health care were already underway, with more than $8 billion in foreign investment. Moreover, for nearly a decade China has invited western officials to visit and tutor them about these systems.

I know this firsthand. In 1999, when I was Under Secretary of Commerce for Economic Affairs, I was invited to Beijing along with Janet Yellen, then chair of President Clinton's Council of Economic Advisors, and Roger Ferguson, the vice chairman of the Federal Reserve. We spent three days shuttling from one gray government building to another, talking about different ways governments can support people when they grow old or lose their jobs. Behind the official Chinese rhetoric of "market-based socialism," one after another official asked many of the basic questions of western public administration and finance. What do you do about pensions for people who can't or won't work? How can you pay for all of it without stunting growth? What role can local governments play? The meetings culminated with an audience with the prime minister in a great, gilded hall in the walled compound where China's leadership lives and works, on the shore of a private lake in the middle of Beijing. As the most senior visiting official, Dr. Yellen sat on a raised platform with the prime minister, while the rest of us sat in chairs arranged by rank. While young women in long silk dresses served tea, the prime minister spoke of the tasks of economic modernization without a murmur about socialism.

In trying to accomplish all this, China has two incalculable advantages.

First, here, too, globalization is working in China's favor. When England, Germany, France, and the United States took on similar challenges, they each largely started from scratch. Either the task had never been attempted before, or they thought they had little to learn from those who had already done it. China is modernizing itself in a time of far-flung worldwide business networks and elaborate global governmental relationships. Over the next fifteen years, China will choose the best of the West's multinational corporations to build and initially manage its new modern systems, while its own domestic companies become their suppliers and eventually their competitors.

China has another inestimable advantage. Unlike most democratic societies, China has not only political leaders utterly committed to radical change, but a unique consensus to carry it out by almost everyone who could affect its success. This goes well beyond the absence of the opposition parties and private interest groups that usually stymie large changes everywhere else. In taking on the enormous tasks involved in its lightning-fast modernization, China so far has evidenced extraordinary unity of purpose and discipline, and an astonishing absence of hesitation, backed by the power and apparatus of the Party.

In 2020, the United States will still be the world's sole military superpower, with no global near peer. But globalization will also support China's geopolitical ascendancy. Not only is it generating the resources for China to build a modern military alongside a modern economy, it also is extending China's influence into the economies of every other major market-based country. The political circles of those countries will always attend respectfully to those who can affect their economies.

China will have another trump card to play in geopolitics over the next fifteen years: A single superpower with no near peer makes everybody else nervous. From Paris and Moscow to Delhi and Tokyo, whenever their interests are implicated by a conflict or problem somewhere else, it will raise the question of what Washington will do and how they might influence it. Only rarely will any country acting alone be able to pressure the United States in meaningful ways. But with America's and China's deepening involvement in each other's economic development, a common position by Europe and China, or Russia and China, or Japan and China, will make Washington at least listen.

This all suggests that the central geopolitical development of the next fifteen years will be the evolving character of the U.S.-Sino relationship. While Europe, Japan, and Russia all will try to influence it as problems dictate, the United States will always be on the lookout for either cooperation with or containment of China's rising power, in managing such inevitable global issues as the world's growing dependence on oil from the world's most unstable region, the threat of nuclear terrorism, and the financial crises that are still part of the DNA of global capital markets. At times, the United States will almost certainly treat China as a threat, an attitude evident in the fast-growing numbers of U.S. military installations on the soil of many of China's neighbors. As America's rival or its partner, China's growing wealth and power will exact the greatest price that a sole superpower can pay—its acknowledgement of China's influence and desires, at least across Asia.

There's no way to opt out of this future. No country has the freedom to decide that they just don't like what its demographics, globalization, and the new shape of geopolitics hold for them. Opting out will be no more of an option this time than it was in the time of mass industrialization and the rise of a militarist Germany. As in that period, every country will eventually figure out how it will adapt to these forces. Nothing on the horizon can substantially alter or offset the fundamental forces poised to slow upward mobility for hundreds of millions of people across Europe, Japan, and even the United States. No decision or event in the next fifteen years can undo the fast-growing pool of educated labor in the world's large, fast-developing countries, armed with the capital and technologies of the global companies of America, Europe and Japan. And over the same period, only the most extreme and historic events could unravel the military and economic superpower of the United States and the ascendancy of Chinese wealth and influence.

CHAPTER TWO

THE DEMOGRAPHIC EARTHQUAKE

THE BIG PICTURE

FOR AS LONG as nearly anyone today can remember, most advanced economies have grown sufficiently to enable most of their people to steadily improve their lot in life. In recent decades, the government of every advanced country also has credibly promised its people that when they retire, their basic living standard won't decline dramatically. In both cases, basic demographics have played a key role, steadily expanding the workforce that creates wealth and pays the taxes that support others in retirement. Yet these demographics have been changing in ways which over the next ten to fifteen years will significantly affect the living standards and retirement security of hundreds of millions of people.

For hundreds of years, in virtually every country, each generation was only modestly larger or smaller in size than the ones that came right before and after it. Until the second half of the twentieth century, history records only a handful of cases in which extraordinary events—a terrible plague or war, or a large wave of immigrants—affected the relative size of one generation in a meaningful way, and then only in one or two countries. Since the Black Death 650 years ago, there were no instances of numerous societies dealing with a significantly smaller generation at the same time. And there were no recorded instances at all of countries on every continent suddenly producing, in sequence, "boomer" and "bust" generations significantly larger and smaller than the ones preceding them.

Modern governments and societies naturally approach many of their basic responsibilities with this enduring pattern in mind. Communities build schools and roads to last for decades, confident that there will be children to fill them and cars to ride on them twenty and forty years from now. Every modern country also has assumed roughly stable population growth in fashioning its basic approaches for promoting jobs and growth and providing social insurance.

This pattern has been broken on a historically unprecedented scale over the last two generations. First, for a variety of reasons in different societies, a boomer generation much larger and longer living than the one that produced it has appeared in country after country, followed by a baby bust generation significantly smaller, not only than the one that produced it but also, in terms of birthrates, smaller than any other in centuries. These boom and bust generations have arisen serially, virtually everywhere in the world except sub-Saharan Africa and parts of the Middle East.

These great demographic changes will help shape the economic and social development of every nation over the next ten to fifteen years. As baby busters age into full adulthood, their reduced numbers will slow the growth of their nations' labor forces or even cause some of them to contract. Almost inevitably, this will reduce their nations' saving and investment rates and, with them, their capacity to grow economically. As the boomers age into retirement over the same years, their disproportionate numbers will strain government finances around the world, producing wrenching political debates and divisions over raising taxes on financially strapped working families, cutting basic pension and health-care benefits for tens of millions of elderly people, and much more. Behind the dry population data, these demographic waves will vitally affect the comfort and security of most people's lives.

These unchangeable developments will especially affect people in Japan and much of Europe, because their baby bust generations are so small that their labor forces will begin to shrink, year after year. In good times, their economies will be able to expand by perhaps 1 to 2 percent a year, leaving tens of millions of working families with little prospect of improving their lot. The social and political effects may be just as great. Optimism and broad public support for consensus politics typically occur in places with growing economies. In many countries in Europe, those critical social attitudes could slowly leach away, leaving the disappointment and anger evident recently among economically marginalized ethnic minorities in France, Denmark, and Germany. And when these economies sputter into recession from time to time, as economies always do, the crosscurrents of globalization could drive down the incomes of the majority of their people with a power not seen since the early postwar years. Over the next ten to fifteen years, millions of average families in Lyons, Munich, Turin, Osaka, and a hundred other places, will likely find they can no longer afford the normal comforts of middle-class life. Millions of their children may question or reject their parents' values of hard work and playing by the rules, even more than many young people do as a matter of course. If history is any guide, these economic strains will also increase the incidence of broken marriages, depression, crime, and a host of other personal and social problems—and there will be little that anyone can do to materially change their circumstances.

The political and social stresses will be particularly severe in Italy, Germany, France, and other countries that will have to struggle with the soaring costs of their boomers retiring into the world's most generous public pension systems. The options for their leaders will come down to raising taxes on strapped and discontented working baby busters, disappointing huge numbers of elderly boomers by cutting their benefits, or trying to protect their political careers by borrowing tens of billions of euros and tens of trillions of yen—and further hamstringing their country's economic growth. Some European countries will escape the worst of it. Ireland will avoid most of these problems and stresses, based on the country's strong immigration, relatively high birthrates through its baby bust, comparatively small public pension benefits, and Europe's best-performing economy. The dissatisfaction of Britons also should be less severe than what awaits Germans, Italians, and French. Not only are Britain's demographics a little less skewed, but the legacy of deep cuts in state pensions put in place a generation ago will reduce the financing and political problems of its boomers' retirement. But those deep cuts in pension payments will cause their own problems, because as British retirement benefits slowly contract over the next ten to fifteen years, British leaders will have to either raise taxes and public debt to sweeten them—the likely course—or consign millions of elderly Britons to retire in poverty.

America's demographic challenges will be less dire than those facing most other advanced countries. America's relatively high birthrates produced a larger baby boom than those seen in Europe and Japan; but a less severe subsequent decline in birthrates and strong immigration also produced a relatively larger baby bust. New immigrants account for more than half of the growth of the U.S. labor force since 1990; and among all the major advanced countries, only the U.S. labor force will continue to grow at decent rates over the next ten to fifteen years. The political strains from the boomers' retirement will also be less severe in the United States, because there are more working-age people to pay for the pensions and because America's elderly claim smaller public pensions than most European seniors. Serious problems for the

U.S. social security system will eventually emerge, but not for a decade or more after 2020. As we will see later, the area where demographics is likely to disrupt U.S. prosperity and politics in coming years is health care for its elderly.

In the developing world, the outlook is most favorable for workers in the East Asian Tigers, where broad economic reforms and investments in education enabled their boom generations to build the richest economies their people have ever known. By contrast, across much of Latin America the boomer bulges could produce profound economic and political strains, because their governments largely failed to reform their economies in ways that could create jobs for their bumper generations and neglected the educational investments that could have made their boomer cohorts into productive assets.

China's population is aging faster than any other in the world, including a baby bust with the sharpest fall in birthrates anywhere; but some unique features still produce a largely rosy outlook. Well before 2020, a boomer generation larger than the combined populations of the United States and Europe, and the most extensive and relentless modernization program ever seen, will make China's economy the second-largest in the world. Moreover, much of the impact of China's severe baby bust should be offset by the country's huge productivity leaps, as its economic reforms move hundreds of millions of people from eighteenth-century agriculture to twenty-first-century industry and commerce. Unless China's social and political cohesion begins to unravel, its extraordinary economic progress will take much of the economic sting out of its plummeting birthrates. But China's demographics waves could still produce considerable social and political stresses. The leadership's decision to unravel most public pension and health-care guarantees in the 1980s and 1990s to help finance modernization will give rise to two great, open questions for the next generation of leaders—namely, will China's boomers accept impoverished retirements as their country grows richer, or will the country's busters accept slower growth to finance the gradual restoration of pensions and health care for their parents and grandparents? With the prospect of nearly 170 million elderly Chinese by 2020—larger than the current populations of all but four countries—China could be forced to rebuild some semblance of those social supports or face social unrest.

THE DEMOGRAPHIC WAVES IN DIFFERENT COUNTRIES

Until the early eighteenth century, the world's population had grown slowly but steadily for hundreds of years. It took the three centuries from 1500 to

1800 for the global population to double. But starting in the early 1800s, population growth accelerated, especially in the industrializing countries; and the number of people on the planet increased by more than 75 percent in one hundred years. In the typical pattern, societies that had been organized around large families and short life spans began to change as improvements in sanitation and public health enabled more children to survive their infancies and childhoods, and enabled more adults to live longer. Facing longer lives, more people began to demand education so they could earn more and save for large purchases, even possibly for retirement. Starting in Europe and United States, this pattern became an essential part of the transition to modernity almost everywhere. By the early twentieth century, life expectancy had risen sharply in the West; but because the changes took several generations to unfold, it didn't initially produce a boomer generation.

In the second half of the twentieth century, other developments piled on to produce those boomer generations not only in the West, but in most countries. In the industrialized nations, birthrates rose sharply in the early post-World War II period, in part an echo of a minor baby boom that had followed World War I and the 1918–1919 global influenza epidemic, and in part a result of decisions by millions of couples to delay starting their families during the Great Depression and World War II. Across much of the developing world, birthrates didn't rise after the Second World War; but like the West in the 1800s, advances in sanitation and public health sharply reduced death rates for infants and children. But in contrast to nineteenth-century Europe and America, where gradual improvements in sewage systems and public health produced gradual increases in population, many developing nations in the second half of the twentieth century replicated modern European and American sanitation and health systems in a decade or less, producing large and abrupt increases in the size of their first and second postwar generations.

For one hundred and fifty years before the late-twentieth-century baby boom, Western scientists and social critics worried about what would happen to a country if its population suddenly grew rapidly. From what we now know, most of them got it wrong. Beginning with the gloom-saying Thomas Malthus in the late eighteenth century, the pessimists figured that fast-expanding populations would send most places into downward spirals, as all the new bodies to feed, clothe, and keep warm consumed its resources or forced everyone to abandon their professions and produce those necessities. This dark view hugely underestimated the ingenuity of people in most societies, especially their ability to figure out new ways to expand harvests, fuel supplies, and most everything else their country might need—or in a global age, their ability to adapt successful strategies from other places. It would be unnecessary to mention the Malthusian view today, but for the fact that as recently as the 1970s, the United Nations Population Council, the U.S. National Academy of Sciences, and blue ribbon commissions in Europe and America solemnly predicted that baby booms would drastically slow growth almost everywhere.

An opposite and much sunnier view was first offered by Nobel Laureate Simon Kuznets and Danish economist Ester Boserup: An expanding population naturally leads to progress and growth, not only because pressures on resources focus ingenious engineers and scientists on the problem, but even more because a larger population means more smart and motivated people generally; and because larger societies can take advantage of more economies of scale. This view didn't turn out to be wrong—it just doesn't always happen. Baby booms hit most of Latin America in the 1960s and 1970s, which by this thinking should have brought economic booms there in the late 1980s and 1990s. In fact, Latin American economies contracted by almost 1 percent a year through the 1980s and grew just 1.6 percent a year in the 1990s. It takes more than a larger number of smart people to deal with the demands of a fast-growing population—they have to become well educated and trained, everybody has to have decent health care, and a country's leaders have to support the changes that can relieve the problems.

The view that makes the most sense of what has actually happened is that a country with a fast-growing population can progress and grow if it takes particular steps to help make it so. As the number of young people increases, for example, social investments in health and education, and a little later, job opportunities, too, have to increase at least as much. Families, businesses, and government have to save more of what they earn or collect in taxes, so they can pay for more teachers, schools, hospitals, clinics, and doctors. The country has to continue to save more as its large generations become old enough to work, so it can finance the public investments in roads and telecommunications, and the business investments in technologies, factories, and office buildings that all those new workers will need to become productive. In countries that favor markets like the United States, or to a lesser degree much of Europe, tax policies changed to encourage the higher savings needed to expand private investment. In places more skeptical of markets, including most of Asia in the 1960s and 1970s, when their baby boomers entered their workforces, much of the investment came in large government subsidies enhanced by industrial policies.

In a big world with countless national differences, some countries are always ahead of others in developing the technologies and business methods that can help turn a baby boom into widespread prosperity. Since most countries now share access to each other, the rest can try to get the companies that own the technologies and the people who know how to use them to invest and locate there—one of the core dynamics of modern globalization. If a developing society has educated its boomers, it can offer foreign companies large numbers of skilled workers willing to work for relatively low wages. It not only has to allow those foreign companies to invest there; it also must protect their property and ideas and create a stable economic and political environment that reduces the risk that the investments will disappear in a revolution, hyperinflation, or deadly epidemic.

Where all that has happened—most notably East Asia since the 1970s—baby booms helped produce economic booms. Where it didn't come to pass—across much of Latin America, parts of Eastern and Central Europe, the Soviet Union, and most of Africa—bumper generations have brought on the economic slowdowns or worse that Malthusians expected. Where it didn't happen at first and then did—most notably in China and India—long periods of economic stagnation have been followed by great spurts of growth.

When baby booms arrived in the world's more advanced countries, they already had the basic social provisions that could turn their bumper generations into national economic assets. In Japan, Germany, France, and the other major European societies, improvements in education and health care were part of the rebuilding process that followed the Second World War, and together with the maturing of their own boomer generations into the workforce, these reforms played a part in their "economic miracles" in the 1960s and 1970s. The United States also played a singular role as a tacit sponsor of boomer-generation prosperity in both other advanced economies and many developing nations. While Europe was rebuilding itself, America developed new technologies and business methods that helped jump-start much of both that rebuilding and the East Asian miracles. In the 1970s and 1980s, Japan and Germany also became sources of growth-enhancing technologies and business methods for the rest of the world—until the 1990s, when Japan's dogged industrial policies and Germany's lumbering welfare state caught up with them. By then, American companies and financial institutions were aggressively expanding their foreign direct and financial investments, making the United States the leading foreign agent of the recent turbocharged industrialization of China, Korea, and others.

In other ways as well, America has been the world's indispensable economy throughout the demographic waves of the last half-century. Successful developing economies with bulging baby booms typically have prodigious saving rates that can fund the additional public education, public health, and native businesses they need to make real progress. The flip side of high national saving is low national consumption, so all those workers in Asia and everywhere else have had to sell their goods to somebody else. From the 1950s, the United States progressively and sometimes unilaterally opened itself to foreign imports, providing the consumers that many other countries needed to keep their investment-led economies growing. As it actually happened, America opened itself to foreign imports not because its economists believe in it, but because U.S. presidents from Truman to Carter offered to provide a rich foreign market to any country that would align itself with the United States in the cold war. For nearly forty years, this geopolitical trade policy largely isolated the Soviet bloc and its allies India, China, Cuba, and North Korea from the wealth and technological advances of the West, denying them critical resources for their boom generations.

While many thinkers have pondered the impact of a fast-growing population, far fewer thought much about why birthrates might suddenly slow down sharply—until it began to happen around the world. It turns out that many of the same factors that drove up the size of the first post–World War II generation drove down the size of the second and third. As infant and childhood mortality rates fell in the developing nations of Asia and the Soviet Union, tens of millions of couples realized they could achieve the family size they wanted by having fewer children—and birthrates began to fall. In 1950, a typical Thai or Korean woman had six children; today, the average number is two. Decent sanitation and public health care not only reduced infant and childhood disease and death, they also extended the average life span of adults. In Asia, life expectancy rose from forty-three years in 1950 to seventy-two years today. In Western Europe, Japan, and the United States, where those basic improvements were ancient history, new medical technologies, income gains, and lifestyle changes also reduced death rates in middle age and early retirement years. From 1960 to 2000, life expectancy moved up seven years in the United States, nine years in France, and thirteen years in Japan; and much of those increases came from the longer lives of those aged fifty and older. Nor is this process over. As pediatrics and childhood immunizations developed fully in the first two decades of the baby boom—bringing down childhood mortality—the boomers' retirement years could see breakthroughs in gerontology, as well as in the conquest of a number of adult diseases. Many scientists believe that within the next twenty years, current research lines in biotechnology and genomics will lengthen average life spans by another decade.

Longer life spans have contributed to the slowdown in birth rates. As most people became accustomed to the likelihood of living into old age, their economic time horizons lengthened. Facing a longer future, people around the world increased their saving and spending on education—and higher savings and more education, along with technological progress, promoted greater business investment. The end result was greater economic opportunities, especially for much of the world's women to work outside their homes. From 1950 to 2000, the share of adult American women working jumped from 37 to 77 percent; and the increases in the number of women working in European countries were nearly comparable. When women have jobs to go to they tend to have fewer children—especially when inexpensive birth control is also available. The first practical oral contraceptive was developed in the 1950s by Dr. Gregory Pincus; and by the 1970s, birth control pills, IUDs, and early term abortions all were widely available and used across Europe and the United States. And with concerns in the 1970s about the "limits to growth" in the absence of population controls, the United Nations spread the gospel and technologies of family planning to much of the rest of the world.

Just as the baby boom wave moved from East Asia and Japan to Western Europe and the United States, and then to China, Latin America, and Eastern and Central Europe, the subsequent baby busts followed the same path. Today, the decline in birthrates is well underway in China, South Central Asia, and Latin America, followed more recently by India and Bangladesh. North Africa and the Middle East achieved longer life spans more recently, and if the rest of the world's experience is a decent guide, their birthrates will begin to fall by the end of this decade.

Two places stand apart from these historic demographic shifts. A modern baby boom has never developed in sub-Saharan Africa, because improvements in basic health conditions and education haven't happened on a wide scale, and famines, wars, and epidemics of AIDS and other deadly diseases continue to keep life expectancy short. The other special case is Russia. The Soviet Union had a fairly typical postwar baby boom, followed by a normal baby bust in the 1970s and early 1980s. A brief increase in birthrates followed in the late 1980s, but the birthrate plummeted again, by half, in the 1990s. In the last decade, the Russian population has fallen by 3 million and is expected to fall by another 3 million in the next decade.[1] Moreover, life expectancy among Russian men began to decline in the 1960s, reaching just fifty-nine years by 2003, lower than male life expectancy in countries such as Egypt, Guatemala, and Vietnam. These trends accelerated in the 1990s, including sharp increases in infant and childhood death rates, as public health spending plummeted to the levels of the world's poorest countries: By 2003, per capita spending on health in Russia was just 3 percent of per capital health spending in the United States.[2] The combination of low birthrates and rising death rates has made Russia the first instance in modern times of a developed country where life expectancy and total population are both falling sharply.

While this boom-and-bust pattern is evident nearly worldwide, many countries have special conditions and attitudes that create notable variations in the intensity of their demographic waves. Those variations will have large consequences for the fates of societies over the next ten to fifteen years.

First, through both baby boom and baby bust, Americans have produced comparatively more children than most Europeans or Japanese. Throughout the boom decades of the 1950s and 1960s, American birthrates were almost 50 percent higher than the Swedes, roughly 30 percent higher than Britons or Germans, some 20 percent higher than the French or Italians, and 16 percent higher than the Japanese. In the bust years of the 1970s and 1980s, birthrates fell nearly as much in America as across Europe—but from its much higher boom rates. (Sweden is an exception to this pattern, starting with Europe's lowest birthrates during the boom and then declining less during its baby bust.) The result is that American birthrates during the baby bust years were still 40 percent higher than the Germans, nearly 25 percent higher than the Italians or Swedes, 15 percent higher than the British, and 6 to 7 percent higher than the French or Japanese. On top of consistently higher birthrates, the United States also has consistently permitted much greater immigration. In recent years, the net migration rate to America has been almost ten times higher than Japan, four times higher than France, double that of Italy and Britain, and 50 percent higher than Germany.

Beyond the numbers, immigration has had much different effects in Europe than in the United States. A significant share of America's recent immigrants are people from quite far away—nearly one-third come from Asia or Europe—and they often bring high levels of education and training. The other two-thirds come from neighboring countries, mainly Mexico and Central America, and generally fill low-level jobs in agriculture, hotels, restaurants, and personal services that many Americans disdain. By contrast, nearly 90 percent of immigrants in Europe come from Europe, many from Islamic countries, and compete for jobs that less-educated young Europeans are happy to take for all the benefits and guarantees they provide. Across Europe, immigration is a deeply passionate issue mixed up with religion and fears of terrorism, and its actual numbers contribute little to most European economies. Immigration is certainly controversial in the United States, too, but the question for most Americans is how best to integrate the millions of foreign-born people living here now—not how best to send them back to where they came from.

The next generation will not bail out the baby busters in Japan and Europe, because their low birthrates already have entered a second generation. In some cases, birthrates are still declining, reducing the numbers of future mothers as well as future children. At current trends, the United Nations projects that Western Europe will have some 2 million fewer children under age fourteen in 2020 than in 2005—a 7 percent decline. There are differences among these nations, but the bottom line is similar. France and the United Kingdom are expected to have 3 to 4 percent fewer children in 2020 than in 2005, while the number of German and Japanese children will slump by about 9 percent. The worst case is Italy, where the number of children will fall by more than 1 million, or over 12 percent, between now and 2020. An exception to this second-generation pattern is Sweden, which should have roughly the same number of children in 2020 as today.

The United States, however, remains unusually fecund. Here, the baby bust has begun to give way to an echo of the baby boom, due in large part to the high birthrates of millions of first-generation immigrants. In 2020, America should have some 5 million more children than it does today, an increase of more than 8 percent. But it is Ireland that stands out most among the advanced countries. Irish birthrates during the baby boom years nearly equaled America's, and then declined much less in the 1970s and 1980s. Moreover, Ireland has attracted immigrants in recent years at a rate 40 percent greater than even the United States. Many are Irish-born, returning now that the country's per capita income is 12 percent higher than France or Germany and 23 percent higher than the EU average. Many others are educated Americans or Britons, moving there to work for foreign companies or start their own businesses. Put it all together, and Ireland's strong immigration and birthrates are projected to produce almost 12 percent more children in 2020 than today.

China represents a national extenuation of many of these patterns. Nowhere was the baby boom larger than in China, because nowhere else was the shift more sudden from a backward country of large families with high infant and childhood death rates to a modernizing nation with decent sanitation, public health care, and the fast-rising life expectancy and childhood survival rates that accompany them. Through the 1950s and 1960s, as China brought basic medical care and sanitation to its agricultural communes and fast-growing cities, its birthrate remained premodern—more than double those of the United States and Europe during their baby booms. The result was a Chinese boomer generation born from the mid-1950s to the mid-1970s that was more than twice the size, relative to total population, of any advanced country.

Moreover, nowhere else did childbearing fall more sharply than in China. Birthrates began to drop in the late 1970s and 1980s in the pattern typical of societies once parents recognize that they can achieve the family size they want with fewer births. China's birthrate fell initially by half, to the rates of the U.S. baby boom of the 1950s and 1960s. By the 1990s, China achieved genuine baby-bust rates, as the economic and social reforms following the Cultural Revolution and Mao's death brought wide-scale education for women, widespread birth control, and new job prospects for women unimaginable a decade earlier. With all these factors in play, official policies penalizing Chinese women for having more than one child probably played a less central role than is usually assumed in the West, since it largely reinforced the choices that tens of millions of Chinese families were making already. By the last half-decade of the twentieth century, China's birthrate fell below that of the United States and Ireland; and today the fertility rate of Chinese women (the average number of children born to a woman over her lifetime) is 1.7. That's lower than America, Japan, and most of Europe; and in Beijing and Shanghai, the fertility rate is barely 1.0. The decline in Chinese births is so sharp that according to UN projections, the country is expected to have 55 million fewer children in 2020 than in 2005, or a decline of more than 13 percent.

China also has undergone far-reaching changes at the other end of the age spectrum. Since 1970, life expectancy has shot up from forty-one years to seventy-two years, so it's now just six years less than the United States and eight years less than Europe. In fact, China will age faster over the next fifteen years, and beyond, than any other country in the world. However, these huge demographic changes will likely have different effects on China's economic and political path—largely more positive ones—than they would in a more advanced country. China's double-sized baby boom so far has well suited the country's turbocharged modernization policies, providing tens of millions of reasonably skilled workers for tens of thousands of new domestic and foreign-owned enterprises. As China becomes a fully industrialized nation with modern agriculture over the next ten to fifteen years, its baby bust should be sufficient to meet the more slow-growing labor needs of a more mature economy. To avert the daunting social and economic problems that would accompany such seismic demographic shifts in an advanced country, however, China will need the continued forbearance of hundreds of millions of Chinese. By 2020, China will have almost 170 million elderly people—nearly 70 percent more than today—and every year from now until then, 10 to 20 million Chinese boomers and busters will lose their jobs to the forces of economic modernization. Yet, no more than 20 percent of Chinese today are eligible for public pensions or health care, and unemployment benefits are unknown. To manage the demographic waves that will wash over China for the next ten to fifteen years and well beyond, China's leaders will have to phase in these social supports at a rate just fast enough to preserve the

regime's popular support, but gradually enough to maintain their investments to modernize the economy.

THE PRICE OF THE BABY BUSTS: THE END OF STRONG GROWTH IN EUROPE AND JAPAN

Nothing that any government could do over the next ten to fifteen years
will materially affect these numbers or their effects on people's lives. The economic crux of the demographic issue, especially for advanced countries, is
how fast their workforces will expand—or whether they will grow at all. In
normal times, most nations' labor forces expand steadily by around 1 percent
a year, so every year there are more people working to produce the country's
goods and services. The demographic shifts of this period change that basic
factor, boosting growth rates as boomers take jobs in the economy, and lowering the rate when baby busters replace the boomers. A country can squander
the potential extra growth from a fast-expanding labor force by failing to
educate its new workers or to provide the conditions and incentives that can
raise their average productivity. When a country's labor force is contracting,
its economy operates under a serious disadvantage; and greater investments in
education, training, and the technologies and business methods that can raise
productivity become virtually the only way to sustain sound growth.

Focusing on this central fact, it becomes clear that Japan's position between now and 2020 is the most dire: By 2020, Japan will have almost 9 million fewer people ages twenty to sixty than it does today, a contraction of
nearly 1 percent a year in the number of people available to generate most of
their nation's wealth. The only cushion is that older Japanese are more prone
to keep working for a few years past age sixty than older people in other advanced countries. Since how fast a country grows is basically the product of
how many new workers it adds and how fast the productivity of its workers
increases, normal productivity increases of 1.5 to 2.5 percent a year over the
next decade will leave Japan perpetually skirting the edge of recession. In
fact, productivity growth in Japan hovered around 1 percent a year through
the 1990s, although it has accelerated in the last few years.

Japan isn't alone in facing this predicament. The number of Italians of
working age will fall by 9 percent between now and 2020, while the size of
the potential labor forces of Germany and France will decline by nearly 4 to 5
percent. The outlook for Sweden and Britain is a little better: The Swedish labor force will stagnate over the next decade or so, while Britain's pool of potential workers will edge up on average by about one-quarter of 1 percent a
year over the next twelve years. A silver lining for Germany, France, and Italy

is that their persistent high jobless rates should come down in the next decade as busters replace boomers at the center of their workforces. But even with lower unemployment, slower growth will mean less economic progress for average families across much of Europe over the next ten to fifteen years.

One reason for this unhappy prospect, and one tied directly to demographic changes, is that working-age people account for most of every country's personal saving; and how much a nation saves affects the financial resources the country has to invest. There's nothing obscure about how aging affects saving and investment—it's simply that people's saving behavior is tied closely to what's happening in the rest of their lives. Younger people spend more than they earn, setting up their households and starting their families, and retired people spend down what they saved in order to maintain their health and lifestyles, while earning little or nothing new (beyond their state pensions). In every country, people in their prime earning years from their midthirties to their midfifties do most of their society's personal saving.

Over the next fifteen years, with tens of millions of Europeans and Japanese aging from those prime earning years into retirement, and fewer younger people moving into their prime-earning years, their nations will have fewer resources to expand their wealth. Japan already demonstrates what these demographic waves can do to a country's savings rate: From 1975 to 2005, as the number of elderly Japanese began to increase sharply and while the number of Japanese in their prime earning years slowed and then began to fall, the country's savings rate plummeted from 25 percent to less than 5 percent—and a recent study by McKinsey and Company projects that it will hit zero by 2020 or soon thereafter.[3] To be sure, demographics are not the only reason for the virtual collapse of Japanese private savings. With the dawn of contemporary globalization, Japanese economic policies that worked for decades became dysfunctional and finally destructive, pumping up and then puncturing financial bubbles in real estate and stocks; and the inept government policies that followed brought Japanese growth to a virtual standstill for more than a decade starting in the early 1990s. One result was a persistent, mild deflation that drove down how much people could earn from savings—so they saved even less. Still, the basic demographic changes not only account directly for at least half of the decline in Japanese savings, they also contribute to the country's low growth rates—so that Japanese families who want to maintain their lifestyle—and who doesn't?—have less left over to save.

Between now and 2020, personal savings rates will fall in most countries, as the share of their populations comprised of older people rises while the share in their thirties and forties falls. Unless there are more sustained productivity breakthroughs—as only America, Ireland, and a few other countries have seen in the last ten to fifteen years—lower rates of personal savings will cut sharply into the wealth that average families will be able to accumulate. In Japan, where the median age will hit fifty by 2020, and where increases in household wealth depend mainly on private savings, experts expect average household wealth to actually decline. In Germany, France, and Italy, the assets of most households will merely stall. In many of the world's richest countries, the lives of most people will stop getting better, and many could find themselves closer to poverty than they ever imagined they might be.

For a nation, a normal aging pattern in which the number of people entering the workforce and moving into their prime earning and saving years increases a little faster than the number of people leaving the workforce and moving into their low-savings retirement years, is akin to picking up a pair of kings in poker. It's not enough to win, but it keeps a country in the economic game. To win a big pot, a society needs to draw additional good cards—for example, a vibrant scientific establishment that develops great new technologies, an entrepreneurial culture that encourages people to commercialize those technologies, and a financial sector that channels resources to promising entrepreneurs. A country with all that can make up for low personal savings by attracting the savings from other societies—as the United States has in this period. Or a country could have thousands of businesses producing goods and services that everyone in the world wants—as do America and Ireland—generating high profits that also can provide financial resources to offset low personal savings. With either of these conditions, a nation can stay in the game even with bad demographics—and if its demographics are decent, it has a real shot at claiming a big pot.

That's one reason why the United States could grow so much richer over the last twenty years, even as its personal savings rate fell sharply (and not just for demographic reasons). Over the past generation, the financial wealth of an average American household grew nearly 4 percent a year, even as private savings rates fell, mainly through the sharply rising value of the stocks and savings accounts they purchased with their fairly meager annual savings. Yet, even with better demographic and growth prospects than most other advanced countries, the aging of America over the next twenty years could still cut by half the rate at which an average family increases its assets. Even in the United States, continuing to do what it has been doing won't be enough to keep large economic gains growing for most people—that would take another leap in technological progress that could lift productivity again on a long-term basis.

Over the next ten to fifteen years, Germany, France, or Japan could do what the United States did over the last generation. Any of them could offset some of the economic drag created by their rapid aging by becoming hot-houses for innovation and by positioning their businesses to be global leaders. As we will see later, doing so could involve changes so deep and far reaching that they might no longer recognize themselves—and changes that seem well beyond the current reformist capacities of their political parties. And the rapid aging of their populations will make it harder still for politicians to sell difficult reforms, as their retiring boomers claim more of their nations' resources for pensions and health care.

In recent years, a number of countries have tried to soften the impact of their demographic crunch in other ways—for example, by trying to convince more of their older people to keep on working. With no popular demand to work longer, this approach has turned out to be well beyond the political capacity of governments everywhere in Europe. Their older people don't work not only because they generally don't want to, but also because in most cases their state pensions are so generous that they don't have to—and their politicians aren't in the business of cutting those benefits so sharply that seniors would have to keep working.

For instance, Germany has tried to keep more of its older workers on the job; but employing jobless young people has higher political priority, especially since older people aren't scrambling to work while young people are. So, while Berlin offers German companies a special subsidy to keep people over age fifty-five working at least part-time, the program also requires that the companies hire jobless young people to make up the difference between the oldsters' part-time work and full-time. In an economy with structural unemployment of more than 10 percent, the two provisions cancel each other out. A few years back, the German government tried a different tack by cutting back the number of months it will provide unemployment benefits. But then it lost its nerve and added a two-year supplemental benefit that most people can claim. The government also tried reducing the normal cuts in income subsidies and free health-care benefits for lower-wage people who go to work, but did it so clumsily that it created an incentive to work only part-time in order to keep more of their benefits. In the end, none of these policies could overcome the laws that allow Germans to work half-time from age fifty-five on without reducing their state pensions, and which encourage older wives to stay at home by heavily taxing second earners and providing free health coverage to spouses only if they don't work. The end result of all these policies, along with most people's preference for working less, is that nearly 60 percent of Germans ages fifty-five to sixty-four do not work at all.

The sharp baby busts create an economic bind from which most of Europe and Japan cannot escape for at least a generation, unless they're willing to

accept larger and more difficult changes that could help create a lot more attractive jobs and make unemployment, part-time work, and early retirement
less appealing. So far, the French, Germans, and Italians have been unwilling
to do anything of the kind. They're all proud of the economic security their
social welfare systems provide—and why shouldn't they be?—but most people who can count on generous government support when they're out of work
lose the urgency to find new jobs. So, while unemployment payments in the
United States last six months—and no months at all in China—Europeans
who lose their jobs collect benefits for eighteen to twenty-four months, and
their benefits are a lot more generous than Americans can claim. No European or Japanese prime minister has been prepared to make the lives of its jobless people significantly less secure.

It's hard to blame elected leaders for lacking the conviction to reform this
part of their welfare states. Politicians rarely ask voters to make sacrifices—at
least when they're not fighting a major war—unless they can plausibly promise that the sacrifice will help make the same voters better off in fairly short
order. Here, that would mean not only making unemployment or early retirement harder to bear, but also undertaking even more difficult and far-
reaching changes that might encourage European and Japanese businesses to
create new, well-paying jobs for the people forced to give up some of their
generous supports. And they can't promise that without deconstructing much
of the social and economic arrangements that their societies have built up
over the last half century.

Instead, as their labor forces and saving rates decline, Europe and Japan
seem set on handicapping their future prosperity with ostensible acts of social
kindness. As we will see later, complex systems of regulations and trade barriers in Japan and Europe create nearly as much artificial economic security
for those who have jobs as the government provides to those without them.
The price of all the regulations and barriers is less competition, so European
and Japanese businesses feel less pressure to come up with new products and
more efficient ways of conducting their business—and, directly and indirectly, that's what drives modern economies to create new jobs.

In many cases, the regulations directly undermine innovation and job creation, which could soften the economic consequences of their demographics. In
France and Italy, a business that wants to reorganize or change the way it operates cannot legally reassign a worker without "cause"—and taking advantage of
new technologies to become more efficient is not sufficient reason. This creates
a dogged form of job security for those who have jobs—but it also discourages
businesses from creating new jobs. And it makes little economic sense in an
open world in which French and Italian companies compete in their own markets

and around the world with American and Asian rivals that can hire, fire, and reassign at will. Such regulations are slowly draining the competitiveness of thousands of European businesses—and that's not all of it. The cradle-to-grave economic security treasured by Europeans costs tens of billions of euros, most of it financed through payroll taxes much higher than those that American and Asian companies live with. Even when a company wants to take steps that would generate new jobs, the taxes on new workers discourage it. German payroll taxes account for 25 percent of labor costs, in Italy they make up 32 percent of labor costs, and in France that level is 38 percent. Even in Japan, payroll taxes account for over 21 percent of labor costs. Only in the United States and Britain, at about 14 percent and 17 percent respectively, are payroll taxes only a modest barrier to creating jobs.

It's no accident that for two decades American businesses have created jobs at three, four, or even ten times the rate of the major economies of Europe and Japan. It's also not serendipity that unemployment in Europe has been about twice U.S. levels for just as long, or that new businesses are created there at much lower rates than in the United States. It's a direct consequence of European social contracts that promised solid security—and delivered it at tolerable economic costs—when baby boomers made up their workforces and foreign markets and foreign rivals were much less important. With globalization and much-less-favorable demographics, the European social project is suffocating the prosperity—and eventually the social allegiance—of those it's designed to protect. Over the next ten to fifteen years, the baby busts of Europe and Japan will make the economic burdens of their social welfare systems irreconcilable with healthy economic growth.

The outlook is much more positive in China, despite demographic changes that dwarf those of the world's more advanced economies. China's huge baby boom—twice the relative size of any advanced nation's—so far has well-suited the country's wide-ranging modernization programs, providing tens of millions of workers for the tens of thousands of new domestic and foreign-owned businesses. As the Chinese economy matures over the next ten to fifteen years, its baby bust generation should be sufficient to meet the more slow-growing labor needs of an industrialized society.

On the crucial question of how long China has before its falling birthrates begin to eat away at the size of its working-age population, China falls in between America (and Ireland) and the rest of the advanced world. The plummeting fertility rates that produced China's seismic baby bust came about a decade or two later than in the West. China's labor force will grow by nearly 90 million between now and 2015, when the effects of its low fertility rates begin to appear, and will begin to contract only around 2020. Looking further ahead, there's no turnaround in sight: Over the next fifteen years, the number of Chinese children is projected to decline by 7 percent and at an accelerating pace.

China has another trump card to help it weather its demographic waves. Even when its labor force begins to shrink, China's development agenda and unique place in the global economy should protect its ability to generate wealth and income for some time. The shift of tens of millions of China's workers from agriculture and the old state-owned enterprises will produce labor surpluses well beyond 2020. These surpluses may be so great that they could lead to, in the words of one of China's leading economists, Hu Angang, "an unemployment war, with people fighting for jobs that don't exist."[4] That competition for jobs will also keep Chinese wages from rising as rapidly as the country's production for a long time. And the reemployment of tens of millions of Chinese in private businesses, armed with western production technologies, should continue to drive up the country's productivity and growth.

China's demographic challenge over the next ten to fifteen years will be more political than economic. Today, only a small fraction of China's hundreds of millions of dislocated workers and elderly people are eligible for any government benefits or supports. Successfully managing the stresses that accompany such vast demographic shifts will require their continued forbearance, even as they watch many of their countrymen grow steadily more prosperous.

THE PRICE OF BABY BOOMS: THE FINANCING CRISIS FACING GOVERNMENT PENSIONS

These demographic waves will create political problems in advanced countries that go beyond people's predictable reactions to disappointing growth. As baby boomers begin to retire by the millions, the politics in many countries will be shaped by acrimonious debates over the terms and solvency of state pension systems. Political battles are almost inevitable in many places, because the relatively smaller cohorts of baby busters will have to come up with most of the trillions of dollars, euros, and yen required to keep benefits coming for the fast-rising numbers of retiring boomers.

The most direct way to grasp this dilemma is to face up to how many potential workers each country will have over the next ten to fifteen years, compared to how many elderly people they will have to help support. Even today, Germany, Italy, and Japan look demographically a lot like Florida (without all the sunshine), where millions of Americans retire. All three of these countries already have just a little over two people ages twenty to sixty for every one person over age sixty. Sweden, France, and Britain are in a little better shape with between two and one-quarter and two and one-half working-age men and women for every older Swede, French, or British person.

These numbers are worse than those from many official sources, because the analyses produced regularly by the European Union and by the Japanese and European governments define "working age" as fifteen to sixty-five, even though relatively few people in advanced countries enter the workforce before age twenty or continue working much after age sixty. By including teenagers and younger seniors in the workforce and excluding younger retirees from the ranks of the elderly, they can claim that the seismic demographic changes shaking their societies somehow won't require hard policy changes—until they actually do.

The retirement systems of a few countries don't face terribly difficult problems, at least in the near future. Ireland sits at the top of the demographic heap in this regard, with nearly four working-age Irish today for every Irish person age sixty and over. The crunch is also less severe in the United States, with about three and one-third workers today for every older person. The fact that working-age people still dominate the American and Irish populations is one of the reasons why U.S. growth over the last decade has been double or more than that of Japan and most of Europe, and why Ireland has grown even faster than the United States.

Yet everywhere, the rapid aging of the population already has driven major changes in retirement programs. In the last fifty years, as the number of Americans of working age relative to the number of elderly fell from more than 16 to 1 to 3.3 to 1, the payroll tax rate that finances U.S. retirement benefits increased more than fourfold. The taxes that finance most other countries' state pensions—value-added taxes (VATs) and often a piece of income taxes, as well as payroll taxes—have risen as much as or more than in America. Many Americans grumble today about paying 12.4 percent of their paychecks to finance other people's social security, but they get off lightly: The cut from the paychecks of Italians is almost 33 percent, the French pay more than 24 percent of their wages, and the Swedes, Britons, and Germans pay between 19 and 22 percent of their paychecks. One result is that most working people in every advanced country now pay more taxes to finance the benefits for their elderly compatriots (including their health care) than they do in direct income taxes.

With such high payroll taxes already in place, it will be very difficult to raise them even more as boomers actually retire. Apart from Japan, the big payroll tax hikes occurred mainly from the mid-1950s to the mid-1980s. Since then, the popular tax revolt that began in America in the late 1970s spread to other advanced countries. As unemployment rose and growth slowed in much of Europe in the 1980s and 1990s, resistance to higher taxes increased. Now, whether the total tax burden of payroll, income, and value-added taxes is 35, 40, or 50 percent, people almost everywhere have said, Enough. American, European, and Japanese economists have long debated the level at which taxes begin to stifle business's willingness to invest, people's readiness to work longer, and their economies' capacity to grow at healthy rates. The answer depends on many national factors—factors that include the tax burden, but are far from limited to that. Whatever the economics, there's little question that over the next ten to fifteen years, as the number of older people collecting public pensions (and health-care benefits) rises faster than the number of working people, political leaders that want to stay in power will find it difficult to raise taxes much more.

This resistance will create thorny political problems across Europe, Japan, and the United States, as the numbers worsen and their consequences become difficult to manage. By 2020 there will be fewer than one and one-half Japanese ages twenty to sixty for every compatriot over age sixty, and just over 1.6 working-age Italians for every elderly person there. The numbers look grim as well for Germany, France, and Sweden, with between 1.8 and 1.9 people of working age for every elderly person in 2020. The Anglo-Saxon countries have a little more breathing space. In 2020, Britain will have about 2.1 potential workers for every elderly person, and there should still be almost 2.4 Americans of working age for every senior. Ireland will remain the major outlier among the more advanced nations, with about 2.9 potential workers for every retiree in 2020.

There's virtually nothing any government can do between now and then that could materially change how many older people it will have to help support and how many working-age people it will have to pay for it. Unless countries figure out how to make their economies grow a lot faster, the policy options flowing from their hard demographics will not be of the win-win sort that politicians look for. Instead, nearly every advanced nation will have to trade off changes that will either make the retirement of most of their boomers a little more nasty and brutish than they expect today, somehow raise taxes on working people—including many whose incomes will be stagnating or falling—or run large, permanent budget deficits that will undermine future growth.

While the choices will be essentially the same everywhere, the urgency and severity of the problem will also depend on the particular features of each country's state pension system. The French, German, and Italian systems are so generous today, it's certain that they cannot be maintained through even the first decade of their boomer generation's retirement. In this respect, the English and Irish systems look more sustainable, mainly because they provide much smaller benefits that, on their present path, will consign large numbers of their boomers to retire into near poverty. On this matter, the United States and Japan fall in between, with public retirement systems much less generous than the continental Europeans, but more so than the British or Irish.

The simplest way to gauge the generosity of a country's retirement program is by what's called the "replacement rate," which measures a person's monthly retirement benefits against his or her average, monthly income over his or her entire working life. Actuaries in different countries calculate these rates in a number of ways, but the most useful is the "net replacement rate." That tells you how much a person's monthly benefits, after he or she has paid any taxes owed on it, will replace his or her average monthly preretirement take-home pay, again after taxes. This is the best measure of both how well off retiring boomers around the world can expect to be in another decade and the financing burden each country will face at that time.

It should be clear to everyone but a politician that once boomers begin to retire, the benefits now promised across most of continental Europe will be simply too large to survive the collision of sharply rising numbers of retirees and the falling numbers of workers able to pay for them. Italians face the worst crisis, with a system that after taxes replaces almost 90 percent of an average Italian's lifetime preretirement take-home monthly wages. The Germans, French, and Swedes are a little less generous, promising monthly benefits that after taxes equal between 68 percent and 72 percent of an average worker's preretirement, monthly take-home pay. This makes no sense, unless you're caught in a time warp from the 1960s, and still believe that the baby boom and unusually fast growth of that time will go on forever. A more secure sense of reality is evident outside continental Europe. The after-tax replacement rate for an average Japanese is 59 percent, and in the United States and Britain, it's roughly 50 percent—although benefits levels in Britain and Japan are set to decline. Finally, Ireland provides the smallest public pensions of the group, replacing just 37 percent of an Irish person's average preretirement, lifetime monthly take-home pay.

One safety valve that could let some of the steam out of this crunch would be greater willingness among people age sixty and older to keep working, along with jobs to employ them. Once again, the differences across countries will make the boomers' retirement more difficult to manage in most of Europe than in the United States or Japan. Unsurprisingly, people who receive retirement checks that pay a good portion of what they made working are not likely to stay on the job. In Italy, less than 40 percent of those ages fifty-five to fifty-nine are still working, compared to 81 percent of Swedes, 76 percent of Japanese, 69 percent of Americans, and between 60 and 65 percent of even the French and Germans of the same age. By their early sixties, however, more than 80 percent of not only Italians, but also Germans and French no longer work—compared to about half of Swedes, Japanese, and Americans the same age. By ages sixty-five to sixty-nine, the problem of the continental Europeans is stark: Only between 3 and 6 percent of Germans, Italians, and French people are still working, compared to roughly one-quarter of Americans and Britons and more than 37 percent of Japanese—all with much-less-generous retirement benefits. It's unsurprising that this problem is most severe in Europe, because governments grappling with high structural unemployment are unlikely to take meaningful steps to keep their oldest workers in jobs that might go to younger, jobless people.

It should be easier to pare back a government retirement system when large numbers of its beneficiaries also hold private pensions. But here, too, when people expect to receive government benefits equal to a good portion of what they earned working, they have little reason to build up private retirement savings. French, German, and Italian families save at higher rates than American, Britons, or Japanese, but not for retirement. The total assets of all private pensions in Germany and Italy are equal to about just 18 to 22 percent of their GDP, respectively—one-fourth the level of Britain and America—and in France, private pensions equal just 7 percent of the country's GDP. Even for the 20 to 25 percent of Europeans with a private pension, it's not large enough to finance more than a year or so of taking it easy—averaging about $25,000 for every German with a pension, $30,000 for an Italian with one, barely $10,000 for every French person holding a pension—and zero for everyone else. The reasonable path when the crunch hits these countries in a few years would be to cut back their generous benefits—and the longer they put off cuts, the more drastic they would have to be to keep their systems solvent. But with most Europeans saving little for their own retirements, the most persuasive politicians won't be able to convince many people that the reforms their systems need won't leave them worse off.

Nowhere are private pensions universal or even nearly so. In the United States and Britain, where public pensions are relatively modest, private pension and retirement accounts hold assets almost as large as each nation's total gross national product. But even there, only one-third to one-half of all working and retired people have them—and they're almost entirely from the better-off half. The large pool of private retirement saving in Britain and America will help shape the debate when the two countries decide to make changes to keep their systems afloat, but it's not enough to produce easy answers. The most obvious change would be cuts in benefits for higher-income people with the private resources to help cushion the impact. But since almost everywhere, older people and more affluent people vote in greater numbers than everyone else, even that change will be politically treacherous.

Major pension reforms can be done, as Margaret Thatcher demonstrated in the 1980s. Under her plan, Britain's public pension now provides a bare-bones payment that's not intended to replace a significant share of anyone's preretirement earnings. To fill the gap, British workers also pay into supplementary personal pensions managed by their employers, financial institutions, or the government itself, and receive tax benefits for contributions to additional private pensions. This approach of pairing a small public payment and a small personal pension with tax benefits to encourage more private retirement savings has become an Anglo-Saxon model; and over the same period, the United States, Ireland, and Australia all expanded tax incentives for private retirement accounts.

Even this approach cannot sidestep a country's underlying demographics. To begin, it can't really relieve the financing crunch, since tax benefits for private pensions cost Britain nearly half as much as its basic public pension payments, and similar tax breaks in America, Ireland, and Australia cost the equivalent of 60 percent or more of their annual spending on public pensions. Moreover, the English approach is gradually creating the least-supportive public pension system in the advanced world. By 2020, an average British retiree will collect monthly checks from the basic benefit and mandatory personal pension that will replace barely one-third of his or her average lifetime monthly earnings. To avoid relegating millions of English seniors to poverty, a future British government will probably have to increase the state pension. That certainly wasn't part of the Thatcher plan. She was convinced, largely by conservative American economists, that the prospect of meager public pensions would drive Britons to build up fat private savings to make up the difference. But everywhere, most people tend to be myopic about saving for anything far off, since most people's drive to live as well as they can afford is more powerful than the discipline required to cut back in order to live better decades later. Today, only about one-third of British people pay into private pension plans. With private pension assets equal to about 80 percent of Britain's GDP, that third should be in reasonably good shape for retirement. But the other two-thirds of British retirees will be left with little more than their near-poverty-level public checks.

In the end, the British reforms have traded the direct financing crisis now .

facing other European societies for a looming social crisis of consigning most British boomers to nearly impoverished retirements. With that social crisis still a good decade off, most British politicians have embraced the same politics of denial as their continental colleagues. In the 2005 elections, the Labour and Conservative parties both offered marginal public-pension proposals of little practical consequence. Only the small Liberal Democratic Party called for serious steps, including larger benefits and mandatory participation in private pensions. Perhaps the Liberal Democrats chanced it, because they knew they'd never have to actually finance the benefit increases or face the opposition of English business to mandatory private plans. In any case, very soon it will be too late for most British boomers to build up large enough private pensions to avoid hard times in retirement.

The outlook for change is still better in Britain than in the rest of Europe, because it's always easier politically to expand benefits than to cut them. With Thatcher's reforms failing slowly, public pressure to sweeten the British state pension payment will likely become irresistible over the next decade. Even if British workers have to swallow higher taxes or large deficits to finance it, millions of grateful elderly and near-elderly voters should take the political sting out of the changes. The UK also has additional leeway to relieve some of the political stress, because compared to most of Europe, its economy is in better shape for the long run. With growth a little stronger and unemployment consistently lower than in France, Germany, or Italy, Britain will have relatively more workers earning wages to finance higher benefits, and those workers' wages should rise faster and generate relatively more revenues for those higher payments.

In the end, future German, French, and Italian politicians will not be able to escape the basic facts that their elderly comprise a larger slice of their populations and have been promised public pensions that are relatively much larger than Americans or Britons. By 2020, the European Union estimates that public pension benefits will claim 15 percent of the GDPs of France and Italy, and 12.6 percent of Germany's. For a sense of how implausible that burden will be to sustain, it's equivalent to the United States earmarking two-thirds to three-fourths of its entire current federal budget to retirement benefits. And the EU projections understate the economic cost of these programs, since its estimates rely on overly optimistic assumptions about how fast Europe's major economies will grow over the next 10 to 15 years.

Lady Thatcher sidestepped the political trap that awaits most European leaders by shifting course long before its effects actually cut into anyone's living standard. By delaying change until their own boomers are about to retire, future German, French, and Italian prime ministers will have to reduce benefits in ways that quickly cut into the pensions of millions of voters, raise taxes on their shrinking workforces, or, as seems most likely, undermine their nations' prospects for growth with large, permanent budget deficits.

These prospects have all of the essential ingredients for a nasty political conflict. In democracies where leaders are accountable to the voters who bear the costs of their decisions, most politicians will find it virtually impossible to deny large numbers of people what they expect and need to maintain their lifestyles. That basic fact of democratic politics makes not only sharp benefit cuts but also large tax increases unlikely, at least until their pension systems actually face bankruptcy. Tax collectors in Germany, France, and Italy already claim between 38 percent and 45 percent of their countrys' GDPs. With poor Europeans paying much smaller shares than that—along with the elderly themselves, who don't pay payroll taxes—these countries' working-class and middle-class voters already pay out about half of what they earn in taxes.

European politicians and the experts who advise them have understood these problems for years and recognized that the sooner they enact their reforms and the taxes to finance them, the smaller the annual pain would be. Yet, nearly every time they've gingerly proposed even modest changes, popular resistance has turned them back. In Germany in 2001, for example, when it appeared that soon there wouldn't be enough marks coming in to pay for the checks going out, the government enacted modest and gradual reductions in the rate of increase in pension spending and the payroll taxes that finance it. Within two years, when weak economic growth threatened another solvency crisis, they borrowed from the future by reducing the system's legally mandated reserves. German businesses complained so much about their payments for workers' private pensions that the government cut them back, too. Berlin did raise the minimum age of early retirement—but only for those citing unemployment as their reason—and then appointed another special commission to propose other reforms. It's still unknown whether any of the original changes will ever take full effect; and even if they do, they will make only a small difference in the system's long-term prospects.

France, too, has debated these issues for twenty years and produced two rounds of marginal reforms, in 1993 and 2003. The 1993 changes raised the statutory retirement age and adjusted how much pensions rise year to year—but not enough to meaningfully affect either the flood of early retirements or the system's high replacement rate. The 2003 reforms were targeted to early retirees, especially civil servants, and could modestly increase the number of middle-age French workers available to help finance the retirement of their elders—but again, not enough to make much difference.

With the window now closing on small and gradual changes that would have large effects by 2015 or 2020, what leader will be able to pull off more painful reforms? Modern politics can be rough, and those willing to endure it are usually those driven to achieve and hold on to power at almost any personal cost. A politician who tried to cut benefits or raise taxes enough to solve this problem not only would likely face a quick and painful end to his or her political life, he or she also would know that the sacrifice could be meaningless, since a successor might well be elected on promises to reverse course. An extraordinary charismatic leader can pull off changes that ordinary politicians find unimaginable. That may happen somewhere between now and 2020—but not in most places. As the crunch comes closer, the choices could come down to a prospect of masses of elderly people, or masses of working people, protesting in Berlin, Paris, and Rome.

To be sure, public pension reform has been a hot topic around the world for more than a decade, and some thirty countries have managed to enact some form of mandatory private retirement saving. But these reforms have happened mainly in places where governments don't have to deal with the legacy costs of a long-standing, pay-as-you-go national system—in developing countries in Asia and Latin America, the new republics carved out of the old Soviet Union, and the Soviets' former Eastern and Central European client states. For example, Bolivia, the Dominican Republic, El Salvador, along with Kazakistan, have followed Chile's model of a system based almost entirely on private saving accounts. Bulgaria, Estonia, Hungary, Poland, Costa Rica, and others have opted for mandatory accounts on top of small, pay-as-you-go basic state benefits. Argentina, Colombia, Peru, and Uruguay let people choose between mandatory savings and small state pensions. Among the advanced economies, Australia, Hong Kong, Sweden, and Denmark have all added mandatory occupational pensions or universal private accounts on top of their traditional state systems.

These approaches work particularly well for developing countries because they can help build up domestic capital pools, especially in places like Latin America, where capital flight is a common problem. They also can provide business for a country's large financial institutions and a political role for national employers and unions. But in the advanced countries with universal, pay-as-you-go systems in place, replacing any significant part of the basic public pension with mandatory private accounts won't be a realistic option between now and 2020. Across Europe and Japan, governments face large, looming shortfalls just to cover future benefits, and such reforms would divert billions of euros and trillions of yen from that task, just as millions of new retirees claim their promised benefits.

Especially across Europe, future leaders are most likely to choose the alternative that democratic governments usually favor when the cost of popular programs collides powerfully and painfully with people's willingness to pay a lot more or receive a lot less: They will create the credit and cash they need and then borrow it. War has been the usual occasion for this strategy, seen currently in the way the United States finances its Iraqi engagement. As an occasion for large deficits, war at least has the saving grace of normally lasting only a few years. The demographic pressures these countries face, however, will persist for probably two generations and will only intensify with time.

Globalization virtually guarantees that if European governments choose this course, ordinary Europeans will pay a large economic price. If government self-financing is the continuing response, year after year, Germany, France, and Italy—along with Sweden, Spain, and others—could face year after year of higher inflation, so that ordinary people pay for the borrowing through higher prices that force them to consume less. Even if globalization's intense competitive pressures dampen this inflation, private investment will slow, as much of Europe's national savings go to finance their public pensions. As that happens, economic growth and people's incomes will slow even more. Eventually, global investors will see real returns in Europe decline, and they will begin to desert the euro. Large continuing deficits also could seriously strain the fabric of the European Union, dimming the prospects of a trans-European capital market that could help its large economies become more successful players in the global economy.

It's almost inevitable that over the next decade or so, the politics of these issues will turn very nasty. Europe has a long tradition of divisive politics and unhappy experience with extreme popular movements. At a minimum, slow growth, high taxes, and pressures on pension systems will force some governments to cut social services—and if recent experience is any guide, those whose services are cut will take their grievances to the streets. Some will blame immigrants, others will blame the generation of their parents or their children. Political outcomes are impossible to predict; but the social, economic, and political conditions for a revival of far-left and far-right politics will be there.

Ireland is the one place in Europe likely to avoid these problems. Most important, Ireland has resisted a large welfare state and avoided mortgaging its economic future to programs for its elderly. And at the least-regulated and fastest-growing economy in Europe, the Celtic Tiger has actually finessed its demographic shifts by sustaining higher birthrates and attracting large numbers of educated immigrants—tens of thousands of foreign professionals along with even more Irish-born, returning to take part in the Celtic boom. Once dismissed by many Europeans as an economic basket case and social curiosity, Ireland has become a new model for European development. So far, the new Baltic and Balkan countries have shown considerable interest. If French, German, and Italian elites could see Ireland as a salutary example, they might avoid some of the painful times that now await their societies.

Halfway across the world lies Japan, with essentially the opposite conditions of Ireland. With the world's most senior-heavy population and the worst-performing advanced economy in recent times, the pension crisis has come to a head earlier in Japan than anywhere else. And unlike Germany, France, Italy, and others, Japan has been willing to cut benefits and raise taxes to keep its pension system afloat. Nevertheless, those changes won't be enough to avoid another crisis in the next ten to fifteen years.

Certain distinctive features of Japanese culture and politics probably made it easier to face up to its pension problems in the 1990s, when the country's fast-deteriorating demographics and years of little economic growth took their toll on the system's finances. For one thing, Japanese seniors are less dependent on state benefits than the elderly in Europe or America. Relatively more workers in Japan have private pensions than anywhere else, a legacy from the halcyon 1970s and 1980s, when the country's large corporations provided comprehensive benefits. Japanese society's attitudes towards old age also allow seniors there to depend less on state pensions, because most Japanese work well into their sixties—nearly a decade longer than most continental Europeans—and most adult Japanese still provide regular support to their elderly parents.

Japan also doesn't have Europe or America's long history of steadily rising benefits, which in other countries has created a sense of entitlement about future benefits and a popular expectation of their untouchability. While Washington mailed out the first U.S. Social Security checks more than sixty-five years ago, and most European countries began much earlier than that, Japan didn't establish broad state pensions until the 1970s. Moreover, significant numbers of Japanese are indifferent to the promises and problems of the system, because even today it isn't truly universal. The Japanese public pension system has two basic parts—a meager, flat benefit for everyone and supplementary benefits based on each person's wages—which together provide retiring Japanese fairly small benefits. But about 10 percent of Japanese are left out entirely, because they work for very small businesses exempt from the system, and another 20 percent or so manage to avoid paying their full premiums.

Japanese politics also are very different from what Europeans and Americans are accustomed to, in ways that can make difficult changes easier to accomplish. For as long as Japan has had democratic government, one party, the LDP, has dominated its politics—even in recent years, as the country's economy stagnated year after year. Japanese voters' patience with the LDP's extraordinary economic mismanagement would be unimaginable in Europe or the United States; and this pattern of one-party government and popular deference to its decisions may help explain how Prime Minister Junichiro Koizumi could cut back pension benefits and raise payroll taxes, while equally talented Western politicians, from Bill Clinton and George W. Bush to Gerhard Schroeder and Jacques Chirac, could not.

So, barely a generation after Japan created its modern public pension system, when benefits began to grow at an unsustainable pace and the system's revenues slowed dramatically, Koizumi could go where no other democracy since Britain's Thatcher has ventured: In 2004, the prime minister persuaded Japan's Diet to raise the payroll tax rate by more than one-third over the next ten years, from 13.6 to 18.3 percent, and to cut basic benefits by 20 percent, based on a set of automatic adjustments tied to demographics and future increases in what is already the world's longest life expectancy.

Even these strong measures will not stave off the storm poised to hit Japan's retirees over the next decade. Japanese demographics will assault the system's financing between now and 2020, as the country's labor force contracts by nearly 12 percent while its elderly increase by nearly 30 percent. Those changes will leave barely 1.4 Japanese working and paying taxes to finance the public pension for each elderly person. Even with the recent reforms, the system will not be able to meet its annual obligations.

Moreover, the national advantages that have insulated many elderly Japanese from financial straits are showing serious wear. Large Japanese employers struggling with globalization are cutting back private pensions. Japan's accelerated aging and anemic growth also have hollowed out the society's once-envied private savings rate, slowly eroding the personal resources of many retirees and their families. Since the mid-1980s, Japanese households with retirees—nonsavers all—have actually outnumbered those in their prime savings years, and that gap will widen substantially over the next ten to fifteen years. And after more than a decade of rock-bottom interest rates, those prime-age savers are putting away much less than their parents did. When the current crop of elderly were thirty-five years old, they saved more than one-quarter of their incomes; today's thirty-five-year-olds save less than 5 percent of their incomes—and hold a lot more debt.

In every way that matters, the financial outlook for retiring Japanese will almost certainly deteriorate. In the 1970s and 1980s, the glory decades for Japan's economy, the financial assets of the average Japanese family grew by more than 5 percent a year. Much of that was put aside for retirement. Those days are gone: The last fifteen years of low saving rates and average returns of

1 percent or less on people's investments have actually driven down the value of a typical Japanese family's financial assets. These trends could worsen over the next decade; and if that happens, the wealth of average Japanese households will keep on declining. The consequence will be falling living standards for Japanese seniors over the next ten to fifteen years, as their assets decline in value while their adult children have less to share.

So, within the next decade, even as the private resources of the Japanese elderly shrink, Japanese politicians will once again have to either cut their benefits or raise taxes on strapped working families. Yet, unlike Germany, France, or Italy, some future Japanese prime minister may well be able to pull it off politically. Even if the LDP finally loses its hold on power, the extraordinary forbearance of Japanese voters may allow whoever runs Japan's government to impose more sacrifice, without suffering the incendiary debates and social protests awaiting most western countries.

Over the same decade, China's political challenge around its retirement system will be entirely different from what Europe and Japan face—but in its own way, equally daunting. China's problem doesn't really begin with its demographics—even though its elderly population will grow faster than Japan or anywhere in Europe. From 2005 to 2020, the number of Chinese age sixty-five or older will increase by nearly 100 million, or about 65 percent—more than Europe and Japan, whose elderly populations will rise 57 percent and 30 percent, respectively. But the baby boom and baby bust came later to China than to the West; so even as the country's labor force begins to contract by 2020, its working-age population will still dwarf its seniors. In that year, China's 167 million elderly will equal less than 20 percent of its working-age people—compared to Europe and Japan, where the numbers of seniors have been growing for generations and by 2020 will equal almost 50 percent of their countries' potential workers.

What sets China apart is that, unlike every western society, it doesn't provide any semblance of universal retirement security. For the twenty years from the mid-1960s to the early 1980s, the country's vast state-owned enterprises and agricultural communes did offer a rudimentary safety net for most people, including retirement supports and health care. During that period, the benefits were a pittance and relatively few Chinese lived long enough to claim them, but even that largely disappeared when, just as the country's life expectancy soared, the new economic program unraveled the communes and many state enterprises. Nor did China's leaders hesitate to do so: The changes were deliberate and an integral part of their modernization strategy, freeing up resources for more pressing economic projects.

Since then, the leadership has given some attention to eventually rebuilding what it had torn down—but always in ways that might help support their economic program. The first step came in 1991 when the State Council issued a "Resolution on Reforming the Pension System." The 1991 changes ostensibly were designed to gradually create a mixed system of basic state benefits and private accounts, but its central purpose was to encourage individual accounts that state banks could tap to help bail out the remaining state-owned enterprises. The way it actually worked was that local governments assumed responsibility for the benefit payments to retirees from local state enterprises, and used the funds from the individual accounts to help finance them.

In 1997, the People's Congress acknowledged the failure of the 1991 reforms and established a new system of pension coverage—but only for those who advance the government's economic plans. The only Chinese eligible for the new state pension benefits are college students, soldiers, and those working in large cities for the government, private businesses, or the remaining state-owned enterprises. By design, this system covers about half of urban workers and one-quarter of all Chinese; in practice, its coverage is even more limited. Tens of millions of private-sector workers outside the large cities are left to their own resources. Those city workers eligible to receive benefits must pay such high taxes—24 percent of an eligible person's paycheck, or twice the cut taken out of American paychecks by social security—that tens of millions evade them. The only consistent contributors thus far—including even the military—are those working for foreign-owned or joint venture companies, because the government withholds business permits from foreign businesses that don't deliver full participation. All told, experts estimate that the new system actually covers only about 6 percent of all Chinese working in private businesses.

For the hundreds of millions of Chinese living in the countryside, the 1997 reforms created a voluntary pension plan that costs the equivalent of 25 cents a month and promises benefits that are equally meager. Even that contribution is too much for typical rural Chinese living on barely subsistence incomes, and just over 10 percent of those eligible have signed on. Nor are private pensions any help, with just 1 percent of the population enrolled—mainly those working for foreign-owned companies. In its current incarnation, China's retirement security program not only fails to cover most Chinese, it's not even financially sustainable at its current scale. Since the program depends on pay-as-you-go taxes that are widely evaded, the cost of providing benefits for those who worked in the state-owned enterprises that still exist already exceeds its contributions.

For the near future and some years beyond, most Chinese who live into old age will have to rely on themselves and their families for support, much as Europeans and Americans did in the nineteenth century. Today, nearly two-thirds of Chinese sixty-five years or older live with their children—and 80 percent of those over age eighty. But attitudes among younger Chinese are changing. With jobs in large cities paying three times as much as those in rural areas, tens of millions of young Chinese have left the countryside and left their parents behind. Some three-quarters of the income of those over age seventy-five still comes from their children—but for retired Chinese ages sixty to sixty-five, the share already has fallen to just one-third. And with birthrates slipping below those in America and much of Europe, China's boomers will have far fewer children to support them when they begin to retire ten to fifteen years from now.

In the last few years, China's government has experimented with a few pilot programs designed to persuade people to put aside more of their prodigious savings for retirement—and let the state off the hook. In 2003, Beijing unveiled a program for personal retirement accounts in Liaoning province, a region of 40 million people in northeast China that's home to the Manchu dynasty's version of the Forbidden City. The program offers substantial subsidies to entice people to participate, tax deductions for businesses that contribute, and stricter efforts to discourage evasion. The program thus far has achieved a participation rate of 28 percent—better than the state system but still a long way from broad-based coverage.

With a life expectancy rivaling the West and the world's fastest-aging population, the financial cost of providing China's seniors a sufficient benefit to stave off poverty cannot be easily reconciled with the demands of modernization. By 2020, such a basic benefit would cost Beijing 10 percent or more of the country's GDP. Such huge sums would have to come out of basic economic projects on which the state's stability could depend, including its massive drives to educate hundreds of millions of people, build a modern military, and construct every manner of modern infrastructure—roads, highways and bridges, airports and seaports, and sanitation and communications systems—across a geographic expanse the size of the United States.

Even if China's leaders determined to provide for the elderly in any way like other major countries, the government today lacks the basic means of collecting the business and employment data it would need to apply and enforce job-based taxes, at least outside the large cities. Creating a government-run universal retirement system in China today would be akin to creating Social Security in 1870s America or in Britain in the 1820s. Not even China could accomplish that today or in the near future.

Nothing stands still, and especially in China. Over the next ten to fifteen years, China will make more attempts to establish new supports for their elderly. By 2020, China's 167 million elderly, most of them poor and with little savings, will live in a prospering society that will raise everyone's expectation of escaping poverty. In this area as in others, China's leaders will continue to seek advice from western experts, private and official. To protect the resources needed to continue modernizing the economy and the military, they will likely experiment with schemes that can channel the high savings of younger Chinese to earmarked private retirement accounts. The most appropriate model could be the Taiwan retirement system; and while Beijing may find it distasteful to adopt arrangements invented in Taipei, the steward of a great state, in Machiavelli's words, "must have a mind ready to turn in any direction as Fortune's winds and the variability of the affair require."

Creating a pension system that depends on private retirement savings will still require difficult changes. For decades, Chinese have held their savings in a handful of huge state-owned banks, which channeled those savings into loans for the country's inefficient and often insolvent state-owned companies. To ensure a range of investment vehicles that could attract savers, China will need a modern banking and financial service industry to take the place of its state-owned banks. That project is doable, since Citigroup, Deutsche Bank, the UFJ-Mitsubishi Group, and other global financial service giants are eager to fill the void. But if the goal is to enable hundreds of millions of Chinese to build up the means to support themselves in their old age, China will have to stop using its people's savings to maintain its remaining state-controlled "vital industries," such as steel, aircraft, and ship building. Without that support, however, the remaining state-owned monopolies could fail quickly, creating a serious problem of dealing with tens of millions more workers who would lose their jobs.

As everywhere else, China ultimately cannot escape its demographics. There aren't enough years between now and 2020 for the 167 million seniors over age 65—244 million over age 60—that China will have by then to build up much security in a system that depends on their own savings. Based on their recent record, China's leaders will likely prefer to leave most elderly Chinese to fend for themselves. If that's the course they choose, the retirement of China's boomers could create political problems that could dwarf what awaits Germany, France, and Italy. In recent years, millions of Chinese have been willing to take their grievances to the streets. China's top police official, Minister of Public Security Zhou Yongkang, has acknowledged publicly that in 2004, more than 3.7 million people took part in some 74,000 large public protests on various topics. A leading Chinese expert, Sun Liping of Tsinghua University, figures the actual incidence of civil unrest is much greater: She estimates that protests involving at least one hundred people occur 120 to 250 times each day in Chinese cities and 90 to 160 times a day in China's villages. Even those estimates are probably conservative, since local officials often try to black out any news of protests in their jurisdictions.

What actually unfolds in China over the next ten to fifteen years will also depend on who wins the next two succession struggles. There hasn't been real dissension about the country's steadily widening embrace of market-based arrangements since the early 1990s. Beneath that consensus, what westerners would call the conservatives have come out on top the last two times—those who see rapid economic and military modernization, and as little political freedom as possible, as the keys to China's future. But the leadership has included a few liberals, or at least what Americans might call a Chinese version of compassionate conservatives, most notably Premier Wen Jibao. Wen and his followers favor a little more emphasis on social projects, including retirement security and unemployment supports, and even a little leeway for ordinary Chinese to express themselves politically. For a number of reasons, no new leadership corps is going to slow down China's military expansion and modernization—not given the power of the People's Army in that leadership and not while the Korean peninsula remains a tinderbox and the U.S. military maintains its huge presence in Asia. If the liberals come to power in 2012, they might finesse the pension problem—or their social and political reforms could end up aggravating public unrest by slowing down the rapid economic gains that hundreds of millions of Chinese now expect. If the conservatives hold on, they might finesse it, too, by generating enough growth to satisfy most Chinese—or face even wider unrest if a global energy or financial meltdown stalls Chinese growth, with little or no public supports to soften the blow.

Finally, there's the United States. Despite alarms raised regularly by some American politicians, in every way that matters the outlook for the U.S. social security system is much more manageable, economically and politically, than what awaits national pension systems almost anywhere else. Compared to Europe and Japan, America's demographics are more favorable (with the exception of Ireland), its promised pension payments are smaller (with the exception of Britain and Ireland), its older people retire a little later (with the exception of Japan), U.S. prospects for the growth and wage gains that finance pension payments are stronger (with the exception, again, of Ireland), and its life expectancy is a few years shorter. Given the right encouragement, tendentious number crunchers have calculated huge shortfalls some seventy-five or more years out, carefully passing over their own unavoidable ignorance of countless intervening developments that can hugely affect such projections. Focusing on the probabilities closer at hand, U.S. payroll tax revenues should be more than enough to cover annual benefit payments until at least 2018 or so; and in 2020, the system's cash flow shortfall should be less than $63 billion, in an economy projected to top $25 trillion. Even by 2030, the cash flow deficit should still be in the range of just 1 percent of America's GDP in that year—and that doesn't include the pension system's income from interest on its trust fund assets. Add that in, and payroll tax revenues plus interest should cover benefits for another quarter-century.

Providing and financing these payments also should cramp normal economic activity much less in the United States than in other advanced countries, because the U.S. system is relatively so much smaller. Social security benefits claim about 4.2 percent of America's annual economic output today, and that's expected to rise to 5.3 percent by 2020. Except in Britain, where basic pension benefits now grow less than the economy year by year, the public pension systems in other advanced countries take two to three times as much out of their national investment pools and budgets. By 2020, payments will range from 11 percent of the GDP of Sweden and 12 to 13 percent of GDP in Germany and Japan, to 15 percent in France and Italy. On the revenue side, the burden of financing the benefits also cuts much less deeply into the investment and hiring plans of American businesses than their European and Japanese counterparts. Today, the taxes that finance state pension benefits amount to 14.2 percent of the labor costs of U.S.-based companies, compared to 17 percent in Britain, more than 21 percent in Japan, nearly 25 percent in Germany, 32 percent in Italy, and a whopping 38 percent in France.

With much more running room, the American public pension system will produce much less divisive political fights than those almost certain to erupt in Europe, since they won't have to involve painful benefit reductions or broad new taxes. Kenneth Apfel, who headed the U.S. Social Security Administration in the late 1990s, recalls President Clinton exploring a bipartisan compromise in secret meetings with Bill Thomas, then chair of the House committee with jurisdiction over public pensions, and in public forums with prominent conservative Republicans such as then Senator Rick Santorum.[5] Bipartisanship dissolved in the impeachment battle of 1997–1998, and President Bush's efforts in 2003 centered on a strictly conservative approach that failed to attract any Democratic support. But a return to a less-partisan, more centrist approach to Social Security reform will have a real chance of success. Even if Americans decided to eliminate any projected shortfall for the lifetime of anyone old enough to work today—a development both unlikely and unnecessary—almost any meaningful package of reforms could take care of it. American social security could "balance its books" for the long term by paring the benefits of the 10 percent richest seniors, tweaking annual cost-of-living adjustments to reflect the average expenses of elderly people, and raising the ceiling on the wages subject to payroll taxes to 90 percent of earnings instead of 85 percent, as it did in the 1980s.

While older Americans can expect to receive the public pensions they're now promised until they die, the United States will not get a free pass. Medical care is much more expensive in America than anywhere else in the world; and as we will see later, colossal financing problems will almost certainly erode publicly financed care for American seniors over the next fifteen years. Moreover, for many years the United States has used payroll tax revenues collected for social security to help fund the rest of government. As retiring boomers claim those excess revenues, U.S. presidents and congresses will have to cut other programs or raise other taxes—with all the political unpleasantness that will entail. Looking beyond 2020, those pressures will intensify as the social security system demands repayment of all the surplus revenues it will have lent to the U.S. Treasury for almost two generations.

So, while Japan and most of Europe face pitched political battles over cutting pensions, raising payroll taxes, or running economically destructive deficits just to keep their pension systems afloat, the United States cannot avoid painful debates over cutting other programs, raising other taxes, or running economically draining deficits. Still, there are enough differences to expect that the political firestorms facing Europeans will be brush fires in the United States. For one, it's more palatable politically to cut back a number of smaller programs with limited beneficiaries, than to slash the promised payments for huge numbers of vulnerable elderly people or raise the payroll taxes of every working person. And if it comes to more deficits, America's relatively stronger economy should make foreign investors more willing to finance them than in Europe of Japan.

It also may be easier for future politicians to raise taxes in America, where the government (at all levels) claims one-third of the nation's income, compared to Europe where states already take 40 to 50 percent of that income. But taxes will be a major battle in the United States of the near future, because the increases necessary to finance health care over the next ten to fifteen years will be substantial. Elderly Americans also face what will be a nasty surprise for many of them: More than 40 percent of Americans have substantial private retirement savings, nearly all of it saved on a tax-free basis. As boomers begin to draw down their retirement saving over the next decade or so, they will discover that everything they take out will be subject to income taxes. This will likely become its own heated political issue over the next decade, as millions of new American retirees call on elected officials to stop taxing the money they need to live on.

In one way or another, the unique demographic shifts of this period will shape much of the next decade's domestic politics in every major country. It could hardly be otherwise, given the real impact these shifts will have around the world on everyone's lives—on how much national economies and family incomes will grow, how high people's taxes could go, and how comfortable or deprived their retirements will be. How well leaders and governments in America, Europe, Japan, and China ultimately respond to these shifts—how successfully they respond to new pressures on their labor forces, saving and investment rates, and public pension programs, and how effectively they can muster a political consensus for reforms—will be a critical test of the quality of their democracies (or in China's case, an autocracy) over the next decade.

CHAPTER THREE

THE PRIMACY OF GLOBALIZATION

MOST PEOPLE WHO TALK or write about globalization use it as shorthand to convey that whatever the subject—jobs, inflation, pop culture, or almost anything else these days—a new and much larger context now affects it. It's imprecise but not really off-base. The landscape and context of the work most people do, the incomes they earn, how they invest and spend their money, and the goods and services they use all have become larger and different in ways that matter.

Nearly one-third of everything produced in the world is now traded from country to country. In little more than a decade's time, the global labor force—the number of workers involved in some way with global production and consumption, directly or indirectly—has soared by 500 million to 750 million. Unprecedented transfers of investment, technologies, and business know-how from the world's most advanced economies to scores of developing countries have unleashed the most rapid modernization in history—moving hundreds of millions of people around the world from poverty to basic comfort or better. Within countries and between them, millions of workers are moving to take advantage of new opportunities. Trillions of dollars in capital move easily across borders, bringing down the trend lines of global interest rates. And a new intensity in competition is holding down inflation in most countries, despite the strong growth of rapid modernization and the low interest rates from global capital.

Each of these features of globalization carries a new and sometimes unexpected price. The late entry and rapid rise of China (and to a much lesser extent, India) as a major manufacturer of the world's goods is squeezing out less efficient producers in scores of smaller developing countries like Thailand and Mexico. In advanced economies like the United States and Germany, manufacturing of all but the most sophisticated or customized goods is on its way to becoming niche activities. The migration of millions of workers across borders is producing political stresses and conflicts, especially in Europe, where influxes of Muslim workers from North Africa and the near Middle East have triggered social unrest in France and Germany and waves of social intolerance in countries like Denmark and the Netherlands.

The globalization of so much production and consumption is also changing the character of large corporations. The easy and cheap availability of global capital and labor have made various forms of intellectual capital the most important and scarce resources for global companies, widening the gap between the incomes of the most highly skilled people and everyone else. By some measures, the top-fifty or so global companies have become bigger than most countries; and the complexities of their operations across scores of national markets have made political capital and skills almost as important as intellectual capital. It's paying off: The needs of the global megacompanies increasingly influence national policies around the world.

In the traditional spheres of labor and capital, globalization also has spurred more offshore outsourcing of jobs to developing countries. While this kind of outsourcing has so far actually cost few Americans, Europeans, or Japanese their jobs, it's entering a new phase in which the piecemeal outsourcing of service tasks could cost millions of service workers in advanced countries their current livelihoods. The new intensity of global competition is slowly and largely indirectly eating away at job creation and wage increases across the United States, Europe, and Japan. As to the new world of global capital markets, while they're dampening inflation and interest rates, those pressures have found new outlets in country after country through boom-and-bust bubbles in stocks and housing. And much as China is the 800-pound panda in global production, the United States uses global capital markets to maintain its place as the world's 2,000-pound customer. To finance the consumption that helps keep much of the rest of the world growing, America now borrows hundreds of billions of dollars every year from foreign savers, perhaps straining global capital and currency markets to nearly a breaking point.

Modern globalization is too pervasive and powerful already for any country to choose the parts that favor it and then wall itself off from those it doesn't care for. It is the largest economic development of our lifetimes and, like it or not, its astonishingly complex and interconnected facets will shape the path and life of every society for the next decade and well beyond.

THE NEW GLOBALIZATION OF PRODUCTION AND CONSUMPTION

Changes in size and dimensions are easy to see. Over the last fifteen years, China, India, the countries of Eastern Europe and the old Soviet Union, and much of Latin America have steadily opened their economies to more foreign investment and greater domestic and foreign competition. These changes took a world comprised of several economic blocs that had little to do with each other economically—the basic arrangements of the previous half-century— and turned them into a genuine global economy.

The data speaks for itself: Over the last fifteen years, the share of every-thing produced in the world that's traded across national borders has in-creased steadily from about 18 percent in 1990 to just under 30 percent today. By 2005, more than one hundred and eighty national economies traded more than $12 trillion worth of goods and services, out of a $42 trillion world GDP. These are the highest levels and largest increases ever recorded. In the United States alone, imports reached $2.2 trillion in 2006—more than the GDP that year of all but five other countries in the world.

For millennia, people have bought foreign-made things because they couldn't make them for themselves or people in those other countries could make them for much less. Much of what we trade today involves the clothes, food, furniture, and the like, or basic commodities like energy and metals, that countries have bought from each other for centuries. The cotton and al-monds that Americans buy today from Indian producers, or the equipment that Germans and Italians buy from companies in China, are not very differ-ent from the garments and tools that ancient Greeks brought back from Egypt, the precious metals and stones that East Africans traded with people in India and China in the Middle Ages, or the chairs and chests that Ameri-can colonists made for the drawing rooms of eighteenth-century Paris and London. Today's levels of global trade are also not wholly unprecedented. Trade among the economies of Europe, plus the United States, in the late nineteenth and early twentieth centuries accounted for 20 to 25 percent of their combined GDP—but then fell sharply over the next three decades of world wars and depression.

For all the similarities, global exchanges today also are different from the 2,500 years of international trade that preceded it. Much of the difference lies not in where things are made, but in how they are made. It's no longer a matter mainly of people in one country selling or buying finished goods or raw materials from people in other countries, or even companies in one nation operating plants and outlets in other nations. The companies that make up the core of every advanced economy today, and many developing ones—the businesses most people work for, directly or indirectly—now operate through global networks that exist outside and beyond the borders and laws of any country. These networks and systems are built on technologies developed in the late twentieth and early twenty-first century, which now enable corporations to break up the production of virtually everything from furniture and clothing to pharmaceuticals and computers into dozens or hundreds of discrete parts, parcel them out to facilities in dozens of different countries, and then assemble and distribute them to scores of different markets. The computer and Web-based systems that track, transfer, amass, and analyze information about all of these far-flung activities make it possible for corporations to manage and coordinate these global operations. But the revolutionary element is the deconstruction of the production process itself, not the way companies deal with the information that flows from it.

All these economic changes also rest on the seismic political shifts of the late 1980s and 1990s, when communism collapsed, China's leaders opted for a gradual, market-based modernization strategy, and the United States with the other advanced economies created the World Trade Organization (WTO). Almost everywhere in the world, the sudden and stunning end of the Soviet Union and its empire left only the American and European models for economic prosperity and political stability. The Asian financial crisis of 1997–1998 reinforced the message by discrediting extensive state industrial policies even in market-based economies. By the late 1990s, from Central and Eastern Europe to East Asia and Latin America, public debate centered more and more on just how broadly and deeply free markets and deregulation should reach. The WTO, shaped largely by the United States, provided the answers. Virtually every developing country in the world accepted the WTO's rules and began to open their state-owned or state-directed economies to domestic markets and much greater foreign investment and ownership. For the first time, American, European, and Japanese companies that long dominated global business could move sophisticated technologies and operations to dozens of countries, large and small, whose role in the international economy had consisted for centuries of mainly providing basic commodities.

These shifts have now created a series of new, world-changing developments. The first and greatest impact has been felt in some of the poorest and largest places on the planet, which went from economic outsiders to global players. In China the number of people working in modern factories and offices has exploded, raising average manufacturing wages there by more than twofold in a decade's time.[1] Since those wages are still much lower than most other places—in 2004, an average manufacturing worker earned seventy cents an hour in China and 40 cents an hour in India, compared to $2.30 in Mexico and $21.00 in the United States—the addition of several hundred million Chinese and Indian workers to the global labor force has produced new pressures on the jobs and incomes of people thousands of miles from the industrial centers outside Shanghai or Delhi. Workers and companies in countries like Mexico and Malaysia—and not America or Europe—feel the most pressure, because they're the ones now competing directly with Chinese and Indian producers. This is one of the reasons why, for the last ten years, incomes have been rising two to three times faster in China, and one and a half to two times faster in India, than almost anywhere else.[2]

In economic development, one good thing often leads to another. So the emergence of far-flung production systems comprised of thousands of operations in many developing countries has accelerated the development of transportation and communications systems—in societies that a decade earlier depended on dirt roads, carts, and irregular mail service—to move people, parts, goods, and the data about them across regions and scores of nations. China has been building 50,000 kilometers of new roads a year, including 5,000 kilometers of superhighways to handle the coming avalanche of private automobiles.[3] Similarly, in early 2006, mobile phone users passed 400 million in China and 90 million in India, with new cell phone subscriptions growing by 600,000 per month in China and 500,000 a month in India.[4]

The rapid spread of sophisticated production operations in poor developing countries has other world-changing effects. The new opportunities to attract lucrative foreign operations to societies, large and small, that had long existed on the margins of the advanced world helped convince leaders from China and India, to Mexico and Bangladesh, to get more serious about modernizing themselves—especially by upgrading education and public health systems to provide the workers that western businesses would need and the basic amenities that foreign managers expect. In Hungary, Mexico, and Guyana, for example, public spending on education doubled in the 1990s, and it rose at least 50 percent in Thailand and Bangladesh.[5] In a developing-world version of supply-side economics, these improvements increase the transfers of new foreign technologies, operations, and capital, which in turn generate more revenues for governments to extend those improvements. The globalization of the production process by the world's large corporations is driving the modernization of many poor societies around the world.

Globalization's world-changing effects are most clear in those countries—China, Taiwan, the Czech Republic, and some others—that have both opened themselves to foreign investment and domestic competition, and upgraded their transportation, communication, education, and public health systems. In one sense—one commonly cited by politicians from Egypt and Venezuela to Malaysia—one hundred or so huge American, European, and Japanese corporations have imposed these changes to serve their desire for wealth and the needs of another several thousand, large and midsize companies that depend on them. If the thirst for riches is the same as every company's need to earn a competitive return, the characterization is true. However, the greatest beneficiaries are the workers (and their families) in those countries with the education and skills to hold down jobs in the modern businesses, foreign-owned and domestic, that are transforming their economies. In the last ten to fifteen years, more of the world's people have moved from poverty to basic comfort and from basic comfort to a middle-class life than in any other comparable period in recorded history.

Gong Jie is a thirty-five-year-old woman who arrived a half decade ago in the northern city of Jinzhou with a young daughter and little else. It took a few years, but she found work tapping into the human transfers in globalization—tutoring Americans stationed there in Chinese. Today, she earns enough to support her daughter in relative comfort and her aging mother back home in the provinces.[6] Even further from Shanghai or Beijing is the world's fastest-growing and fastest-modernizing city, Chongqing on the Yangtze River in western China. At the bottom are once destitute farmers like Yu Lebo, who can now support his family as a "bangbang" porter carrying goods up the steep hills of the mountain city.[7] Further up the ladder are people like Li Zhiguan, who started as the son of a poor farmer, left the countryside for basic factory work near Chongqing, and now earns a decent living cleaning windows of the city's new office towers. Finally, there are nine thousand new workers, many of them migrants from rural villages, working in the new Lifan Sedan auto factory founded by a sixty-eight-year-old entrepreneur named Yin Mingshan. Yin spent much of Mao's reign on a state work farm and set up his first business, a small motorcycle repair shop, in 1992. He's building Lifan Sedan into a major domestic auto producer by buying foreign factories. To create the Chongqing facility, for example, Yin bought a BMW factory in Brazil, broke it down, shipped it back, and reassembled it in the Chongqing Economic Zone. So far, these stories and hundreds of millions like them are the single most important and powerful effect of globalization

To be sure, these leaps in progress are far from universal. Most Africans remain beyond the reach of all the economic progress, especially in the sub-Saharan countries. Closer to many people's experience, much of Latin America has remained largely on the sidelines of globalization. While China has grown for two decades at supercharged rates of 8 to 10 percent a year and the economies of Southeast Asia expanded 5 to 7 percent a year for the same time, the GDP of Latin America has grown less than 2 percent a year. There's nothing in globalization that in some subtle way favors Asians. In fact, for both historic and geographical reasons, Latin America is a more natural place than Asia for foreign operations by American and European businesses. The difference lies mainly in politics, not economics. Foreign investors and foreign companies could hardly fail to notice the regular bent of Latin American leaders for policies that led to hyperinflations, sovereign debt defaults, and the periodic appropriation of foreign assets, or policies that tightly restricted foreign ownership and encouraged widespread piracy of foreign patents and copyrights, and that left most of their potential workers with few marketable skills. Economic instability has hit Asian countries, too, most notably in the 1997–1998 currency crises. But Korea, Thailand, Taiwan, and the rest of the region emerged from their crises with reforms that increased the market-based arrangements favored by foreign investors and foreign businesses; and by 2000, China and India had followed and even surpassed their example.

Where modern globalization takes hold, it often assumes a dynamic quality that wasn't evident the last time international trade expanded sharply. In the period of globalization in the late nineteenth and early twentieth centuries, American and European international companies used cheap labor in developing countries mainly to export their resources back to America or Europe where they were used to produce goods for their own markets and those of other industrialized countries. This time, developing countries whose economies stagnated for generations have seen year after year of growth and income gains of 5 percent, 7 percent, or even 10 percent. This isn't happening simply because U.S., European, and Japanese global companies have established more foreign subsidiaries. This time, the actual production is taking place, piece by piece, in hundreds of places across the developing world. On top of that, tens of thousands of native entrepreneurs have set up their own businesses both to provide goods and services to the companies producing the pieces for large Western corporations and to satisfy the burgeoning, new domestic markets created by the income gains that their production helps drive.

The emergence of new small-business classes in countries such as Thailand and Taiwan, and later China, India, and Mexico, is as much a part of globalization as the worldwide networks of Citigroup and Siemens. Their emergence and growth also creates a powerful new political constituency in developing countries that sees its future aligned with modern globalization. Much like small-business owners in America, Britain, or France, they resist any return to the old ways of government-sanctioned monopolies and hostility to foreign investment and foreign businesses. Instead, they press for freer markets and more government spending for education, infrastructure, and public health. The growth of small and medium-size businesses in many major developing countries—more than the economic and political power of the United States—will make globalization a permanent feature of most people's future.

Latin America is a good example of how this works in some places and not in others. Among the larger Latin countries, Mexico and Brazil have accepted the market-based terms of globalization more extensively and enthusiastically than Venezuela and Argentina. The centrist Vincente Fox's 2000 defeat of Mexico's Institutional Revolutionary Party (PRI), which had held power for more than seven decades, followed in 2006 by the election of the business-oriented Felipe Calderon and his conservative National Action Party (PAN)—along with the reelection of the fiscally disciplined Lula government in Brazil—represent the first time that popularly elected centrist or even conservative governments have come to power and successfully retained office in major Latin American countries. Argentina and Venezuela have accepted fewer of the requirements of globalization, their small-business classes are smaller, and their politics are more traditionally populist, conventionally left wing and anti-American. This may be a pattern that could apply eventually to Europe, where countries such as France and Italy, that resist the deregulatory demands of globalization and encounter serious economic problem for both those and demographic reasons, develop their own versions of populist, antiglobalization politics.

GLOBALIZATION AND THE NEW, IDEA-BASED MEGACORPORATION

The development of lucrative new markets in fast-growing developing countries also is changing the character of global corporations. For the first time in economic history, there are genuine global markets for both the production and the consumption of everything from cell phones and pharmaceuticals to machine tools and information services. This new aspect of globalization vastly enlarges and complicates the business of business, and it's produced a new kind of global company.

The most obvious change is just how big large corporations have now become. The vast expansion of potential markets and the ability of firms to serve huge and complex world markets have produced a new class of megacorporations. In 2006, 60 countries had a national income or GDP of $50 billion or more—121 global companies had revenues at least as great, including 30 companies with revenues exceeding $100 billion.[8] At the top, the combined revenues of the world's top-ten corporations—WalMart, ExxonMobil, Royal Dutch Shell, British Petroleum, General Motors, Toyota, Chevron, DaimlerChrysler, ConocoPhillips, and Total—came to $2,436 billion in 2006. That was more than the national income of all but four nations (the United States, Japan, Germany, and China). In America alone, the combined 2006 revenues of the top-ten U.S. corporations, $1,984 billion, were greater than the GDP of all but six countries. No matter these revenues, corporations do not have the power of countries. But most of those huge revenues are paid out to workers and governments and other businesses for countless goods and services. So, when the revenues are huge and the country is fairly small, that alignment inevitably creates political and social power.

Unsurprisingly, most megacorporations come from the most advanced economies. Seventy-one of the world's hundred-largest corporations are from the United States, Japan, Germany, France, and Britain. For a variety of reasons, British and French corporations, and to a lesser degree America's, are overrepresented among the top hundred, relative to the size of their economies. Yet even American, British, and French companies dominate only a handful of global sectors. U.S. companies, for example, are strongest in pharmaceuticals, with about half of all the world's large drug companies, and software with all four of the world's big four. U.S. computer makers also share domination of that global market with Japanese firms. European and Japanese companies account for most of the world's major auto makers, and European firms dominate the world's telecommunications giants.

Globalization changes what it means for a company to be an American, French, Swiss, or Chinese enterprise, but not in the ways that many people believe. A decade ago, the pundit politician Robert Reich attracted popular attention by pronouncing that the sharp increase in foreign investment and foreign operations by multinational companies divorced their decision making and allegiance from the interests of their headquarter countries and their workers, all to enrich a small number of global shareholders. It's a seductive and easy-to-grasp formulation, but it is off the mark. As we will see, more than 70 percent of the people employed by U.S. global megacompanies live and work not in outposts in Europe or developing nations but in the United States; and the shares of home-based workers in European and Japanese multinationals are higher. Reich also didn't get the shareholder issue right. Some of the profits from foreign operations certainly go to shareholders—but again, for U.S. multinationals, about 40 percent of the profits distributed to all of their shareholders, American and foreign, go to pension funds and retirement accounts held by ordinary Americans.

We also will see presently that globalization shifts the strategic focus of large companies in ways that reinforce the American or European character of large corporations headquartered there. One important reason why Exxon Mobil doesn't decamp to Bahrain or Lichtenstein, and why the British pharmaceutical giant Astra-Zeneca hasn't moved to India or Poland, is that globalization has shifted their critical strategic focus from securing cheap capital and labor to the more central task of developing and implementing new, economically powerful ideas. The data show that large companies today invest and care most about their "intangible" assets, including not only the intellectual property embodied in newly developed products and processes, but also their branding, databases, and new ways of operating and organizing themselves, and training and using the highly skilled employees who work with these intangibles. It's the intangibles, and not the location of less than 30 percent of their workers, that makes a company American or European; and almost everything associated with those intangibles happens in the countries that these companies call home.

Stated more concretely, the executives, managers, and other critical personnel who develop and use the intangible assets of large American, European, and Japanese companies, as well as their shareholders, are still very predominantly American, European, or Japanese, respectively—as are most of the relationships for the critical development, financing, marketing, distribution, accounting, and legal functions of a global company. Even huge companies founded in small countries, with most of their production and sales distributed across other countries, hold on to their original national character. Tetra-Laval, the Swedish packaging giant, does most of its business in markets outside Sweden. It's privately owned by the Rausing family, who now live in England, which does not tax gifts and inheritances within hugely wealthy foreign families. Yet in every way that matters, Tetra-Laval remains Swedish—it's run out of Sweden, and the central engineering, marketing, and distribution operations that determine its success remain distinctively Swedish. And when two global businesses from different countries merge, the stronger company becomes dominant—DaimlerChrysler, after all, was not called Chrysler-Daimler.

The more subtle ways in which globalization does make megacorporations less "national" in character mainly involve an increasing cosmopolitanism in tastes, business methods, and standards. A global company like IKEA or Wal-Mart has to work with the conditions and consumer preferences of dozens of distinct national markets. They not only bring products, services, management approaches, and expertise developed in Scandinavia and America to all those other countries, they also bring back the lessons and sometimes new products developed in the other places they operate. American companies learned just-in-time inventory practices from observing the way that their Japanese rivals operated in both Japan and the United States.

McDonald's, the leading global fast-food brand (though the company earns more profits from the underlying real estate), operates more than half of its restaurants outside the United States, including more than seven hundred outlets in more than one hundred Chinese cities.[9] Much of foreign menus came from the American operations, with lots of singular dishes based on local specialties—the McCalabresa in Brazil (a garlic sausage patty seasoned with vinaigrette), beer in Germany, wild strawberry and vanilla pie in Sweden, the Shogun Burger in Hong Kong (a pork bun with cabbage and teriyaki sauce—without the cabbage, it's also the Samurai Pork Burger in Thailand), the McKofta meatball sandwich in Pakistan, and red bean sherbet served in Korea. But the McLobster rolls on many New England McDonald's menus were first developed in Canada, and the deli sandwiches started in the U.K. Perhaps sometime soon, McDonald's in Los Angeles will offer home delivery like the outlets in Egypt, Turkey, and much of Southeast Asia, or McDonald's in the southwest and northwestern United States will adopt the "McCafe" coffee-and-dessert operation seen first in the Australian branches and now popular across the company's Latin America outlets.

The same internationalization often extends to other aspects of business. U.S., European, and Japanese manufacturers aren't as safety and pollution conscious in their Indian or Mexican operations as they are at home; but they're more so than the native companies because economies of scale dictate that they use much the same equipment and operating methods everywhere. And it can work the other way when a developing country's standards are stricter than those in a global company's home country. China's fuel-efficiency standards for cars and trucks are higher than those in America; and if General Motors or Ford plan to compete successfully in what will soon be one of the world's largest auto and truck markets, they will develop more energy-efficient technologies and vehicles that will make their way back to the American market.

Sweden is not the only small country which, despite its small domestic market, has produced some of the world's largest companies. The other standouts in Europe are the Netherlands, with 14 of the world's 500 largest firms, in banking, oil and gas, insurance, aerospace, and electronics; and Switzerland with 11, including five world-class financial service or banking firms and two global drug makers.[10] Canada is also home to 16 of the world's 500 largest companies, spanning finance, insurance, metals, aerospace, and auto parts. Perhaps most striking, 14 of the world's 500 largest companies are from a developing country—South Korea, where smart government policies helped create the global auto maker Hyundai and a vibrant electronics sector led by Samsung and LG Electronics. Over the next two decades, Korea is on track to duplicate Japan's singular achievement in the twentieth century, as the only nation in the world that's moved from being a low-income to a high-income country.

Other developing nations, even the largest ones, are still represented only modestly on the list, with just 56—or barely 11 percent of the top 500 companies. The impediment to this kind of success is not access to capital or technology, which they have in abundance, but shortages of the entrepreneurs and experienced managers that could build and operate exceedingly complex and innovative global organizations. Companies like Toyota, IKEA, or Exxon Mobil that span global markets have to create, operate, and manage not only huge, international networks of plants and offices using workers and materials from scores of developing and advanced countries, but they also must manage distribution and marketing systems to reach customers in scores of countries with varying needs and tastes. And these markets don't stand still, but shift and change all the time, gradually or sometimes suddenly. So, for all its wealth and growth, only 24 Chinese companies are among the top 500— not far ahead of Canada or the tiny Netherlands—and they all principally serve China's large domestic market. Similarly, India accounts for just 6 out of 500, fewer than Spain or Australia; and once-mighty Russia has only 4— fewer than little Belgium and Sweden, and tied with Brazil.

The shortages of entrepreneurs and world-class managers in most developing countries often reflect government policies that throw up innumerable obstacles to both entrepreneurs who want to start their own businesses and existing companies that want to expand and challenge large, old-line native firms often controlled by the country's most influential families. What South Korea achieved is within the reach of any developing country, given several decades of pursuing the policies to actively promote it.

No country exemplifies more starkly than Russia the costs that globalization can impose on an economy and its people if its government doggedly pursues policies that drive away foreign investors and expertise. In the early 1990s, Russia took enormous steps to introduce markets, privatizing almost everything and decontrolling most prices. But in every other way, it created one of the world's most distorted playing fields for market competition, one that systematically favored the newly privatized old state companies, all dismally inefficient, and penalized new companies, domestic and foreign. The old enterprises received such extensive special subsidies and tax breaks, government contracts, and debt forgiveness that new companies couldn't compete. Foreign companies backed away: In recent years, just 6 percent of all business investment in Russia is foreign, compared to 42 percent in Poland.[11] The result is that the productivity of Russian core industries, and the nation's total output, collapsed. Russia is the world's second-largest oil exporter, but it gets little benefit from it because the productivity of its oil sector is just 30 percent that of American oil companies.[12] Productivity in big Russian steel, which once rivaled U.S. steel makers, fell to 28 percent of U.S. levels—and 40 percent of Brazilian levels.[13] Similarly, Russian productivity in housing construction is just 10 percent of U.S. levels and two-thirds less than Brazilian levels, largely because the privatized state banks don't offer mortgage financing. So 90 percent of new single-family houses are built in the most inefficient way imaginable—by the homeowners themselves, brick by brick, after they collect the hundred or so permits they need to take on the job. Russia has become one of the world's minor or dispensable economies: Its 2006 GDP of $987 billion was barely 7.5 percent of America's economy, about 42 percent of Britain's GDP, smaller than Spain or Brazil, and just 10 percent larger than South Korea.

Globalization is also changing the character and basic needs of modern corporations, and everyone who works for them will feel the effects. For centuries, large national and international companies used their heft to get sweet deals on their most basic resources, capital and labor. But modern globalization makes labor and capital easily and relatively cheaply available to global companies. So their basic business strategies no longer focus on securing those resources; instead, the truly scarce and critical resources for most global businesses today are the intellectual and political capital—the intellectual capital of their patents, brands, distinctive business methods, and the knowledge and relationships of their professionals and managers; and the political capital of favorable regulation, tax treatment and government subsidies, and the connections and influence in their home countries and major markets required to secure them.

The "idea-based" economy has been a metaphor for years. Globalization has made it a concrete reality. Since the mid-1990s, for example, U.S. companies have invested as much in intangibles—mainly the intellectual property of patents and trademarks, but also databases, branding, organizational changes, and the training or human capital to use these ideas—as in all physical assets, from equipment to land and buildings. This shift to intangible assets of course coincides with the rapid progress and spread of computer, software, and Internet technologies, which play large roles in the development and application of the products and processes that win patents and trademarks, the databases and branding, and even organizational changes and high-level training. The shift is also evident in the way investors value public companies. As noted earlier, twenty years ago, the market value of the physical assets of the top 150 U.S. public companies—their "book value"—accounted for 75 percent of the total value of their stocks. A firm was roughly worth nearly what its plant, equipment, and real estate could be sold for. By 2004, the book value of the top 150 corporations accounted for just 36 percent of the total value of their shares.[14] Today, nearly two-thirds of the value of a large company comes from what it knows and the ideas and relationships that it owns.

Creating and applying valuable new ideas, especially those built on the best old ones, has always been the most important factor determining how much economic progress a country and its people make. Since the pioneering work of Nobel laureate Robert Solow in the late 1950s, economists of every school and stripe have recognized that the development and application of economic innovations has more impact on how fast a country grows and the incomes of its people rises, than how much capital they invest or even how much they improve their education and skills. For example, 30 to 40 percent of all the gains in wealth and productivity made by the United States during the twentieth century can be traced to innovation in its various forms—not only the development of new products and technologies, but new materials and processes, new ways of financing, marketing, and distributing things, and new ways of organizing and managing a business. All the increases in the U.S. capital stock account for just 10 to 15 percent of that progress, while improvements in the education and skills of American workers explain 20 to 25 percent.

Judging by how much businesses now invest in ideas and how much investors now value them, the economic role of innovation is still expanding. For a number of reasons, the central economic innovation of our period, information technologies, affects every aspect of the economic lives of most people on the planet in unusually powerful ways. To begin, compared to previous sweeping innovations such as electrification, information technologies continue to advance at very rapid rates while their prices continue to decline at equally and unusually fast rates. The rapidly falling price of computing power, storage and transmission, and most recently software, has become an important feature of globalization itself, because it makes it possible for the innovations to spread quickly across not only the world's advanced countries, but almost everywhere else as well. It's not surprising that elaborate information systems should pervade finance or manufacturing in the world's richest and most highly developed countries. What's unique is that the same systems have so quickly become part of not only every corner restaurant and repair shop in small-town America and Europe, but also most large and medium-size businesses in places like China, India, Egypt, and Peru. In 2006, an estimated 103 million Latin Americans were online; and in the world's least-developed region, sub-Saharan Africa, nearly 90 million people had mobile phones and more than 22 million were online.[15]

For the first time, globalization and information technologies give companies virtually everywhere in the world real access to ideas and innovations generated not only a few miles away but also several continents away. For developing countries, the ability to host, buy, rent, borrow, or steal the most productive technologies and business methods of companies from more advanced economies has opened the door to lightning-fast progress. For all its cheap labor and land, China would look a lot more like rural Russia today if China's resources weren't being combined with the intellectual property of American, European, and Japanese companies.

To keep their progress going over the next ten to fifteen years, China and other fast-developing countries will have to keep on attracting and absorbing the new ideas being developed today and in the near future by western companies. To do that, they will have to accept the strategic bottom line of modern business. To begin, China, India, Brazil, and others will grudgingly adopt American and European intellectual property protections. They're not there yet: Much of the Chinese government and People's Army has long run on pirated versions of Windows, while the Chinese automaker Chery produces clones of GM cars; and India, Brazil, Argentina, and others look the other way as local businesses make and sell millions of doses of counterfeit western drugs and tens of millions of pirated DVDs and CDs. Even harder, those developing nations that are most determined to keep their progress going will gradually let go of most remaining restrictions on foreign investment by western companies, along with layers of traditional regulation, especially about whom a company can hire and fire and where it locates its operations. In these respects at least, China, India, and much of Eastern Europe in 2020 will look a lot like America, at least from the vantage point of corporate headquarters in London, Tokyo, and New York.

In the end, globalization offers countries like China, India, the Czech Republic, and Brazil no other real option. Their governments know—or will soon recognize—that the rapid economic progress on which much of their own legitimacy now depends requires constant infusions of new western technologies, investment, and expertise. The global businesses that provide it similarly know that their own rapid growth equally depends on their ability to claim their share of the value of the goods and services produced with their ideas in the world's successful low-cost economies.

Global businesses won't always get their way in developing nations. Western companies cohabiting with rapidly changing developing societies have to find ways to respect and accommodate local traditions and laws. This has proved quite difficult and sometimes impossible in Islamic countries, where western methods and managers are sometimes unable to operate under Shira law—one of several reasons why most strictly Islamic states have not moved past the fringes of globalization. But much of the non-Islamic developing world is adapting quickly to many western tastes and manners, from western work hours, teamwork, and IP rights, to jeans and rock and roll. And when the issues are big enough—especially when they could affect the entire world—even the megacorporations' strategic bottom lines will give way. In perhaps the very near future, a deadly, global avian flu pandemic could quickly trump the intellectual property rights of large pharmaceutical companies, and the prospect of worldwide climate change could force big oil, big steel, and big autos to revise their carefully laid investment plans in large developing countries.

Businesses that operate in so many different ways in so many countries need political capital and political facility, almost as much as they need intellectual capital. Megacorporations, especially as a group, exert real political as well as economic clout. It's not simply a matter of a particular politician or party caving to the demands of Wal-Mart or Total, the French petrochemical giant—though that certainly happens. And no single foreign business is indispensable to the life of any country or the stability of any government. But no country or government—at least none but the most autocratic or populist—can afford to create conditions that would be inhospitable to the world's largest companies. As a group, huge companies operating in numerous countries have certain common needs, from basic health care and training for their workers, and capital markets that aggregate large tranches of capital for easy access, to standardized global telecommunications and low taxes on their profits. The upshot is that across most advanced and developing nations, across societies that see themselves as traditionalist, progressive, or socialist, a roughly common approach to the government's role in economic life is arising out of globalization. Across societies with little else in common politically or culturally, deregulation and training programs have gathered strength, health-care coverage and educational access are becoming broader, telecommunications and other technological standards have largely become global, and, especially in developing countries, corporate taxes are generally falling.

In a commonly quoted passage from *The Communist Manifesto*, Marx and Engels wrote, "The need of a constantly expanding market for its products chases the bourgeoisie over the whole surface of the globe. It must nestle everywhere, settle everywhere, establish connections everywhere. . . . It compels all nations, on pain of extinction, to adopt the bourgeois mode of production." Among all their overheated rhetoric and plain errors—not the least their dismal failure to anticipate how the process would lift hundreds of millions of people out of poverty—Marx and Engels got that right long before modern globalization.

The ability of a megacompany like Pfizer or Nokia to use its global systems to develop, produce, and market products on a worldwide scale depends on its capacity to successfully navigate countless legal requirements, health and safety regulations, tax and labor laws, and much more, across scores of national markets. In places like China, India, Russia, and Argentina—where it's hard to do business at all without the cooperation of innumerable local and national bureaucrats, and where the legal system can't be relied on to hand down consistent rulings—networks of political connections are especially important. And in their own countries as well as their foreign markets, the megacorporations' ability to compete successfully with rivals across much of the world also often depends on their skill in extracting special tax and spending subsidies.

The shift to intellectual and political capital as the central assets of large companies is also changing the mix of industries that dominate global big business. In 2005, 39 percent of the world's 150 largest corporations were in financial services and health care—sectors that employ large numbers of professionals and managers who create value through their knowledge and interactions, and are subject to extensive government regulation and oversight. That's more than three times the 12 percent share of the 150 biggest companies just twenty-one years earlier, in 1984. Even in manufacturing, the number of professionals and managers has been rising sharply. From 1984 to 2005, the share of General Electric employees in professional or managerial positions more than doubled, from 25 percent to over 50 percent, even as the total number of GE employees contracted. Corporations like GE are changing in this way because it makes business sense in an idea-based global economy—over the same period, the inflation-adjusted, net income per employee at GE soared from $13,000 to $54,000.

There's no way of knowing which companies will be global market leaders in 2010, 2015, or 2020; but we can reasonably conjecture about which sectors

are likely to be relatively more or less profitable or important over the next
ten to fifteen years. Among the major players in the global economy today,
the outlook may be most bearish for the big oil companies. For more than
a half-century, the world's large oil and gas companies have been powerful
movers in their national and global economies; and today, six of the world's
top ten companies and eleven of the top fifty are in the oil and gas business.
Over the next ten to fifteen years, however, the sector will face increasing government regulation from climate change, more intense competition from alternative energy sources driven by climate change and high energy prices, and
new political constraints, as the struggle over energy sources becomes a critical part of most countries' national security policies and a source of regular
tension among them.

Likely winners over the next ten to fifteen years include the world's big
pharmaceutical, biotech, and new genomic companies. Burgeoning numbers
of elderly people almost everywhere will increase demand for their products,
financing crises in many national health-care systems will make more of
their products the low-cost treatment alternatives and, for the biotech sector, rising incomes in many less-developed nations will increase demand for
genetically modified foods. Yuri Verlinsky, who directs the Reproductive
Genetics Institute in Chicago, sees cures for children with leukemia and anemia within ten years using stem cell lines from their parents' embryos; and
Norman Borlaug, who pioneered genetically modified crops for Latin America, expects genetically enhanced wheat strains to vastly increase its production in India soon. It should also be boom times for global auto makers that
can best adapt to the energy and environmental constraints of the next decade, as fast-growing incomes and vast road-building programs in China,
India, Bangladesh, and other large developing nations drive up the demand
for cars. China alone forecasts that its national fleet will soar from 25 million
cars today to 140 million by 2020.[16] The next decade also looks bullish for
global information technology and telecommunications companies. As the
prices for computers and Internet access have steadily declined and their uses
have steadily expanded, every advanced country has seen these technologies
spread from large corporations and affluent early adopters to the smallest
businesses and almost everyone else. Over the next decade, the same process
will almost surely repeat itself across much of the developing world, starting
already with cell phones and the $176 laptop with open-source codes re-
cently unveiled by the MIT media lab.

As these technologies reach across most of the world, they will almost certainly further accelerate the integration of national markets and global networks in almost every industry. At a minimum, they should produce strong,

sustained productivity gains in many places, which should mean higher profits and perhaps faster-rising incomes in countries that take best advantage of them. More highly integrated national and global industries could also force all but the largest companies to become more specialized, increasing the pressures on workers to develop even more specialized skills. And much like Cisco Systems, most megacorporations could slowly withdraw from the business of directly producing anything at all and focus instead on designing, managing, adjusting, and enforcing the global networks of other companies that will do the development, production, marketing, sales, and distribution of the goods and services—from automobiles or clothing to IT or financial services—that the public will know them for.

There may well be natural limits to just how large megacorporations can grow, and still remain profitable. Global markets are already unimaginably large and complex. For a single company like Toyota or Wal-Mart, global networks of developers, producers, suppliers, and distributors are involved to provide a large range of products and services for customers in scores of different national markets. Carrying out those activities requires layer upon layer of middle- and upper-level managers trying to communicate and coordinate with each other as well those working for them. In one recent survey of large companies, more than half of executives said they spend as much as two days a week on unproductive e-mails, voicemails, and meetings. Even with the most sophisticated information and telecommunications systems, billions of details can swamp the strategic plans of the best-run companies.

Already, a number of megacompanies have tried to make their lives simpler (and more profitable) by divesting themselves of various sidelines and noncritical operations. Citigroup, for example, sold its life insurance subsidiary to MetLife and its transportation-financing line to General Electric, so it could focus on the higher profits it earns from banking. Dell Computer is one possible model for the near future, especially for manufacturers. Dell made itself the world's most profitable computer maker (measured per employee) by focusing on one thing—providing customized products using rigorously standardized processes carried out by subcontractors, and avoiding the temptation of expanding into higher-end products or taking on related lines of business such as systems consulting.

WORKERS IN A GLOBAL ECONOMY: THE SPECTER OF OUTSOURCING

As globalization shifts the strategic focus of large businesses from labor and capital to ideas and politics, the implications for everyone who works for a living are enormous. Low unemployment rates and healthy income gains for most workers in the 1960s and 1970s were unmistakable signs that labor in those days was generally a scarce resource in the United States, Europe, and Japan— just as the high unemployment and stagnating incomes in most developing nations in those years were reliable evidence that labor was anything but scarce. And since it takes real sacrifice and resources for workers to move from country to country, the labor surpluses in developing countries couldn't relieve the tight labor markets in the advanced economies. Modern globalization burrows under those geographical barriers. Chinese, Czech, or Mexican workers can join the potential labor pool of European, Japanese, and American employers by simply moving to the industrial and business centers of their own countries, where those employers have set up their foreign operations. With fast-rising wages to draw them to those areas and new transportation networks to get them there, these internal migrations have already helped create megacities of 10, 15, or 20 million people in Mexico, Brazil, India, China, Indonesia, Bangladesh, Egypt, Argentina, South Korea, and Russia. Facing them across oceans or continents are tens of millions of working people in every advanced country, worried that waves of offshore outsourcing over the next decade will cost them their jobs. They have good reason to worry, even though direct outsourcing to workers in developing countries is unlikely to touch most of them.

There's no doubt that the global labor pool suddenly became much larger. Richard Freeman, a Harvard University economist, calculates that globalization has effectively doubled the size of the world's connected workforce: In 2000, China, India, and the countries of the former Soviet Union and its Eastern European empire had roughly 1.5 billion workers among them, or about the same number of workers as all the advanced countries, Latin America, North Africa, and Southeast and South Asia. Professor Freeman's arithmetic is correct, though his conclusion seems exaggerated. While China, India, and the rest have joined the global economic system, hundreds of millions of their subsistence farmers and village workers are still isolated from the world of trade, foreign investment, and outsourcing. Nevertheless, in barely a decade, globalization has probably expanded the effective global labor pool—people who take part directly or indirectly in producing or providing things that are sold in other countries or that compete with goods and services produced in other countries—by at least one-third and perhaps as much as 50 percent. That comes to an increase of 500 to 750 million workers.

According to some politicians and pundits, this could mean that hundreds of millions of Americans, Europeans, and Japanese will lose their jobs to Chinese, Indians, Central Europeans, and Brazilians over the next ten or twenty years. In the 2004 U.S. presidential campaign, John Kerry—who knows better—called outsourcing businesses "Benedict Arnold corporations"; and a usually reasonable writer for *The Wall Street Journal* wrote in mid-2005, "if you can describe a job precisely or write rules for doing it, it's unlikely to survive. Either we'll program a computer to do it, or we'll teach a foreigner to do it."[17] Theoretically, outsourcing to developing countries could affect any job that doesn't involve direct physical contact with either customers or a company's other domestic employees. The practice of offshore outsourcing and its potential for the near future, however, are both more complicated than that and much more limited.

Outsourcing itself is a normal and fairly pervasive feature of advanced, diversified economies, and it has certainly increased greatly as corporations have become larger and more complex. But almost all outsourcing happens inside a country, when a Citigroup, Mitsubishi, or any of thousands of much smaller companies take something it used to do itself—its payroll systems or routine legal services, call centers or IT servicing—and begin buying it from another company. A survey of four thousand companies in the Northeast, most of them with less than one hundred or five hundred employees, found that 83 percent outsourced a wide range of activities to other American companies.[18] More than half use other companies to manage their retirement or health benefits, 30 percent outsource training, and nearly 20 percent their call centers and IT services. In IT services, companies such as CIBER set up centers in lower-cost locations in the United States, including Oklahoma City and Tampa, that provide remote IT services for East and West coast companies. And it's not limited to America and other advanced economies: Eastern Software Systems handles not only the routine maintenance of IT systems for medium-size Indian companies, but also hiring for domestic IT businesses.[19] By one good estimate, this kind of domestic outsourcing accounts for more than 87 percent of the overall phenomenon, and has little to do with globalization.[20]

Three other kinds of outsourcing do involve workers in other countries and are part of globalization. One form that has gone on for a long time involves multinationals creating foreign subsidiaries to produce parts or entire products, or provide or manage certain services as part of their global networks. American companies do more of this kind of outsourcing, by far, than European or Japanese companies, mainly because labor laws in Japan and most of Europe make it much harder for companies to fire anybody at home. Moreover, a fair share of those jobs involve services: Some 40 percent of all U.S. imports of services—some $125 billion worth in 2005—occur between a U.S. parent company and its foreign subsidiaries and affiliates. In areas of business that are very IT-intensive and involve advanced skills, it's even higher—some 60 percent of all imports of financial services and 70 percent of all imports of business and professional services.

Yet this still involves a small share of jobs. The U.S. Bureau of Labor Statistics figures that offshore outsourcing cost about 13,000 Americans their jobs in 2003 and another 52,000 in 2005. In 2004, after decades of gradual outsourcing of this kind, U.S. multinationals had almost 30 million people on their worldwide payrolls, of which 8.3 million or about 28 percent were employed abroad.[21] But foreign multinationals also employed about 5.1 million workers in America, so American firms outsourced 3.2 million more jobs abroad than other countries outsourced to the United States. In recent years, for example, many of Europe's leading pharmaceutical companies have shifted much of their R & D to the United States, to tap into a market that is both the world's largest and the only major one with largely unregulated prices. The 3.2 million difference equals about 2 percent of all American jobs. And those numbers have barely increased in recent years. The number of foreign workers employed in the foreign subsidiaries of U.S. companies increased by about 100,000 in 2004, following a decline of about 60,000 in 2003. If every one of those 100,000 jobs outsourced in 2004 cost an American his or her job, the job losses came to less than one-tenth of 1 percent of all American jobs that year.

Offshore outsourcing cost a lot more American jobs than that in the past. The United States lost 2.3 million manufacturing jobs from 1979 to 1993, during the so-called period of deindustrialization, and a fair share of them went to Mexico, Taiwan, and other developing countries. As globalization took off, the United States lost even more manufacturing jobs—2.7 million jobs from just 2001 to 2004. But over those years, the number of people employed by the subsidiaries of U.S. companies in developing nations didn't increase at all. Millions of American, European, and Japanese manufacturing workers have been losing their jobs because they're either being replaced by technology or outsourced to other businesses in their own countries.

That's not the end of it, because globalization is giving rise to a new and distinctive form of offshore outsourcing. For the first time, American and European firms (Japanese firms do this much less) have begun to outsource certain services—call center operations, software development, a piece of their R & D—not to their own foreign subsidiaries, but to separate firms in places like India, Ireland, Israel and, inevitably, China. This new kind of offshore outsourcing will grow more important over the next ten to fifteen years and pose a genuine threat to millions of workers in advanced countries.

———

It is certain that some highly skilled Americans and Europeans employed by large multinationals are losing their service jobs to other firms in other countries. But low wages are not driving this development, because most of this offshore outsourcing involves jobs moving from one advanced country to another. The leading country for offshore outsourcing today, as a share of its labor force, is not India or China but Ireland, which redid its tax and regulatory regimes especially to induce British, European, and American software companies to set up shop there. And Israeli companies, many of them spinoffs from Israeli military developers, now perform a lot of R & D for American and European firms.

Unlike manufacturing, service outsourcing doesn't happen simply because a developing country modernizes itself, at least not yet. India has become a center for some forms of software and IT service outsourcing, after a decade of committing significant resources to educating young Indians in a few major urban areas. Even so, foreign firms cost only an estimated 100,000 American software and IT service jobs from 1983 to 2003.[22] And the United States and Great Britain totally dominate the realm of offshore outsourcing of highly skilled service jobs to developing countries. U.S. global corporations alone account for more than half of all offshore outsourcing of engineering jobs, for example, with British firms accounting for another 16 percent.[23] In continental Europe and Japan it's still very difficult to shift jobs to other countries. All of Europe outside Britain accounts for 16 percent of all engineering jobs outsourced offshore, and Japan's global companies are responsible for just 2 percent.

What about the great push of recent years in China and India to educate hundreds of thousands of their younger boomers to be business analysts and managers, doctors and accountants? These investments will produce a new corps of skilled professionals there eager to work for a fraction of what their American or European counterparts earn. But most of them won't end up taking the jobs of Americans or Europeans. Instead, most will be snapped up by domestic businesses or provide their professional services to western companies selling into the fast-growing Chinese and Indian markets. Moreover, despite the push to educate and train millions of young Chinese and Indians as service professionals, it's very hard to significantly expand a developing country's system of higher education in a few years, especially to provide the skills required by most sophisticated western companies. In many places in the developing world, shortages of skilled service workers have appeared already. Moreover, the experts at McKinsey calculate that within four years, given current trends, there won't be enough workers skilled in IT and engineering-based services in China, India, and the Philippines to meet the needs of just the American and British operating there today.[24] In theory, migration could ease some of the shortages—but for a long time, those and other developing nations have been even more resistant to foreign immigration than places like France and Germany.

Again in theory, this kind of outsourcing could spread to countries that get little of the business today, especially as wages rise in China and India. In the five years from 1998 to 2003 for example, wages of IT managers in India more than doubled. By 2015, shortages of engineers and software and financial professionals in China and India should drive their wages to the levels of countries like Mexico and Brazil; and in places like the Czech Republic and Poland, those wages will probably reach the levels of Portugal and even Spain. If governments in places like Malaysia, Nigeria, Peru, and Morocco were to start upgrading their training and educational efforts now—including intensive language training in English, French, German, or Japanese—they could become small new centers for offshore outsourcing of highly skilled jobs.

Still, the McKinsey analysts figure that over the next thirty years, less than 5 percent of U.S. service jobs will be outsourced to other countries.[25] That would mean a loss of 225,000 service positions per year in an economy today with more than 108 million private-sector service jobs.

In some cases, Americans—as well as Europeans and Japanese—effectively outsource medical services by picking themselves up and going to a foreign source. This practice is no longer mainly the very wealthy searching for the best cosmetic surgeon in Switzerland. More typically today, it's seniors crossing to Mexico or Canada for expensive dental work not covered by Medicare, Americans going to the Middle East and Asia for transplants, or middle-class Japanese coming to the United States for in vitro fertilization treatments. There are no data on precise volume, but the demand is sufficient to support private services to "accredit" foreign hospitals for medical tourists, such as Joint Commission International, Inc., that certifies eighty medical facilities in Asia, Europe, and the Middle East (and sells consulting services to foreign facilities on how to get accredited). Ann Lombardi, an Atlanta travel agent, discovered she could save $4,000 by going to Germany for an outpatient procedure for a cataract—and when she returned, she set up European Medical Tourist, Inc., in North Carolina, to arrange similar savings for Americans willing to go to Germany, Belgium, and France for treatment. Similar services have a considerable history in Britain, where a number of firms market surgical procedures in France, Hungary, Malta, India, Latvia, and elsewhere to Britons who want to avoid long waits at the National Health Service and the high prices for private treatment in Britain.[26]

Some Americans businesses trying to cut health-care costs are looking in the same direction.[27] For example, Blue Ridge Paper Products in North Carolina, which self-insures, offers its workers a share of the saving from sending them to New Delhi for surgery; and a Florida health-care manager, United Group Programs, includes a Thai hospital as a preferred provider. Moreover, several Fortune 500 companies have recently investigated the possibility of large-scale contracting with foreign hospitals.

Moreover, globalization is not standing still; and over the next ten to fifteen years, western service workers will be hit by an entirely new form of offshore outsourcing. The future of offshore outsourcing lies in piecemeal tasks, in which workers in China, India, Mexico (and perhaps Malaysia, Nigeria, Peru, and Morocco) handle discrete parts of much larger and complex services, such as reading an X-ray, running a tax program, or analyzing a set of business data, networked through global information systems. A few years from now, when there are shortages of Chinese, Indians, and Mexicans with western-style degrees in engineering, computer science, and business administration, those countries will be able to train millions of their young people to carry out discrete subsets of those jobs. Moreover, the western customers for piecemeal outsourcing will probably reach well beyond the relative handful of huge global corporations that dominate it today. Over the next decade, thousands of midsize and large America and European hospitals, accountancies, and financial institutions will find it possible and practical to network through their information systems with workers trained in sets of discrete skills, from almost anywhere.

Piecemeal outsourcing began a while ago with data processing and call centers, but it proceeded more slowly than the outsourcing of manufacturing jobs in the 1970s and 1980s. One of the main reasons was language barriers, because most of the service tasks that have shifted to lower-wage countries involve communicating some kind of standard information. And that usually requires that the outsourcer and outsourcee (or its customers) speak the same language. That's why General Electric, American Express, British Airlines, and Microsoft set up offshore data processing and call centers not in China or Brazil, but India—along with the U.S. and British subsidiaries of European and Japanese companies. Given America's fast-growing Spanish-speaking market, Mexico has become another favorite place for the call centers of U.S. companies and the American subsidiaries of European and Japanese businesses.

For the same reason, companies such as Siemens, Lufthansa, and the French electronic giant Thomson, along with European subsidiaries of dozens of U.S. companies, have set up call centers in the Czech Republic.

The new push to piecemeal outsourcing of service tasks, however, rests on recent advances in software, and will reach far beyond call centers and data processing. Through the 1980s and 1990s, while computing power, transmission, and storage progressed by almost constant leaps, software development lagged behind. This began to change in the late 1990s, when software advances enabled companies to separate a program's design from its implementation in computer code.[28] In tech-talk, software programming became "modularized." Without this advance, the programming centers in Bangalore and Hyderabad would hardly exist today, regardless of the wages that Indian programmers will gladly accept. Moreover, since about the year 2000, breakthroughs in financial, accounting, medical, and engineering software have enabled companies for the first time to take apart other, very complicated tasks, such as the complex analyses required to value a company for a possible takeover or the tax liability of a global company under a range of scenarios. With these new software, a company can divvy out the pieces of large service jobs to any number of professional staffs, connected through Internet networks, and then assemble the results in one place.

Over the next decade, this software-driven piecemeal outsourcing could cost the jobs of an estimated eight to ten million service workers in advanced countries. The first stage won't destroy many jobs in the United States, Europe, or Japan. As the software technology advances, thousands of large and midsize Western businesses will be able to outsource parts of their services, and they'll start with other companies in the same city or their own country. In this, as in other forms of outsourcing, American businesses will likely lead, if only because this kind of outsourcing reduces their health insurance and pension costs.

By 2010 or sooner, American and British global companies will likely begin to outsource service tasks on a piecemeal basis to the developing countries where they're already well established. The third and major stage should come a few years later. In India, Mexico, China, and other developing countries, thousands of native entrepreneurs will set up firms to handle the piecemeal service outsourcing of the megacorporations. Once they're in place, Web-based middlemen companies will be able to make the same service and savings available to the thousands of American and British companies already adept in piecemeal outsourcing in their own countries. By 2015 and beyond, when Japan and much of continental Europe face baby bust labor shortages, many businesses there, too, will probably join in.

Complex manufacturing went through a similar process twenty and thirty years ago. But the economic dislocations and other changes will be much greater this time, because the demand and opportunities for piecemeal out-sourcing will be much larger in services than they ever were in manufacturing. While manufacturing makes up about 20 percent of every advanced economy, services account for three times as much; and every business, whether in man-ufacturing or services, uses enormous quantities of services. And Americans spend more on medical care and personal services than on housing and food— both of which also include large quantities of services. Moreover, interna-tional demand for services will inevitably grow over the next ten and fifteen years as developing nations follow the same pattern. As countries develop and people's incomes increase, without exception they spend larger shares of those rising incomes on services rather than things. As it is, global trade in services has grown twice as fast as global trade in goods for the last fifteen years. With piecemeal outsourcing of services, this is almost certain to expand. The silver lining for the advanced countries is that the rapid development of China, In-dia, and Central Europe will also increase their need for high-end business and professional services; and there, the United States and Europe will hold the trump cards for the foreseeable future.

While piecemeal service outsourcing is likely to cost several million Ameri-cans their jobs in the next decade, the software advances that make it possible could have other and perhaps more powerful effects that have little to do with globalization. In at least the United States, the impressive productivity gains racked up by firms in manufacturing, finance, retail and other sectors have come substantially from the way they've used new information technologies. IT has spread as far and fast as it has across the American economy because technological advances, cheap foreign production, and competition all dra-matically reduced the real price of IT hardware. These same dynamics now appear to be taking hold in software.

Over the next five to ten years, as software becomes more powerful and versatile, foreign software production expands, and (perhaps) competition increases, the real price of software should fall sharply, too. The inevitable result is that businesses of every sort will become even more IT intensive, es-pecially in sectors such as health, education, and personal services, which so far have adapted IT more slowly and reluctantly than manufacturing, finance, and retail.[29] If the spread of IT hardware is a reliable guide, this could pro-duce even greater productivity gains over the next decade than we saw in the last one. And that should mean higher profits, larger incomes, and ultimately more jobs—at least for people with the education and skills required to work with the technologies.

Software advances will certainly spread quickly across the world's advanced countries. But as we will see, Japan and much of Europe may not reap as many economic benefits as the United States and a few European countries. For more than a decade, businesses in Europe and Japan have invested in IT as eagerly as American companies, yet they've reaped much smaller productivity benefits for their money. For all their power and potential, information technologies don't make any economic difference at all unless companies also change the way they operate to take good advantage of them. In Japan and across much of Europe, countless laws and regulations create thousands of small obstacles and additional costs to precisely the kinds of changes—relocating facilities, firing or even reassigning workers without the appropriate experience and skills, shifting hours of operation, and much more—required to wring higher productivity and profits out of IT investments.

HOW GLOBALIZATION REALLY AFFECTS JOBS AND WAGES IN ADVANCED COUNTRIES

Globalization has another nasty surprise for working people in America, Europe, and Japan. It begins with the most basic dynamic of modern globalization, the waves of technology, investment, and expertise that have come to China, and to a lesser extent India, from companies in America, Europe, and Japan. Across much of the world, the labels of everything people use are silent testimony to the results. Chinese and Indian companies now produce some sophisticated standard goods, from laptops and semiconductors to video games. They also still make thousands of less-sophisticated things that they've been producing for decades, from steel and concrete to furniture and toys—and thanks to all the transfers from companies in more economically advanced places, they're producing them in much greater quantities, and of much higher quality.

Workers in advanced countries haven't made toys, steel, and concrete in great quantities for a long time; and before China and globalization, most of those jobs had gone to Mexico, Brazil, Taiwan, and Korea decades ago. U.S., European, and Japanese companies continue to produce most of the world's most advanced goods; and once again, many of the jobs producing the parts that go into those goods also had shifted to developing countries some time ago—to the Asian Tigers, and as their wages and other costs steadily increased,

to countries like Mexico, Thailand, and Malaysia. This makes a critical difference today, because the thousands of jobs now lost every year to China and India are mainly coming not from America, Europe, or Japan, but from Mexico,
Thailand, Malaysia, and other countries like it.

It couldn't be otherwise. The International Labor Organization and U.S.
Bureau of Labor Statistics report that an average manufacturing worker
earns $23.65 per hour in the United States, $24.63 in France, and $21.76 in
Japan, compared to $6.38 in Taiwan, $13.56 in Korea, and $2.63 in Mexico.[30]
Manufacturing workers earn about 70 cents in China and 40 cents an hour
in India. Chinese and Indian workers employed in the modern facilities
set up by foreign companies earn more—as they do in Korea, Mexico, and
Thailand. But the big labor-cost differences of operating in China and India, compared to Taiwan and Mexico, are the main reason why China is now
claiming hundreds of thousands of jobs every year from other developing
countries.

The World Bank reports that China's merchandise exports soared almost
1,300 percent in the last fourteen years, shooting up from $68 billion in
1990 to $837 billion in 2005. And they're still growing by 20 to 25 percent
a year. At those levels, China's exports swamp those of its manufacturing rivals in other developing countries—they're almost two-thirds more than all
the rest of East Asia, including four times that of Malaysia, five times that of
Thailand, and eight times that of Indonesia. Chinese exports are also more
than twice those of Mexico, more than five times those of Brazil, and 30 percent greater than all of Latin America and the Caribbean nations.

Size alone matters a lot here. Chinese companies, both domestic and
foreign-owned, armed with technologies and business methods comparable
to or better than those used by producers in scores of other developing countries, can now produce much of what their rivals in Thailand, Mexico, or
Malaysia have been exporting—in much greater quantities and at lower cost.
China's modernization and globalization are producing waves of new competitive pressures for scores of industries in dozens of other developing economies.
The good news for them is that the fast-growing demand for skilled Chinese
labor is pushing up Chinese wages. From 1997 to 2003, real manufacturing
wages in China doubled, compared to increases of about 30 percent in most
other places in Asia (and 10 percent or less in most advanced countries).[31]
The downside is that it will take years before Chinese wages catch up with
those in Thailand and Mexico, much less Korea and Taiwan. In the meantime, Chinese businesses are squeezing their companies out of many industries and markets.

The businesses and workers in those countries feel the squeeze most directly, but it doesn't end there. Global capital markets transmit some of the competitive pressures from industry to industry and country to country, until they reach businesses and workers across the United States, Europe, and Japan.

Here's how it happens. Zhejiang Linhai Guohai Forging Co., Ltd., a Chinese manufacturer of forged metal parts, expands its production of automobile and motorcycle parts, at prices that undercut producers in Mexico and Thailand, while Dongguan Sunpower Enterprise, Ltd., a large producer of hotel furniture, and the Chinese subsidiary of Perry Ellis International, ramp up their operations and undercut rival producers in Malaysia and Egypt. As customers in Munich, Seattle, and Seoul learn about it, some of the less productive makers of metal parts, hotel furniture, and apparel in Mexico, Thailand, Malaysia, and Egypt are squeezed out of business—and on the margin, capital and expertise in those countries shifts from those industries. It migrates either down the value scale to agriculture or other commodities, or up the scale to, say, basic electronics or more sophisticated equipment. In either case, the infusion of capital and expertise makes those industries in Mexico, Thailand, Malaysia, or Egypt a little more competitive. When the resources shift down the value chain, the result squeezes producers in poorer countries. When it shifts up to more advanced products, the additional capital and expertise for Mexican and Thai electronics manufacturers or Malaysian and Egyptian equipment makers puts new pressures on rival producers in those industries in, say, Korea and Brazil. This process repeats itself; and on the margin, capital and expertise shifts in those economies as well, often up the value chain again to, say, LCD makers in Korea and auto producers in Brazil. This time, the new competitive pressures may begin to squeeze LCD producers in Germany and auto makers in the United States.

China's manufacturing sector is so big and diversified that fast-growing exports are intensifying competition across scores of industries in scores of countries, ultimately ratcheting up competitive pressures everywhere in the world that's connected through global capital markets and trade. On balance, this dynamic is accelerating modernization in many other developing economies. But once these competitive pressures reach American, European, and Japanese businesses, they have a very different effect.

When the competitive pressures that begin in China and are transmitted from country to country finally reach America, Europe, and Japan, they have the same impact as increased competition has within any advanced economy:

The competition makes it harder for companies to raise their prices. For the last several years, thousands of companies in the United States, Europe, and Japan have found that they have, using the economist's term for it, less "pricing leverage." And what happens in companies with little pricing leverage when their costs go up? In the United States, for example, health insurance costs of business have risen by about 90 percent since 2001, their energy costs have more than doubled, and pension costs for many of them have gone up sharply as well. When a firm's costs increase and competitive pressures prevent it from raising its prices enough to cover those increases, it finds other costs to cut. In recent years, most of them have turned to jobs and wages.

One more worldwide factor intensifies the squeeze on companies and their workers, even though it's one that the companies ultimately welcome: The global supply of capital has expanded less in the last decade than the global supply of labor, with the result that the returns on capital that global investors look for have risen. These overarching factors, again all arising from current globalization, mean that financial markets expect companies to earn greater profits—the profits of U.S. companies have broken all records—so companies can't use all those profits to protect their workers' wages or jobs without driving down their stock price, and with it their investment levels and future profits.

This squeeze on the ability of companies to raise their prices when their costs increase is rapidly changing two basic dynamics in advanced economies. First, it changes the relationship between how fast an economy grows and how many jobs it creates. The United States provides the clearest example. Here, the squeeze from globalization is measurably weakening the vaunted American job-creating machine. The first evidence came in the 2001 recession, when job losses relative to the actual decline in economic growth were six times greater than in previous postwar recessions. It didn't stop there: Once the U.S. recovery took hold, it took four years to get back to prerecession job levels, compared to eighteen months in the previous recovery. Five years into the current American expansion, job creation was still running at half the rate of the preceding expansion. Despite this unprecedented slowdown in U.S. job creation, the official unemployment rate has remained low—but only because the number of working-age Americans looking for jobs has declined even as the economy has grown.

In much the same way, globalization is weakening the long-standing connection between increases in the productivity of workers and the wages they earn. Since 2001, labor productivity in the United States has grown on average by more than 3 percent a year. That's the best U.S. performance in decades. Yet, for the first time on record, the average real wages of American workers declined through five years of strong productivity growth. Even

counting the value of the employer-paid health insurance and pension contributions whose costs have gone up so much, the average American worker's total compensation has increased little despite the five years of strong economic
growth and productivity gains. Andy Stern, the reform-minded leader of the
Service Employees International Union, recently put it this way:

> The challenge in America is not to stop globalization. The real
> question is how, in the long term, can the jobs that remain in
> America become decent-paying jobs . . . If we don't find [the an-
> swer], then Alan Greenspan is going to be proved right—the gap
> between the rich and the rest of the population is growing so wide
> and so fast that it's going to threaten democratic capitalism.[32]

Nor are these effects limited to the United States. Across most of Europe and
Japan, job creation has been weak even by their standards, and wages there
have declined or stagnated. "Globalization is a reality, but one with the ability
to mutate into a monster," according to President Horst Kohler of Germany,
who holds a Ph.D. in economics. "For this reason, we have to "constrain it
with carefully considered rules of governance."[33] These new pressures on jobs
and wages will grow stronger in coming years. The competitive pressures emanating from China are not going away. China's exports should continue to
grow by leaps until at least the next global recession—which will also cost
jobs and reduce wages in America, Europe, and Japan. As China's lightning
economic development continues to push up wages and other costs there in
coming years, it should ease some of the initial pressures on rivals in other developing countries. But India's large and fast-developing manufacturing sector, with wages today that are still a fraction of China's levels, may take its
place in this new global process.

America, Europe, and Japan cannot escape this hidden cost of globalization; instead they'll have to outrun it. They'll have to figure out how to relieve some of the cost pressures on their businesses. We saw earlier how hard
it will be to reduce the costs of pensions, especially as tens of millions of western baby boomers begin to retire in the next five to ten years. As we will see a
little later, containing rising costs in health care and energy will prove just as
difficult, in both practical and political terms. It won't matter much for the
top 20 percent or so of each country's people, the professionals and managers
whose advanced education and experience give them something valuable to
bargain with in global markets. Ironically, it also may not matter much for
the bottom 20 percent or so of workers, whose physical labor is much less
subject to global pressures and who, in the United States at least, often don't

receive private health insurance or pensions anyway. But if governments and businesses in America, Europe, and Japan don't meet this problem head-on, the great middle classes in these countries could face stagnating incomes or worse for the next generation.

THE MONEY THAT MAKES THE WORLD GO ROUND: THE GLOBALIZATION OF CAPITAL

Intellectual and political capital may be the critical, scarce resources for most large corporations, but much of the current stage of globalization is still about money. The same might have been said of money and finance in other periods of economic development, but not in the same way. As capital markets have gone truly global, the sheer volume and variety of financial assets flowing through the world's economies dwarf that in any previous time. The volume of capital moving around the world, coupled with the price-squeezing effects of competition under globalization, have driven down global interest rates and created the conditions for modern boom-and-bust cycles in housing and stock markets in the advanced countries. Finally, modern globalization has substantially reversed the traditional direction of these flows, with hundreds of billions of dollars going from developing nations to the world's most advanced country. All of these features make the current operations of global capital markets a critical leading edge of modern globalization.

For most people, countries, and now the world as a whole, the most important feature of money is how much of it there is. The world is awash in fast-growing stocks of financial assets—money. For some time, global capital has been growing faster than world GDP, faster than global trade, and faster than worldwide savings.[34] A fair estimate of the global capital pool today is more than $150 trillion, more than three times world GDP and more than three times its size less than fifteen years ago.[35] By 2010 it will reach $200 trillion and almost certainly double again by 2020 to more than $400 trillion. Moreover, the rate at which dollars, yen, and euros move from one country to another (and often one currency to another) is accelerating even faster than their quantities, tripling in just the last ten years and reaching more than $5 trillion a year in 2006.

The fact that global financial assets are growing faster than global GDP is meaningful. Since these assets are claims on the future, this rapid growth signals that, overall, the world's rich people and rich businesses that hold most of them are bullish about the future—certainly more so than in 1980, when the world's financial assets were growing much more slowly and totaled just

10 percent more than world GDP. But that optimism doesn't apply to every country. With 4 percent of world GDP, Latin America holds just 2 percent of the world's financial assets, because Latin Americans as well as foreigners are largely uninterested in investing there. Much of Latin America remains off the track of globalization.

The historic size of both the global capital pool and capital flows from country to country ultimately reflect the unprecedented prosperity of much of the developing world, along with the revolution in information technologies. After the last fifteen years of massive transfers of western investment, technologies, and expertise to many countries that had largely stagnated for decades or centuries—China, India, Malaysia, and Mexico, for instance— their businesses and people are amassing enormous amounts of new savings and wealth. Moreover, modern finance exchanges much of this wealth for corporate bonds, bank deposits, stocks, and other kinds of financial assets— economists call this process "securitization"—so that much of this new prosperity ends up in local or national capital pools.

Information technologies play a special role in moving these local and national pools of financial assets into the global capital system, because most of these financial assets now exist in the form of the bytes created, stored, and transmitted by those technologies. And no sector has more thoroughly globalized itself than banking and finance—Citigroup has operations in ninety countries, for example, and the London banking giant HSBC does business in seventy-six different nations. These technologies allow them to not only link up and manage their global operations, but also turn physical wealth into securities and financial deposits that, unlike paper or gold, can move from account to account and country to country in a nanosecond, with no shipping costs. So, while there may be relatively limited numbers of businesses in Chile or Indonesia—or even China—that can profitably use all the capital they create and save, firms and wealthy people in Santiago and Jakarta can easily and seamlessly invest their profits and savings in companies in Shanghai and San Jose, or lend it to the government of the United States. This whole, complicated process ends up bringing much more of the real wealth created through globalization into the monetary or financial base of scores of nations, which gives rise to yet more credit or money.

China, India, and a few other developing nations may be great beneficiaries of the globalization of capital, but most of the world's financial assets are still held in western financial institutions. The United States, Europe, and Japan dominate this sphere even more than they dominate GDP, with more than 80 percent of worldwide financial assets held in their institutions. Where assets are held is different from who owns them. With about 27 percent of world

GDP, America holds 37 percent of the world's financial assets—some $44 trillion—but non-Americans hold 12 percent of U.S. equities, 25 percent of U.S. corporate bonds, and 44 percent of U.S. government securities. In July 2007, the Chinese government held $408 billion in U.S. Treasury securities, and $611 billion were held in Japan. Together, they account for more than 20 percent of the U.S. federal debt held outside America's federal, state, and local governments. It's a window into how bullish global investors are about the prospects of the United States.

The currencies that dominate global capital markets testify to the same judgment, and to America's general preeminence in the global economy. In 2004, 45 percent of all global equity trades and 43 percent of bond trades were conducted in dollars—and since dollars are the currency for buying and selling U.S. stocks and bonds, these proportions tell us whose stocks and bonds are being traded through global markets. They also suggest how bearish global investors have been about Europe and Japan: Only 15 percent of global equity trades in 2004 were conducted in euros and 9 percent in yen, even though the Euro economies are almost as large as the United States and Japan's economy is one-third the size of America's. Britain's prospects look a little better to global investors, with about 8 percent of world equity trades conducted in pounds in 2004.

The sources of the growth of these assets in different countries provide another window into what drives different economies. America is the center of global business life, and the growth of U.S. financial assets comes mainly from corporate equities, fueled by IPOs and the high price–to-earning ratios of American stock markets. In Japan, where equity markets only recently came back from years of decline or stagnation, government debt fuels the growth of that nation's financial assets. Europe, still struggling to figure out how to stimulate private business while preserving unsustainable welfare systems, derives most of its new financial assets from the privatization of government enterprises and, again, government debt. And in China, where equity and bond markets are still largely undeveloped, its fast-rising financial assets come mainly from personal savings held in bank accounts as a hedge against illness and retirement in a society without broad-based health insurance or public pensions.

America's financial system will still dominate in 2020, though less than today. Alan Greenspan recently told a committee of U.S. senators that today and for as long he can imagine, "people find U.S. Treasury securities the safest in the world."[36] U.S. government securities will likely still be the world's safest financial assets, and American stocks and bonds will likely continue to attract more foreign investors than those anywhere else. Business writers sometimes attribute this to U.S. "political stability," which is mainly a euphemism for America's conservative economic culture. Whether American politics leans left or right, businesses almost always have a freer hand than in most other advanced economies. Over the last seventy-five years, there are almost no counterparts in the United States to the nationalized industries in Europe—many recently privatized—much less to the state-owned enterprises in China. For as long as anyone can remember, Americans have admired business more than Europeans or Asians, even trusting it with providing such public goods as health care and pension coverage. Moreover, while European and Japanese economic arrangements have long tended to focus on preserving wealth, Americans have concentrated more on building it. These cultural dispositions lead to certain real differences that tend to produce higher returns on investment in the United States, most notably the much lower legal and regulatory hurdles for American entrepreneurs. In a global economy, the innovation (and jobs) created by young businesses produce enormous wealth that quickly becomes financial assets that investors want. And perhaps most important for global investors over the next decade or so, business taxes are almost certain to remain lower in the United States than in Japan or most of Europe, because the public burden of America's social commitments (apart from health care) will grow more slowly there than in Japan or most of Europe.

The biggest change for the world of capital in the near future will be the growing importance of China's financial markets and assets and the relative decline of Japan's. Today, China holds about $5 trillion in financial assets, or less than one-third those of Japan, one-seventh those of Europe, and one-ninth those of the United States. But China's financial assets are growing faster than those anywhere else except the United States. Most of the growth is in the deposits of Chinese savers in the country's state-owned "banks." Today, China's state-owned banks are still mainly pass-through institutions that channel their depositors' savings into credit for local governments and often insolvent state-owned enterprises. In short, they have little in common with European or American banks that create credit for private businesses and consumers based on reasonably objective criteria. More broadly, China still lacks most of the independent public and private institutions and rules that comprise a modern financial market, and that turn a nation's growing wealth into paper (or bytes) that can be traded around the world. Most of the domestic businesses listed on China's young stock markets—the "red chips"—got there not by meeting capital standards and providing transparent financials, but by political favor. Nor are the brokers that carry out stock trades in China bound by the rules that create confidence in western exchanges. Perhaps most important for global capital markets, China still lacks a convertible currency.

All that will almost certainly change over the next decade, though gradually and very painfully. Already, more than one hundred western banks maintain branches in China, principally to serve foreign-owned businesses. However, recently China has allowed Citigroup, Société Générale, and Bank of America to purchase stakes in state banks, and they will bring modern banking practices with them. The challenge for China in allowing a modern banking system to take hold is that, without the credit from state-owned banks, most remaining state-owned enterprises will be bankrupt in fairly short order, throwing millions of Chinese out of work. Similarly, China has taken a few small steps to bring the remembi a little more in line with other currencies. The challenge here is reconciling a truly convertible currency with Beijing's basic strategy of driving growth through cheap exports. When foreign investors know they can take home their profits in their own currencies, they'll rush into the world's fastest growing economy—and that rush will drive up the value of the remembi and increase the prices of Chinese exports everywhere else in the world. The result could be the ruin of thousands of Chinese companies and, again, millions of Chinese thrown out of work. The only way China can maintain strong growth while its exports falter is to become a more consumption-based economy, like other major countries. And that will require that the Chinese government restore some semblance of broad health-care and pension coverage; because until that happens, most Chinese will still prefer to save rather than consume.

China's economic modernization has largely been a story of shocks executed carefully and ruthlessly. It will take the Chinese government at least another decade to create a modern financial system—still a small fraction of the time it took Europe and America to build theirs. If they carry it off, by 2020 China could have the world's second-largest stock of financial assets and the world's second-most-important financial market.

Whether or not China can make this leap, global capital and its flows will still double or triple again over the next ten to fifteen years. Directly or indirectly, these markets will finance tens of millions of new industrial and commercial jobs in developing countries, transforming the lives and prospects of hundreds of millions of people. The effects for people living and working in the advanced countries, however, are more subtle and mixed.

Most fundamentally, the huge increases in the sheer size of the global capital pool mean that, for nations as for corporations, capital is no longer scarce. One hugely important consequence is that interest rates have been historically low since globalization took off. According to the IMF, since 2000 the world's eight largest advanced economies all recorded lower real long-term interest rates than in any comparable period of the last forty-five years, in both absolute terms and relative to how fast those countries grew. Even countries with dismal national savings rates and high levels of foreign borrowing, most notably the United States and Britain, have maintained these low interest rates. The only precedent for this is the last period of accelerating globalization, from the mid-1890s to World War I, when interest rates were also unusually low relative to growth.[37]

A second, hugely important consequence is that global capital markets are becoming more powerful than central banks. In recent years, the U.S. Federal Reserve and the European Central Bank both discovered that their decisions to raise short-term interest rates had little effect on the long-term rates that drive most borrowing and, ultimately economic activity. From June 2004 to June 2006, for example, the Federal Reserve raised short-term rates from 1.25 percent to 5.25 percent in seventeen separate steps. Yet the rate for AAA corporate bonds was about 6 percent in both June 2004 and June 2006, and a conventional mortgage cost just four-tenths of 1 percent more in June 2006 than in June 2004. In both cases, the long-term rates actually fell through the first year of the Fed's "tightening" cycle, and then slowly returned to roughly their original levels when the Fed policy change began.

In this instance at least, the financial markets have just been smarter about globalization than central bankers. According to the traditional rules of central banking, when capital, credit, and consumption all grow as quickly as they have recently—and are likely to do in future years—fast-rising inflation usually follows. But globalization is changing those rules. Since globalization took off in the mid-1990s, inflation has decelerated sharply almost everywhere in the world. In the advanced countries, inflation slowed to just 2 to 3 percent a year from 1995 to 2005. Similarly, price levels in Asia (outside Japan, where deflation was the rule in those years) rose about 3 to 6 percent a year, or half the rate of the previous decade. Price increases also fell by half in Latin America, which has enjoyed relatively mild inflation (for them) of 10 to 20 percent a year.[38] A senior member of the Bank of England, Dr. Deanne Julius, took this wide view and concluded, "Intensified international competition—what some call globalization—and the spread of new technologies may be thought of as the current drivers . . . towards the old norm of low global inflation."[39] The IMF is still generally skeptical about how deeply globalization is dampening inflation—it's part of the IMF's business, after all, to monitor inflationary forces in developing countries, and nobody likes to acknowledge that part of its mission has become less important. Yet even the IMF says now that, "globalization has undoubtedly provided some break on inflation."[40]

The intensification of competition that makes it harder for businesses to increase their prices, even when their costs go up, isn't the only reason inflation is now rising more slowly. Nearly as important, all the flows of western capital, technologies, and expertise have made large economies like China, India, and Mexico much more productive, dampening inflation there and creating floods of cheap exports that also lessen the inflation in Europe, Japan, and America. Finally, the Asian financial crisis of 1997–1998—itself a phenomenon of globalization—has taught governments in most developing countries that choosing economic policies that risk more inflation can carry a very large price.

Globalization won't end inflation. Most notably, price shocks in energy and other important commodities still drive up prices around the world. But given the underlying trend of lower interest rates and lower inflation, the people in most advanced and fast-growing developing nations will be able to borrow and consume more over the next ten to fifteen years than they can probably imagine today. A 2005 survey of 1,656 top business executives by *The Economist* estimated that worldwide consumption spending will increase 60 percent faster than worldwide GDP, rising from around $27 trillion a year today to $62 trillion in 2020. The clothing, electronics, and personal services that affluent Americans and Europeans enjoy today will be available to both almost everyone there, and to hundreds of millions of households in the rest of the world. Consumption will especially soar in fast-growing developing countries. By 2020, China will be the world's second-biggest consumer market, with as many privately owned passenger cars than America in 1980—an estimated 130 million vehicles—and total consumer spending in India will match France. All that credit also will be available at low interest rates to businesses around the world, but business investment will concentrate itself where the prospective returns are most promising. The countries best situated to take advantage of globalization will attract most of it, which is why the executives in *The Economist* survey see much stronger business growth between 2005 and 2020 in the United States, China, and India than in most of Europe or Latin America.

In economic life, everything comes at a cost. All that liquidity sloshing around the world is producing a new and painful downside in the emergence of new asset bubbles that eventually burst. In the last decade, housing prices in most advanced countries rose faster than during any comparable period since World War II. From 1997 to 2005, housing prices marched up 69 percent in Italy and 73 percent in the United States, 84 percent in France and 87 percent in Sweden, 114 percent in Australia, 145 percent in Spain, 154 percent in Britain, and 192 percent in Ireland.[41] (The only exceptions have been the two weakest major economies, Japan and Germany.) By 2006, the housing bubbles in the most overheated markets, such as Australia and Britain, were beginning to burst or at least leak. In 2007, America's bull market in housing also stumbled. Japan offers a grim example of what the near future may hold for homeowners in some countries: Housing prices soared there in the 1980s; and when the bubble burst in the 1990s, average prices fell 40 percent.

These asset bubbles are not limited to housing. From 1997 to 2004, stock markets around the world rose and fell more sharply than during any comparable period since World War II—up or down by an average of more than 16 percent a year in Britain, 18 percent a year in Austria, 20 percent a year in the United States, and 24 to 27 percent a year in France, Italy, and Germany.[42] This volatility was even greater in emerging markets such as Russia and Turkey, where stock prices have risen or fallen by an average of 81 percent and 92 percent a year respectively, since the late 1990s. Globalization will also change stock markets themselves, and in ways that could further increase their volatility. With cross-border capital flows likely to top $10 trillion a year within five years, the need for the traditional "specialists"—selected brokerages that intervene when buyers disappear in the New York Stock Exchange and a few other major markets—will melt away. Nations may still retain their own national stock and bond markets, but the mighty New York and London exchanges will be indistinguishable from the NASDAQ and the major continental European and Asian markets. Listing standards will be internationalized, and almost all trades everywhere will be handled through fiber-connected computer networks matching buyers and sellers around the world.

With the globalization of capital flows almost certain to accelerate, and inflation (apart form energy prices) likely to remain moderate, boom-and-bust markets will become more common. The shrewd and the lucky will get rich, while the best advice for most stockowners and homeowners is to do your best to get through the down times.

HOW GLOBAL CAPITAL CAN LOSE ITS BALANCE

Global capital markets have certain implicit rules, and nobody likes to play by the rules all the time. China invites foreign investment but won't allow its foreign investors to convert their Chinese profits to their own currencies. But the largest transgressor is the world's most indispensable economy, the United States. Since the turn of this century, the United States has been consuming trillions of dollars a year in foreign imports, supporting growth or booms in countries around the world. To pay for those imports and support its own investments, America has used global capital markets, year after year, to borrow most of the world's excess savings. If those markets finally apply their own rules to the United States, the current global boom will become a global bust.

Even America can't break the rules all by itself. For more than a half decade now, the United States has sustained its strong growth by saving little and allowing Americans to consume much more than they produce—some $600 billion to $800 billion more—through imports. It's not hard to get Americans or anyone else to consume beyond their means—large budget deficits, low interest rates, easy credit, and arrangements that have allowed tens of millions of homeowners to cash in more than $3 trillion in housing bubble equity have done the trick. And every year since 2001, Americans have spent between 4 and 5 percent more than they earn, while their government has spent 2 to 4 percent more than it takes in. With America intent on boosting consumption in an economy five times China's size and larger than Japan, Germany, the UK, and France combined, much of the demand for European, Chinese, and other Asian exports now comes from the United States. For their part, Asia, most of Europe, and the oil exporting countries maintain their own growth and jobs by consuming much less than they produce and exporting the rest, mainly to America.

The hitch is, how can the United States, with a dismally low national savings rate, pay for all its consumption and budget deficits, plus the investment needs of its growing businesses? The funds are out there in the world—most other advanced nations along with China and the oil exporting countries have been consuming so much less than they produce, that they're saving at levels far beyond their own economies' investment needs. The answer thus far is that America has used global capital markets, year after year, to sell U.S. government securities, stocks, bonds, and real estate to everyone else, to pay for imports and help fund its deficits and the capital needs of its businesses. Global capital markets are managing a unique and shaky symbiosis among the world's major economies.

These arrangements are rewriting the relationships between rich, not so rich, and poor nations. For decades, capital moved from the rich countries to developing markets, where opportunities for fast growth and high returns are usually most abundant. Now emerging countries and the oil exporting nations are financing two-thirds of the world's richest nation's capital shortfall, with other advanced countries making up the rest. These arrangements also are materially changing who owns America. By the end of 2006, non-Americans owned $15.1 trillion in U.S. assets, equal to 20 to 25 percent of the value of all physical and financial assets in the American economy.[43] Moreover, all those foreign owners can take home the interest, dividends, capital gains, and rents that their U.S. assets generate every year. Since Americans own about $12.5 trillion in foreign assets, the annual return on the $2.6 trillion difference—about $140 billion—is what it costs Americans each year to be such large debtors to the rest of the world. That's equivalent to the cost of providing medical insurance to every uninsured American.

No household or corporation could indefinitely finance its consumption and shortfalls by selling off more and more pieces of itself, and neither can a nation. As it is, the large foreign financial institutions and central banks that dominate global capital markets now have huge stakes in U.S. securities and other dollar assets; and like all good investors, they know that it's just common sense to diversify their holdings. To begin, their trillions of dollars in U.S. assets has left them hostage to any downturn in America's economy. They also know that history is full of examples of countries with fast-rising foreign debt reducing their liabilities by pumping up their money supply and letting their inflation rates rise. If the United States were to start printing lots of money, it would reduce the value of the dollar and so erode the value of the U.S. holdings of foreign investors in their own currencies.

If the world's megaconsumer and megaborrower were any country but the United States, foreign central banks and global investors would have stopped their huge lending and investing there long ago. But U.S. assets are still generally more desirable than those of other countries, because the political and economic prospects of the world's sole political and economic superpower are still, on balance, better than those of almost anywhere else. The U.S. dollar is also still the world's reserve currency, so foreign central banks have been willing to hold enormous amounts of U.S. Treasury securities as insurance against a future crisis of confidence that could threaten their own currencies.

Even so, the willingness of China and other emerging economies to finance much of the U.S. current account deficit costs them a great deal of money every year. To start, U.S. Treasury securities produce relatively little income. In March 2006, Lawrence Summers, then president of Harvard University, told the Reserve Bank of India that if it invested India's excess reserves in relatively safe private financial assets, it could generate returns greater than India spends on health care—and the current excess dollar reserves of China, Taiwan, Thailand, Russia, and even Malaysia and Algeria are relatively larger than India's. All told, if America's various national financiers used their excess savings to finance more national infrastructure, or even just invested them in a diversified portfolio of global securities, their return would be hundreds of billions of dollars higher a year—as much as all global foreign aid.

These imbalances will not go on indefinitely because, ultimately, they make no economic sense for everyone involved. The crisis could start in a variety of ways: At some point, foreign private investors—who stopped expanding their U.S. dollar holdings in 2004—will begin selling some of them; and even higher purchases by the major foreign central banks won't be enough. Or China's economy could simply slow down, cutting into its excess savings. Or domestic investment in Japan and Asia's emerging economies could finally recover, cutting their excess savings. Or oil prices could fall, driving down Middle Eastern savings. Or perhaps the United States could trigger the crisis on its own, by letting its frustration with huge trade deficits turn America more protectionist.

Even if none of these scenarios comes to pass, globalization has created a trap door under the U.S. and world economies. Next month, next year, or later, one or more foreign central banks will begin to get cold feet about sinking more of their wealth in U.S. assets. When one shifts course, others will follow to cut their losses; and global capital markets will extract a swift and large cost for all the years of American overspending and undersaving. Without infusions of $50 billion to $70 billion every month in foreign funds, the meager U.S. savings won't be enough to fund its deficits and normal business investment. The only option for the U.S. Treasury and American businesses looking for capital will be to offer higher returns to placate the anxieties of foreign investors. When you borrow money, higher returns take the form of higher interest rates, and a sharp rise in interest rates required to slow down this kind of crisis would also sharply slow down the U.S. economy and send U.S. stock and bond prices into a swoon.

In a time of globalization, that crisis won't end there. As the downturn drives down U.S. stock and bond markets, foreign investors will join the sell-off and a vicious cycle will begin. In the worst case—one that many other countries have gone through—everybody gets cold feet at once and dumps U.S. stocks and bonds en masse. American markets plummet along with the dollar; U.S. interest rates soar to attract buyers for the government's continuing debt, and the American economy stalls out. If this happens, everybody will lose a lot of money, because a recession will cut into America's demand for foreign imports. As European and Asian exports to the United States fall back, growth will slow sharply across the world.

This is not a fantastic scenario. Something much like it began to unfold in 2004, when private foreign investors largely stopped buying more U.S. stocks and bonds. With less demand for dollars, global capital markets began to work their will: The dollar fell against the euro, European exports to the United States slowed down, growth and job creation declined in much of

Europe, and U.S. interest rates began to rise. The crisis faded when the Chinese, Japanese, and Saudi governments decided to buck the markets. China and Japan were determined to keep their currencies relatively cheap, so they could continue to depend on exports for their domestic growth; and the Saudis joined in because most oil sales are denominated in dollars. The three governments entered the capital markets and used their own reserves to buy tens of billions of dollars, which they promptly converted mainly to U.S. government securities. In a real crisis, they won't have enough remembi, yen, and rials to turn the tide.

As it is, the globalization of capital has not only left the economies of China, Japan, and much of Europe hostage to U.S. consumer demand, it also has made the United States hostage politically to the investment decisions of foreign governments. Everyone knows how much damage a foreign government could inflict on the United States by simply selling off U.S. assets, and that knowledge has already conferred new political power on America's big foreign funders. If the risk of China withholding its purchases of U.S. assets didn't hang over America's head, perhaps the American administration would press Beijing about its wholesale violations of U.S. software copyrights and pharmaceutical patents. In at least one recent U.S.-Japanese trade dispute, these concerns apparently did affect U.S. policy. The story goes that in 2005, President Bush called Prime Minister Koizumi to press him about Japanese import of U.S. beef. That afternoon, the Japanese finance minister told the press that the Bank of Japan should consider "rebalancing" its foreign currency holdings, and the next day, U.S. trade negotiators quietly shelved their complaints about beef. As it happens during the Suez crisis a half century earlier, when Britain and France depended on U.S. purchases of pound and franc assets, President Eisenhower had used the same threat to force their hands on support for Israel. The power that America's major financers could wield over it today could prove especially troubling in the Middle East, where high crude oil prices have already swollen the dollar holdings of Saudi Arabia and other large oil exporters. As the Saudis' dollar holdings expand and U.S. borrowing increases, their potential influence over U.S. policy there inevitably will rise.

With or without a dollar crisis, however, this will probably be the last of the great global capital imbalances. Global capital flows are growing even faster than the U.S. current account deficit and global savings, so by 2015 or earlier, even the United States and its funders will not have the resources to hold off the discipline of global capital markets. In a decade or sooner, the American dollar will move up and down much like other currencies, subject to painful devaluation if U.S. administrations keep on running large budget deficits and U.S. consumers keep on spending more than they earn and produce. Since devaluations will increase prices for foreign imports in the United States, U.S. inflation and interest rates could trend a little higher, despite all the forces keeping them down. With government spending certain to rise with the retirement of American boomers, global capital markets will probably force the United States to raise national savings, and perhaps by doing what seems politically unthinkable today, raise taxes. At least that should be possible in America, where—in sharp contrast to much of Europe—taxes are still sufficiently low that raising them a bit shouldn't damage the economy or trigger a popular revolt.

The power of global capital markets will present difficult choices for other countries as well. If Americans end up consuming less of what the rest of the world produces for export, China, Europe, and Japan will have to rely more on their own consumption to keep their jobs and incomes growing. That adjustment won't be easy anywhere. China will have the great advantage of fast-growing incomes—but convincing hundreds of millions of Chinese to spend rather than save more of their growing incomes will, again, require major expansions in health-care and pension coverage. In Japan and much of Europe, where dismal demographics and intense global competition are already producing stagnating incomes, it may prove just as difficult to convince people to consume more. Cutting taxes won't be an option, given the enormous financing burdens that their pension and health-care systems face. In fact, these pressures will likely force some of those countries to cut benefits and, in a real crisis, perhaps even raise taxes. In either case, it will leave millions of Europeans and Japanese with even less to spend than today.

As global capital markets force Americans to consume fewer European and Japanese exports, their real choice may come down to either enduring years of economic stagnation or trying to slowly boost incomes by rolling back decades of cherished labor market restrictions and industry regulations. Here, it mainly comes down to politics, especially a country's willingness to jump off a cliff of big political changes. By all the recent evidence in this regard, another long period of hard times is the most likely outcome for Japan, where people stuck with the LDP through more than a decade of economic stagnation, rather than shake up the most inefficient service sector in the advanced world by reforming regulations that protect thousands of mom-and-pop and medium-size businesses. By most recent evidence, many of the smaller countries in Europe—Sweden, Finland, Switzerland, Spain, and perhaps a few more—could well go down a big-reform route with more commitment than the major countries. And Ireland just has to keep on doing what it's doing now. But Italy seems almost destined for real economic decline in the near future, given the scale of its problems and its fragmented politics. Germany and France periodically take small steps to reform themselves and then usually lose their political nerve when powerful businesses, unions, or public opinion objects. It will take an extraordinarily talented new leader—perhaps Nicolas Sarkozy—plus a real shake-up in their political parties to make a real difference there. Without that, tens of millions of German and French people are also likely headed for a slow-motion decline. Among Europe's large countries, only Britain has shown a recent tolerance or taste for big changes. Britain's political parties, like America's, have shown a modern tendency to reinvent themselves—Thatcher's Conservatives and Blair's Labourites, like Reagan's Republicans and Clinton's Democrats, faced up to some of the reality facing their countries and unraveled a number of cherished regulations and outdated social programs. The work Britain has done, especially in its once-nationalized sectors and public pensions, should be enough for the English to muddle through more successfully than their continental counterparts. But a return to real prosperity will take another Thatcher or Blair willing to lead Britons to shake up their way of life even more.

THE OUTLOOK FOR GLOBALIZATION

Globalization encompasses powerful forces, but its development is not universal and, even where it's taken a strong hold, its progress is far from uniform. Markets may be global or nearly so; but the world is still organized around sovereign nations with the authority and power to set many of the terms of those markets, both within and across their own borders. As we will see, countries large and small sometimes can resist the market forces at the heart of globalization with considerable success, although usually at considerable cost to themselves. Some popular resistance to globalization seems quite reasonable, because it intensifies market forces, and those markets may provide very little reward for the majority of workers with few or even average skills and abilities. Since the demographic changes sweeping the world also produce new economic pressures for most of those workers, popular resistance to at least the outcomes of globalization will likely intensify in many places. In the United States, Europe, and Japan—as well as places like Korea, Thailand, Brazil, and Mexico—governments may revive protectionism for sectors like autos, electronics, and agriculture, at least for a time. Another wrenching financial calamity, perhaps triggered by a dollar crisis, could bring back stricter government controls on foreign investment and capital flows, again at least for a time. And mounting job losses from piecemeal service outsourcing could convince governments in many advanced countries to slap new taxes and regulations on companies that import their services, even when they do so through the World Wide Web.

Still, the consequences of globalization also include so much new wealth for so many people and countries that its basic hold on at least the near future is secure. Once before, powerful political and economic forces came together and tore down the institutions of a previous period of globalization, through the two world wars and worldwide depression of the last century. A reprise of that period in our own time is very unlikely, at least for the next few decades. Despite the optimism of its most ardent defenders, globalization cannot banish conflict and war from our near future. But the rise of the United States as a sole military superpower with no near peer, along with the political commitment of most nations to market-based arrangements, should prevent a continental or worldwide conflict involving major players in the global economy. The world also has gone more than sixty years without a global depression, although recessions have and will always develop periodically. Globalization can moderate or intensify a normal recession, depending on conditions at the time; and certain elements of globalization can transmit recessionary forces from country to country. But when stock and bond markets plummet today, governments usually know enough to take steps that can turn them around, at least where the real economy is otherwise sound and strong. And in countries prepared for the commitment—happily, the majority today—globalization ultimately makes their real economies stronger and sounder. So globalization makes another worldwide depression even more remote.

The largest threats to globalization and its core promise of social and economic progress lie in the wild cards stalking the world today, a few frightening possibilities that could possibly arise outside the realm of globalization, with deeply disruptive worldwide effects. Nuclear terrorism, especially in the United States or China, could shut down much of global trade and investment for a time and roll back political and economic freedoms for a much longer time. The U.S. National Intelligence Council's outlook for 2020 warns that "terrorist attacks that killed tens or hundreds of thousands in several U.S. cities or in Europe" could lead to draconian government controls "on the flow of capital, goods, people and technology that stalled economic growth."[44] Similarly, the political disintegration of Saudi Arabia, especially at the hands of radical Islamists, or of China, would send the global economy into an extended downspin. So might a deadly global pandemic on the scale of the 1918–1919 Spanish influenza pandemic that stopped much of global trade and travel for many months or a few years.[45] Thankfully, these prospects remain remote. Moreover, other equally unpredictable wild cards—for example, the development of new, clean, and inexpensive energy technologies—could accelerate and deepen globalization. Either way, unless we assume the utterly unpredictable, the deep trends in global markets, global politics and global technologies should keep the basic process of globalization on track for at last another generation.

CHAPTER FOUR

THE TWO POLES OF GLOBALIZATION: CHINA AND THE UNITED STATES

THE SINGULAR STRENGTHS OF THE U.S. and Chinese economies, combined with their sheer size, will set much of the course of globalization for the next ten to fifteen years, and with that the economic path of many other countries. Now and for the next generation, globalization will have an eastern and western pole, akin to the earth's North and South poles. The eastern pole is unequivocally China: the world's leading platform for manufacturing, the world's largest source of personal savings used in global capital markets, and, within a few years, the world's second-largest market for everything produced nearly everywhere. America's position as globalization's western pole is equally secure for at least the next decade. The United States is and will continue to be the world's largest source of new products, especially those employing and produced by advanced technologies. It will also be the leading source of the advanced business and financial services tied to these technologies and which everyone needs to engage in the global economy. The U.S. will remain the center of world capital markets, and the single largest buyer of everything produced domestically in China and most other places.

Globalization will shape the paths of these two huge economies as much or more than they will affect its course. For instance, unless some unforeseen international conflict seriously disrupts worldwide trade and investment, that trade and investment will continue to grow much faster than worldwide output—as they have for the last two decades. These two countries will dominate much of that trade and investment. Already, China passed Japan in

2005 as the world's second-largest manufacturer. With another decade of
strong foreign and domestic investment in China—and, if they're lucky, another decade of social stability—its share of worldwide manufacturing will
top the United States. At a first glance, the U.S.- China pairing may seem to
closely resemble the pairing of Great Britain and the United States from 1880
to 1900, when America's share of global manufacturing rose from less than
15 percent to almost 24 percent and Britain's fell from nearly 23 percent to
18.5 percent. But China's rise does not presage America's decline. For one,
China's expanding share of worldwide production will come mainly from
other developing nations. For its part, America's share of global production
has been remarkably stable for more than a century—nearly 23 percent in
2003, compared to just under 24 percent a century earlier—even as substantial shares of it have moved offshore. Moreover, its shares of both high-tech
manufacturing and advanced services, both in growing global demand, have
been rising steadily.

There's another critical difference between America's rise over a century ago
and China's in this period. While American industries in the late nineteenth
century developed behind high protectionist walls and so were almost entirely
homegrown, China today owes about half of its share of global manufacturing
to foreign rather than native companies, including most of its high-end and
high-technology production. China's domestic producers mostly serve its own
domestic market; and with some exceptions, the native Chinese companies
competing successfully today for consumers in other countries produce food,
apparel, toys, and other relatively low-value-added goods. By all accounts,
young Chinese managers and MBAs are quickly learning the best practices of
the western companies now operating in their country; and they surely will
adapt them to the next generation of native businesses. With another decade of
practice and investment, Chinese auto, electronics, and machinery producers
may be competitive in world markets with western companies also producing
in China or in other low-wage, developing countries. Chery Automotive is already producing models that resemble GM-Daewoo cars enough to move the
auto giant to sue for patent infringement, and has introduced units selling for
as little as $2,000 in the Middle East. Chery plans to offer other models to the
United States—George Soros has been a leading investor in Chery's U.S.
distributor—but whether Chery autos can make it in the U.S., European, or
Japanese markets is still unknown. Industry insiders say that Chery's plans to
enter the U.S. market in 2008 were put off when crash tests repeatedly decap-
itated the dummy drivers. Whatever the outcome for Chery, it is all but certain
that by 2020 an even greater share of the world's non-Chinese manufacturers
will be producing there, and China's own domestic producers will successfully

challenge them for much of the domestic market. Regardless of how long it will take for native Chinese companies producing high-value products to be broadly competitive in advanced countries, China's position already as the number-one or number-two trading partner with most advanced and developing nations will be more solid in 2010 and 2020, and a greater share of its exports will come from its own domestic companies.

Even so, the demands of globalization in other areas should temper China's economic ascendancy. For example, China's current financial arrangements are hopelessly outdated for even a much smaller and simpler economy. The country's vaunted personal savings all flow into state-owned banks—there's virtually nowhere else to bank them—and flow out quickly as loans to the state's own remaining, wildly inefficient enterprises that employ tens of millions of Chinese. This primitive arrangement hardly constitutes a real financial system and plays no real role in the country's modernization. Instead, most investment has come through foreign transfers and transplants. This system cannot go on much longer if China intends to create modern domestic companies that can help drive its growth and further development. Accomplishing that will require a working financial system that can channel private Chinese funds to the most promising Chinese prospects. China's leaders will have little choice over the next decade but to wind down the state-owned banks and let western financial institutions take firm hold. Inescapably, this will mean much less financial support for its state-owned enterprises and the tens of millions of Chinese employed by them.

As continued globalization helps raise the incomes and expectations of tens of millions of Chinese over the next decade, its leaders also are likely to face popular demands for broad-based old-age pensions and health care. For the leadership, that could mean choosing between a prospect of rising social unrest—already a serious problem—or shifting tens of billions of dollars from future economic development to new retirement and health insurance systems. Further, and perhaps most important for everyone outside China, globalization also will force China to be much more engaged politically with the countries that will be its important markets or sources of resources or investment, especially in Asia, the Middle East, and the United States. This intensified engagement could for the first time limit some of the leadership's freedom of action in sensitive matters such as intellectual property rights and state preferences for native Chinese companies, and produce tensions with the United States, its most important investor, trading partner, and borrower.

Globalization will also affect America's economic and political course over the next ten to fifteen years. Its complex forces already are dampening the U.S. economy's vaunted capacity to create new jobs and deliver higher wages, especially for workers in the center of the economy. Globalization also has stimulated America's high-technology industries by vastly expanding their potential markets, and the continuing development and spread of new technologies will heighten the strains on job creation and wage increases for tens of millions of ordinary American workers. These technologies have been particularly apt at replacing jobs involving repetitive and routine mental tasks—secretaries, inventory controllers, bank tellers, and many others in the center of the economy—and studies now show that for the last fifteen years, those kinds of jobs have grown much more slowly in good times and disappeared much faster in bad times than others. In this, America is not alone. Every other large, advanced economy is experiencing these same, slow-motion shifts and shocks for their workers. As we will see later, most European governments are responding not by thinking through the causes, but by providing additional public benefits to cushion the impact. But the looming demographic problems in most advanced societies will unravel this strategy within the next decade, even if the pressures building from globalization and technology were somehow to ease. The burden of paying for even greater benefits for jobless people, while pension and health care costs for the growing numbers of elderly are rising quickly and labor forces in many countries slowly contract, will prove unsustainable.

The United States will be better positioned to deal with these pressures because its current pension and jobless benefits provide much less security and its demographics will be more favorable for at least the next generation. But globalization is aggravating America's economic inequalities, which are harsher to begin with than those in Europe and Japan. Inequality in itself has never been the political flashpoint for Americans that it often has been for Europeans. American voters rarely object to the rich getting richer, but the 2006 congressional elections showed that they will not quietly tolerate their own incomes stagnating year after year. If America's two political parties fail to address the forces holding down most people's wage gains—the paucity of real opportunities for most workers to train themselves for new technologies, for example, and the cost pressures on businesses from rising health insurance and energy costs—these forces could produce a new a political agenda focused on economic security. That agenda would likely include measures such as wage insurance and trade protections, which ultimately will reduce some of America's current advantages in globalization.

Over the next decade, America will face several problems that will weigh on its economic progress. It is virtually inevitable that pressures from globalization will continue to unravel America's current system of job-based medical insurance, until the number of uninsured working people reaches a political tipping point and produces a full-blown political battle over universal health care. Much of these cost pressures come from the relentlessly increasing availability of new medical technologies that are evermore expensive to develop and use, especially with many more people turning the age that makes them avid, even frantic, consumers of high-tech health care. Whatever plan and cost-cutting measures American politicians adopt to ultimately achieve universal care, it will be highly expensive; which means it will almost certainly involve substantial tax increases that will also slow American growth. U.S. companies also are less energy-efficient than their counterparts in Europe and Japan, because America's lower energy taxes make energy much less expensive there. Every year of strong growth in China, India, Central Europe, and other developing nations will put upward pressures on oil prices, exerting a competitive drag on American companies in their domestic operations. Finally, America's rock-bottom savings rates and consequent dependence on foreign capital could produce the dollar crisis described earlier, which would stall the American economy for a time, and with it growth in much of the rest of the world.

With all these problems, and inevitably more, the United States still has powerful economic advantages that should solidify its preeminence among advanced economies for the next ten to fifteen years. Unlike Europe and Japan, America's labor force will continue to expand, mainly because it attracts and accepts millions more workers from other countries. The American workforce, alone among the large, advanced countries, is itself increasingly globalized, now including millions of highly skilled and educated workers born in Europe and Asia. Sergey Brin, one of Google's two cofounders, who was born in Moscow and emigrated to the United States with his parents in 1979, is only one of the thousands of spectacularly successful American technology entrepreneurs born somewhere else. Moreover, as China's rising preeminence in most areas of basic manufacturing leaves more advanced economies more dependent on their businesses' capacity to innovate and their workers' ability to operate effectively in technologically advanced workplaces, America's edge in these areas is likely to become even greater. One reason is that despite the countless studies documenting the real deficiencies of American education, a larger share of Americans are college educated; and U.S. workers on average have completed two to three more years of schooling than the average European or Japanese worker.[1]

For these and other reasons, American workers and businesses consistently rack up higher productivity gains and business returns, on average, than European or Japanese workers and businesses. Those advantages have made the United States a magnet for global capital, and there's no reason yet evident to expect these advantages to diminish in the next decade. America's trading patterns also are much more global than other advanced economies, importing and exporting to and from developing nations much more intensively than the major countries of Europe. U.S. corporations, as a group, also direct more of their foreign investments to developing nations, especially China, while German, French, Italian, and British multinationals continue to invest mainly in other advanced economies. Large U.S. corporations, on average, also are more thoroughly digital than large businesses in Europe and Japan, so their global networks are more highly developed and efficient.

Finally, America's political conditions produce other advantages in globalization that seem secure for the foreseeable future. Assuming that the widespread hostility in Europe and much of Asia to the Bush administration's foreign policies recedes with the election of his successor in 2008, the United States will be able to use its unrivalled superpower status to extract special consideration for U.S. companies and their particular interests in many foreign markets. Even if pressures on U.S. jobs and wages produce a leftward tilt in American domestic politics, the United States will continue to take a more conservative approach than most other advanced countries, especially regarding regulation and taxes. The nearly unescapable economic result is that America's economy will continue to be more efficient. These differences have made European societies more equal and socially compassionate, and will continue to do so; but they also will make the American economy relatively even stronger than today.

CHINA

China is the singular illustration of how, in the space of a single generation, globalization can help make an economically backward country a genuine economic powerhouse. What it apparently requires, apart from ample human and financial resources, is a sustained national commitment to redo itself to accommodate the demands of globalization, along with an authoritarian political system willing to inflict enormous hardship on much of the population in the transition. China is already an economic powerhouse, but not yet a genuine economic power. The impact of globalization as embraced by China's leadership is clearly evident in its stunning growth. Over the last fifteen years, China's economy more than tripled in size (and that's adjusted for inflation), growing in just the last five years by more than 55 percent.[2] Over the next ten to fifteen years, its economy will grow much bigger, as will its impact on the world. Yet, in 2020, China will still be a developing country, and its radical strategy for rapid development will produce enormous strains that could limit the height and extent of its economic ascendancy.

Just twenty years ago, China was a marginal player in the world economy; today, it is the world's most important developing nation. Its economic advantages rest on the economic basics of labor and capital. China's workforce is the world's largest, but its real labor advantage lies in the fact that its workers are still lower-paid than those in most other developing nations, yet generally at least as skilled as most of those workers. Moreover, every year, its people save so much of their incomes and its private businesses retain so much of their earnings—it comes to 40 percent of China's annual GDP—that it's enough to finance vast public works improvements, subsidize the remaining state-owned enterprises, and lend America enough to finance its own exports to the U.S.

The capital that matters most for China's extraordinary economic progress comes not from Chinese savers, but from the world's leading western corporations. Just a generation ago, China had no use for western investment or expertise, and its economy was profoundly backward. Since the early 1990s, however, China's leadership has successfully used the attractions of its ample low-wage workforce, general social discipline, the prospect of a huge domestic market, zero duties on imports of any equipment, and numerous special subsidies and tax breaks to attract some $850 billion in direct foreign investment, mostly in modern manufacturing operations. Only America and the United Kingdom contain more direct foreign investment today, and they've been accumulating it over the last century. In just the years from 2002 to 2005, China (with Hong Kong) attracted almost $333 billion in new direct foreign investments—as much as the United States in those years and more than Germany, France, and Japan combined.[3] Starting from nearly zero twenty years ago, foreign firms today account for about the same share of manufacturing in China as in the European Union, and more than in the United States. That's how almost overnight China has become one of the world's largest production and assembly platforms.

All that advanced, direct investment has enabled China to grow faster than any other country in the world every year for two decades running; and its annual output is now larger than every other country except the United States, Japan, and Germany. Within a few years, China's GDP will top Germany's, too, and well before the end of the next decade, Japan's as well. From almost nowhere, China also has become a great trading nation. Twenty years ago its factories produced almost nothing that anyone else in the world would buy; today, companies in China sell abroad so much of what they produce there that China is now the world's third-largest exporting country, behind only

America and Germany and ahead of Japan. The numbers tell the story: Twenty years ago, China exported less than $35 billion, mainly in basic goods and commodities to North Korea and the Soviet Union. By 2006, China's exports to virtually every country in the world totaled more than $981 billion, greater than the entire economies of all but eleven other countries.[4]

Like the United States and like no other place, China has become a country that matters for every other significant economy in the world. Its global economic importance rests on its sheer size and rapid growth, transmitted through its exports, capital, and the foreign investment it attracts. Across much of Asia, the Chinese economy is clearly what matters most—in many ways it's more important economically than the United States. Two-thirds of all foreign direct investment to China comes from Taiwan, Japan, Korea, and Hong Kong; and China's imports from those countries, principally parts for products assembled and finished in China, are the largest factor driving their own exports.

Scan the world, and China is now a key trading partner nearly everywhere. It is Japan's number-one trading partner, and on the other side of the world, the second-largest trading partner for the European Union, Canada, and Mexico, behind only the United States. America, too, imports more from China than from any other country except neighboring Canada, and exports more to China than to all but three other countries.

Despite its trading prowess, China is nothing like the economies of its trading partners, including most other large developing countries. Its economy is certainly very large, producing almost $2.7 trillion in goods and services in 2006. But with a population of more than 1.3 billion, its people are still very poor: Its current per capita national income, measured in dollars at current exchange rates, ranks not number 10 or even number 50 in the world, but 128th, behind countries such as Swaziland, El Salvador, and Tonga.[5] Measured in what economists call "purchasing-power parity," which adjusts the income in each country for its cost of living, China's per capita national income still ranks 107th in the world, behind such nations as Namibia, Belize, and the Ukraine. And with the average Chinese family saving more than 30 percent of its paltry income, the quality of life of most Chinese is even lower. Most of China's people are probably a little better off than they were a decade ago; and behind the averages, tens of millions of them live well by the current standards of many developing nations. Several million Chinese households, principally headed by skilled workers in coastal cities on or near the Pacific Ocean, live a Chinese version of a middle-class life. But they constitute less than 1 percent of all Chinese. More than half of their countrymen and women live in desperate poverty in the vast rural provinces of central and western

China, on the losing side of the starkest economic inequality of any large country in the world.

Behind these profound inequalities lies China's unique approach to modernization. No society can do everything at once. But China is adapting to globalization in very carefully selective ways that not only will leave most of its people poor for a long time to come, but also are creating terrible economic stresses and anomalies that will produce large problems over the next decade. How long will China's families support an industrial modernization program that unraveled their free health care to help pay for it? And how will China's leaders promote the creation of thousands of competitive native companies, so long as all domestic business credit comes from the network of state-owned banks committed to keeping afloat hundreds of inefficient state-owned enterprises?

In sharp contrast to the path of economic development in every other major country, China's economic success rests on globalization itself. The United States, Japan, the countries of Europe, and the Asian Tigers all built their economies the only way they could, largely on their own and over many decades. Through the nineteenth and much of the twentieth centuries, even while countries were trading huge volumes of goods and commodities, most investment and new technologies stayed put in the countries and industries that created them, often walled off from foreign competition by high tariffs. It took decades for America, Germany, Japan, or Korea to figure out what industries and business lines would thrive best there, given each country's particular natural and human resources, habits, norms, and laws. By trial and error on a vast scale—which is the usual way that markets work—each society gradually put together its own set of political and social arrangements to help it build up the capital that businesses needed for investment, the educational systems people need to be productive in those businesses, and the public goods a society needs to keep it all going, from roads and sanitation systems to legal rights and law enforcement. Everywhere, all of this took a long time, even centuries in some cases, because no one knew how to embed the requirements of a modern economy in their society until they figured it out. And the lessons and national wealth created by this process and progress spread slowly across the sectors of their economies and the classes of their peoples, so that everywhere but China, modernization proceeded through the gradual rise of middle-class societies.

China alone owes its stunning growth and progress not to this normal, slow process of economic development, but to a series of seismic political decisions. In this respect, China's greatest advantages are not strictly economic, but political. The country's tight circle of leaders maintains a monopoly on power, policy making, and popular allegiance that's unimaginable in other large developing countries, such as India or Brazil. Its unique brand of popularly based authoritarianism provides the political capacity and social discipline, unique among large nations, to press forward with modernization regardless of the wrenching dislocations it causes for hundreds of millions of ordinary people and what in freer countries would constitute powerful interest groups.

The decisive first step came in 1978 at the Third Plenary Session of the People's Eleventh Central Committee in Beijing. China had been at bitter odds for a decade with the Soviet's domination of the world communist bloc and had itself just emerged from the chaos of the Cultural Revolution. With a new détente with the United States taking hold, Deng Xiaoping announced a new era of economic reforms, beginning with easing price controls on grains and allowing rural Chinese to start small businesses. A leading Japanese historian of modern China, Ryosei Kokubun, writes that from that moment forward, every move towards western market mechanisms was part of a political strategy by China's leaders. The goal was to maintain the power and legitimacy of the Communist Party of China as its leaders distanced themselves from the palpable failures of both Maoism and the Soviet-led communist movement, by driving economic growth and the national power it would confer. From its start, China's tentative embrace of capitalism was fundamentally a political choice.

Every important development in China's modernization follows this pattern. Those first steps didn't come about because Chinese agriculture was evolving naturally from land-intensive to more capital-intensive farming, or because a budding entrepreneurial middle class was forming in the hinterlands. They were decisions made at the top, to force up agricultural productivity and drive tens of millions of people from the land and into small enterprises that could help supply China's state-owned enterprises. With a series of decrees issued from the leadership compound near the Forbidden City, China's leaders bankrupted some one hundred million subsistence farmers and, in less than a decade, shifted tens of millions of them to other, more productive parts of the economy. In every other country, a similar process proceeded through the normal dynamics of economic development and took generations.

The collapse of the USSR a decade later unleashed a series of even more radical political edicts that have transfigured the core of China's economy and left the name of the ruling party the only thing still truly communist in the country. To create a new domestic market space for private businesses, Jiang Zemin began withholding capital from many of the less politically important state-owned industries, from apparel and furniture to toys and appliances. Even more revolutionary for China, the leadership opened those industries to foreign trade and investment, and unleashed a new workforce for them by allowing ordinary Chinese for the first time to move from their villages to larger towns and cities. By the force of these decrees—and by holding out the prospect to western multinationals of getting in on the ground floor of what might soon be one of the world's largest markets—China didn't simply make itself an irresistible lure for foreign companies. Its leaders maintained control over the process by initially requiring that all transfers of western technologies and expertise occur through joint ventures, forcing the world's richest and most sophisticated companies to become training grounds for the next generation of Chinese managers and entrepreneurs. That's another stage that takes generations when it happens through the normal dynamics of economic development.

This strategy and its results are unimaginable apart from modern globalization. China's prodigious progress is being achieved almost entirely through the capital, technologies, and business expertise transplanted from America, Europe, and much of the rest of Asia. As a result, almost everything modern and advanced in China today is either foreign-owned or based on foreign models. The nation's prodigious growth comes from deploying a slice of its labor force to enterprises owned and run by the leading corporations of the United States, Europe, Japan, Taiwan, Korea, and a few other more advanced developing countries—organ donors for China's economy—and allowing another slice to start their own modern businesses modeled on those foreign companies or more backward ones that are more traditionally Chinese. Through it all, the vast majority of its people still eke out impoverished lives in backward forms of agriculture or hand-to-mouth small businesses.

China's economic progress, selective and dependent on others as it is, remains its own political achievement. More than any other large developing nation, its leaders took the decisions and risks entailed in opening their society to massive foreign transfers and offering up tens of millions of low-wage workers and a huge potential market. But those risks are large and real, because the leadership's strategy has created a manufacturing sector and large cities that are much more developed than the rest of the country and its other social and economic arrangements. In effect, part of Chinese society has developed far ahead of the rest of it, creating disparities and anomalies that almost anywhere else could tear apart a country. Unless China's leaders address these disparities and work to resolve the anomalies over the next decade, they will threaten China's continued growth and development, and perhaps its prized economic and social discipline.

The central disparity in China's modernization is the huge gap between the great manufacturing capacity that makes it an economic powerhouse, and the rest of China's backward agricultural economy and society. With so much of its growth and progress based on transplants from more advanced economies, most of what makes China modern exists in a kind of economic vacuum without the strong, natural ties to everything else that help maintain intricate forms of economic balance—what economists call "equilibria." This not only leaves China's manufacturing powerhouse hostage to foreign producers who someday could move on to India and Bangladesh; more broadly, this vast disparity makes it much harder for China to develop the kind of integrated national economy that large countries need to maintain their growth and development.

In economies that develop normally, as the manufacturing sector grows more advanced, all the other parts of the economy that manufacturers depend on—workers and managers, companies that provide the materials and build the parts, distributors and financers—grow more sophisticated along with it, especially in response to each other's requirements. Societies like the United States or Germany that developed their core industries more naturally and slowly also develop a second tier of companies that make the parts and provide the services for those manufacturers, and a third tier of businesses that produce products and services for the workers earning rising incomes in all three tiers.

In China, where the manufacturing transplants are advanced but everything else is decades behind, the gaps between the first tier and the rest are as great as the differences in economic development between Japan and Namibia or Italy and Haiti. For instance, most of the sophisticated parts used in China's advanced manufacturing plants are still produced outside China, principally in other Asian economies. China's own companies and workers are moving up the value chain every year, but they're still a long way from the sophistication of Korea or Taiwan, much less America or Germany.

IBM faced some of the problems produced by these gaps in its early years in China. Arvinder Surdhar, who headed up IBM's first Chinese joint venture to produce PCs in the early 1990s, has recalled, "When we opened up those first shipments from China, there was more dust in the boxes than anything else." Most of the units were damaged in transit over China's rutted roads and antiquated rail systems, in their rough unloading and reloading by border agents as the boxes moved from one province to another, and from sitting for weeks in humid containers in crowded ports waiting to be shipped overseas.[6] Those kinds of problems still plague modern companies in China. For example, when Pacific Cycle moved the last of its bicycle manufacturing to China

from Mexico and the United States in 2000, it found that its new joint partner couldn't use Cycle's central tracking systems. Pacific Cycle eventually
bought out its Chinese partners and yet still found in 2005 that the lead time
to produce one of its custom bicycles in China was 270 days, compared to
60 days when they had been made in America or Mexico.[7]

Most native Chinese manufacturing remains distinctly underdeveloped
even today. David Hale is president of International Smart Sourcing, an
American firm that locates Chinese manufacturers for medium-size American
producers of plastic, metal, and electronic products. He works with sixty-two
Chinese companies, but he had to investigate more than six hundred producers to locate 10 percent that could do acceptable work. "We say, 'no deal' to
90 percent of them, because they don't have the equipment, enough knowledge of the process, or quality controls to meet our basic standards."[8] Since
most managers and directors of Chinese companies get their jobs through political connections, normal economic incentives to improve their technologies
and business methods are blunted.

A larger problem is the dismally inefficient performance of China's remaining state-owned enterprises, which still dominate major industries such
as autos, chemicals, and metals. Their technologies and business methods
generally are as outdated as the large Soviet enterprises in the early 1990s,
with most stuck in a rut of shoddy production, large annual losses, and regular loans from state banks to pay their suppliers and workers. They have little
spare capital or organizational capacities to improve or innovate and feel no
competitive pressures to develop them. If China's manufacturing sector had
evolved more naturally and normally, these large insolvent enterprises would
have failed, and their resources would have shifted to private firms that could
make better use of them. That's not possible in China, because there are no
modern bankruptcy laws for the orderly liquidation of a company's assets,
or modern financial arrangements to lend funds to promising private enterprises. Instead, the state-owned banks keep the monopoly companies afloat,
or the state simply closes them down, forcing the banks to write off decades
of bad loans that drain the working capital derived from the deposits of ordinary Chinese.

Another great disparity has emerged between what most Chinese people
and businesses can use or afford and most of what the country's advanced
manufacturing operations produce. The result is that China has to export nearly
half of what the core of its economy produces. This anomaly makes China's
development strategy highly vulnerable to the global business cycle. The
world has managed to avoid a serious recession for more than a decade—the
same period in which China has become a world-scale economy—but it can't

avoid it indefinitely. Nor does China seem prepared for what lies ahead. A leading Chinese economic commentator, Lau Nai-keung, sees a political upside for China from any global recession that starts in America: "To us, the good news is that when the country [United States] is in deep trouble, the U.S. will not have the energy to pick on China. . . . This will provide a . . . 'period of strategic opportunity' for the country to [focus entirely on economic progress and] pass through a turbulent zone [when] per capita income is between U.S. $1,000–$3,000."[9]

Professor Lau is whistling in the dark. When the next global recession comes, Chinese growth will be hit very hard, because the two lynchpins of its modernization strategy, exports and foreign investment, will both decline sharply. If China's growth stalls and incomes decline for a time, the real question is whether the domestic "turbulent zone" that Lau mentions will become the nightmare scenario for China's leaders, with demands across the country for the personal and political freedoms that came to a head in Tiananmen Square. The leadership's rule will probably weather almost any demonstrations and dissent that a global recession might trigger; but if their predecessors' response in 1989 is a guide, the upshot will include tighter political controls that could slow or stall the next stages of economic liberalization.

Building a world-class manufacturing platform that produces things that most of China's businesses and people can neither use nor afford is creating another pitfall for stable economic progress, since China has to use its high savings to finance a good part of the foreign demand for its own exports, especially to the indispensable American market. On this matter, China's Lau echoes the warnings of many western economists: "The U.S. economy is dependent on the central banks of Japan, China and other nations to invest in U.S. Treasuries and keep American interest rates down. The low rates keep American consumers snapping up imported goods. Any economist worth his salt knows that this situation is unsustainable."[10] Virtually every salty economist expects that global capital markets eventually will force the United States to reduce its huge trade and current account imbalance, even by simply driving down the dollar. When that happens, those capital markets will also drive up the relative value of China's currency, making its exports less competitive everywhere. For a country that exports half of what it produces, currency shifts that raise the price of its exports in all of its markets, and especially in its largest one, will be very serious business. Foreign investment could slow dramatically, young native businesses could go under, and unemployment would soar. Even without a full-blown recession, these developments will slow China's growth and probably the next stages of its modernization.

The weak domestic consumer spending that makes China's growth so dependent on exports reflects the exceptionally high savings rates of average Chinese households, and those savings rates in turn mainly reflect China's unique status as the world's only major economy in which the vast majority of people have no guaranteed health-care or pension coverage. This anomaly, created deliberately by the leadership's decisions to unravel the old safety net arrangements tied to the farm collectives and state-owned enterprises, could impinge seriously on their entire modernization strategy. In a society where life expectancy now tops seventy, everyone expects to get sick sometime and to grow old; and since the government won't help out with either, most people put aside what they can. China is not the only country that provides little medical or old age coverage. But among countries with little or no safety net, China alone has a constantly expanding modern manufacturing sector that produces goods which most of its people cannot buy because they're poor and have no medical or old-age coverage.

So long as China's leadership refuses to extend these basic social benefits to more of their people, most will continue to save a good part of what they earn rather than consume what their country produces. Moreover, most of those personal savings are held in those state banks that lend them to the country's hundreds of inefficient and often technically insolvent state-owned businesses. So extending basic social benefits to large numbers of Chinese will not only involve shifting hundreds of billions of dollars from public works, thus slowing modernization, it also will force many of those state-owned enterprises to go belly up, throwing millions of Chinese workers out of work. Here, too, a deep anomaly in China's modernization path could provide the seeds of popular unrest sometime in the next decade or so, from an army of newly jobless Chinese potentially joined by millions of other desperately poor people who see a small slice of their fellow citizens grow comfortable in some of the world's most advanced cities. Such a chain of events is sufficiently worrisome that Chinese officials occasionally speak about it publicly. The former director of the State Council's Development Research Council, Wu Jinglian, warned recently of the "possibility of an economic and social crisis," as "income gaps among the people and between urban and rural dwellers have grown to such a dangerous degree that they might cause social instability."[11] Those concerns are rooted in real events: The Cultural Revolution, with its chaotic yet uniquely managed forms of mob violence, must be a vivid memory for most Chinese adults, and especially the leadership. And much more recently, China's top police official, Zhou Yongkang, admitted that even in the midst of the greatest boom in China's history, more than 74,000 protests involving almost 4 million people occurred in 2004—and that almost certainly involves a large undercount.[12]

One sensible course for China over the next decade would be to begin to close the gap between its foreign-owned or foreign-based manufacturing sector and the rest of the economy. But China's other economic and social arrangements are so underdeveloped that it will be difficult to pull it off. China has plenty of ambitious people eager to start their own businesses, for example, but no modern banking system to provide them financing. An interbank market that could transfer funds to where they're needed doesn't exist, its corporate bond markets are fragmented and short of funds, IPOs are highly restricted, and most of the funds held by the big state banks are dedicated to the insolvent state-owned enterprises. That's why despite China's status as the world's highest domestic saver, the OECD reports that a perpetual credit crunch continues to stifle its small- and medium-size business sector.[13]

Even as the world's largest multinationals carry out complex deals in China's hypermodern cities and industrial development areas, other huge gaps in the basic elements of modern commercial life further hold back the normal development of native businesses. With modern intellectual property rights routinely violated or ignored—government departments in Beijing and most provinces still used pirated copies of Windows until very recently—only the most well-connected native companies invest in technology development. Lax intellectual property protections even deter foreign companies such as Nintendo from transferring cutting-edge operations and technologies to China, since they assume that their most valuable patents and copyrights will be quickly ripped off. Even more basic, China's legal system still applies different rights and different rules to private, collective, and state-owned forms of property, and every judicial ruling is politically correct. These antiquated distinctions make it harder for native private companies to plan and operate in a normal business fashion.

Nor does China yet have many of the other modern arrangements and forms of infrastructure that come together to create genuine national markets. At least thus far, the leadership's strategy of using progress in the Pacific coastal metropolitan areas to catalyze development in the vast interior has largely failed. Without a modern financial system, capital cannot move easily or often enough from province to province; and that's only a small part of the problem. Because China's interior highway and rail systems are out of date and out of shape, basic materials, parts, and finished goods all move slowly from place to place and often are damaged along the way. Even when goods can move undamaged, hundreds of cities and dozens of provinces block the transit of goods from other provinces in order to protect their native companies. Further, while tens of millions of young Chinese have left their rural homes for the cities, China is still a long way from developing a national labor market. Apart from the young and ambitious, everyone else is tied to their land by traditional rural land tenure rules that dictate that anyone who leaves a rural residence for an extended time loses his land use rights—and for most rural Chinese, those rights are the best insurance for their old age.

China can't keep up with its own progress in other very damaging ways. For example, the country is now renown for the world's worst air and water pollution, with more than 600 million Chinese drinking contaminated water on a daily basis and acid rain covering at least one-third of the country.[14] At current rates and thanks to weak regulations, these conditions will worsen significantly. Zhang Lijun, the number-two official at China's state environmental protection agency, recently reported that under current growth paths, China's pollution "will increase by four to five times" between now and 2020.[15] That could cost hundreds of thousands of lives and make significant parts of the country unlivable. Even today, the World Bank estimates that pollution costs China's economy more than $50 billion a year in health costs, time lost from work, and damage to crops, farmland, and infrastructure—that's equal to some 8 percent of the country's GDP. And cleaning up China's environment will not be simply a matter of setting strict new standards and enforcing them, as advanced countries do. Where such a direct approach is possible, China's leaders have already done it. Notwithstanding the choking traffic gridlocks in Beijing and Shanghai, for example, China's current auto fuel–efficiency standards are the equal of Europe and stricter than America's. But most of China's serious pollution is linked closely to the modernization program, coming as it does from the state's own energy-inefficient heavy industrial plants, the rapid growth of dozens of cities that have drawn tens of millions of rural young people, the country's reliance on its most plentiful and inexpensive energy source, soft coal, and the intense use of fertilizers that has made China self-sufficient in food for the first time in its history. Applying modern air and water quality standards to the sources of China's world-class environmental problems would set back much of its basic development strategy.

The chasm in Chinese development between its manufacturing prowess and everything else in the economy was plain to consumers around the world in 2007, when one after another, reports surfaced of Chinese exports found to be dangerous or deadly. Chinese eel and shrimp imported into the United States were tainted by banned veterinary chemicals, children's clothes were soaked in formaldehyde nine hundred times higher than safe levels, imported Chinese toys were decorated with hazardous lead-based paint, dog and cat food sent to Canada and America were contaminated by a potentially deadly chemical called melamine, Chinese toothpaste exports were tainted with a poisonous compound found in antifreeze and, finally, Chinese-made cough syrup containing a toxic solvent killed more than three hundred Panamanians. While China's unique, globalization-based approach to development has enabled it to go from being a minor exporting nation to the number-three exporter in the world in less than two decades, its regulatory and enforcement capacities are still primitive. China has no equivalent of a western consumer-product-safety agency. Nor does Beijing yet have the administrative resources to create a central food and drug regulatory system—in this critical area, six agencies share overlapping and often conflicting authority, and none has the systems or manpower to enforce anything.

The gap between China's manufacturing capacity and its ability to ensure the quality and safety of what it produces could pose serious problems for China's economy in the next few years. For example, the country already produces a substantial share of the world's generic drugs, including by some estimates a majority of the world's penicillin, aspirin, and over-the-counter vitamins. Recently, EU regulators found a lethal bacterium that can cause meningitis in infants in Chinese-produced vitamin A, and American authorities have found traces of arsenic, lead, and iron in products containing vitamin C from China. Before China's capacity for regulation catches up with its capacity to manufacture goods that need regulation, tainted and unsafe drugs and other products from China will kill Americans and Europeans. That could well make "Made in China" a shunned brand throughout the advanced world.

How China deals with its profound economic anomalies and pitfalls over the next ten to fifteen years will depend largely on how much change its tightly held leadership will accept or tolerate, beyond the current fast-forward path for manufacturing. The future course of China's economic development, like its beginnings, will depend mostly on politics.

In a fundamental respect, these problems follow directly from China's singular approach to power and policy. Contemporary Chinese politics has none of the give-and-take or competition among parties, branches of government, and interest groups that almost everywhere else force policy makers to take account of how their decisions affect businesses and ordinary people. The United States government cannot open parts of its Alaskan wilderness for oil drilling without hundreds of compromises and accommodations among the departments of its executive branch, between the White House and Congress, among factions within the Republican and Democratic parties of Congress as well as between them, between dozens of environmental and business groups, and then between them and the two parties—at which point independent judges have their say, too. All of this wrangling and positioning is also covered by the media, so the American public forms its own views that pollsters measure and report—shifting or solidifying the positions and negotiations of everyone involved in the administration, Congress, and beyond. The process is much the same in European countries, where even though parliamentary systems generally ensure that the executive and legislature move together, parliamentary majorities often are comprised of multiple parties and factions within parties. In the end, the result is the same: Economic decisions in democracies almost always take a long time, and the final results are usually fractional measures. The benefit of allowing every group and individual with even a little power or influence to weigh in is that most decisions of consequence take some account of the capacity of the country's major groups and businesses to live with the consequences—or at least with those that can be known.

That's not how politics works in China, where the nine men of the Politburo Standing Committee make all key national decisions on economics—along with social policy, foreign affairs, defense, science and technology, education, and national cultural policy. No other individuals, groups, or institutions can overrule, contradict, or modify those decisions. Those nine men are not always of one mind in their decisions about the path of modernization. In a brief spell of Glasnost, Chinese style, in 2005, the Politburo released files that clearly portray the views of some of the major players on the Standing Committee. Four of the nine members count most in economic decisions: Hu Jintao, the top man and party general secretary; Wen Jiabao, the premier; Zeng Qinghong, the head of the party secretariat; and Luo Gan, the security chief. All but Wen appear committed to promoting more foreign transfers to drive the next decade's technological progress, and expanding exports, as the best ways to ensure rapid domestic growth and increase China's global influence. They're also determined to avoid what is universally seen in China as the "Russian trap" of quickly privatizing the state-owned enterprises. If this thinking holds through the next succession—and the Standing Committee chooses its own successors—the stresses and contradictions in China's current development will almost certainly worsen over the next ten years.

Among China's senior leaders, only Wen favors a different strategy. With a few younger officials in the next lower level of the leadership, Wen seems convinced that measures to reduce the country's harsh and growing inequalities can lay a basis for more sustainable, consumer-led economic growth. So he has argued for shifting resources from industrial and high-tech investments to infrastructure and agricultural modernization, and establishing minimum incomes and more pension and health-care coverage run by local governments. The more conservative leaders thus far have rejected such basic changes, but they have made some concessions to Wen and his allies. In 2005, for example, Hu announced his version of a "New Deal" that includes tax cuts for farmers, more infrastructure projects for western and inland provinces, a few provincial pilot programs in old age pensions and health care, and a massive project to transform the inland city of Chongqing into the next Shanghai.

This debate among China's current leaders, like those of their predecessors, questions not the goal of building a modern, western-style economy as quickly as possible, but only the politically safest way to do so. Whether it takes five years or twenty to dismantle most of the remaining state-owned enterprises, to bring China's farm sector into at least the early twentieth century, and to create modern arrangements for bankruptcy, property rights, and patent and copyright protection; and whether China establishes a modern financial system and broader old-age and health-care coverage in this generation or the next one, there is no dissent among China's leaders about opening the economy to more foreign competition. And that is the essential element that will drive Chinese exports and competitiveness and make China an increasingly powerful force for every other economy to reckon with between now and 2020.

Anywhere but China, the profound shortcomings and pitfalls bound up in its politically directed and economically anomalous strategy would virtually ensure a path of merely modest growth for at least another generation. But China is not like anywhere else. Its sheer size provides the world's largest supply of low-wage workers. The exceptional saving habits of its people and its turbocharged industrial growth should continue to create the financial resources to take on, at once, thousands of new infrastructure projects, gradual agricultural modernization, and the carefully measured shift from state-owned industries to private ones. Its unique political system and single-minded leadership make rapid, society-wide changes possible, regardless of the problems and pain they cause millions of their people. And those changes have brought sufficient prosperity or its prospect for enough Chinese to maintain popular allegiance and social discipline, at least so far. With all these inducements, almost every major global corporation has been, and will continue to be, intent on doing whatever it takes to get rich in China.

The next ten to fifteen years will likely be a period of both major economic and social progress for China, and wrenching economic and social turmoil. Its manufacturing prowess will continue to expand. In 2004, 75 percent of the world's toys, 58 percent of its clothing, and 29 percent of its mobile phones were made in China.[16] By 2020, tens of millions of people in other countries will drive Chinese-made cars, operate Chinese-produced machines, and use computers and electronics made in China. China's manufacturing prowess will continue to eclipse every other developing nation, forcing them to either modernize in specialized areas or slide backward. And the direct and indirect effects of China's expanding competitiveness in manufacturing will continue to reverberate across the world's economies, ultimately reaching the advanced economies of Europe, Japan, and the United States, where they will continue to dampen job creation and wage gains.

All these achievements will not hollow out the world's advanced economies. Over the next decade and beyond, American, European, and Japanese corporations will continue to produce sophisticated goods and services that will still be beyond the capacity of China's native companies. The expansion of domestic Chinese demand over the same period, principally from the fast-growing numbers of skilled urban workers at the top of its economic heap, will make China an indispensable market for western businesses. And when it comes to sophisticated consumer goods, the preferences of China's elites already mirror those of wealthy westerners. *China Daily* reported recently on a survey of six hundred Chinese millionaires, and found that their favorite cars are Rolls-Royce, Mercedes-Benz, and BMWs, their preferred brands in watches are Cartier and Rolex, and they favor Chivas Regal and Dom Perignon when they drink, Christies for auctions, and HSBC for their personal wealth management.[17] China's middle-class households today number some 2 to 3 million, counting everyone earning as much as $12,500 a year—taking into account China's cost of living, that's equivalent to an American household income of about $35,000.[18] By 2020, those urban middle-class households could number 30 to 45 million. By then, they will constitute one of the world's great consumer markets, with spending power close to all Japanese households today. Given China's demographics, most of them will be relatively young, which may make them big spenders. In another decade's time, China's market will be as important to the United States and Europe as the American market is today to Europe and China.

Under the current program, all that won't be enough to enable China to maintain its long record of 9 percent annual growth over the next decade. To this point, China has achieved its stupendous growth and great leap forward by mobilizing domestic labor and foreign investment in manufacturing. But this strategy has failed to develop its own ways of raising productivity even in manufacturing, and has left most other parts of the economy underdeveloped and without the legal and economic mechanisms to develop quickly on their own. As the economist Paul Krugman has pointed out, this approach makes slower growth almost inescapable, even apart from the inevitable shocks from a global recession or the possible shocks from serious political unrest.

A leadership program that uses rapid growth as an essential foundation for its own legitimacy and popular allegiance could find those foundations shaken when growth falters. Every year, the program destroys the livelihoods of millions of rural Chinese and millions of workers in state-owned businesses—and on top of that, millions of young Chinese reach working age every year. One expert recently ran the numbers and found that over the next decade, "China needs to create 300 million new jobs . . . to absorb or re-employ those who lost their jobs in the agricultural sector or former state-owned enterprises and provide work for the new members of the labor force."[19] Turbocharged growth is the only way to do that, and the current program will not produce it indefinitely.

China's greatest challenges over the next decade, therefore, are likely to be political. The Cultural Revolution saw mass political action that was often harsh and cruel. More than three decades later, popular demonstrations of at least a few hundred people occur by the hundreds every week across China. The political disintegration of China is very unlikely. But serious political upheavals over the next decade are a real possibility. Whether such political shocks will produce a more liberal political system to accompany China's liberalized economic arrangements, or, as seems more likely, even more authoritarian politics that slows the next stage of Chinese modernization, we cannot know. But the world will almost certainly watch it unfold between now and 2020.

THE UNITED STATES

America's status as the other indispensable country in globalization is as secure as China's. Most fundamentally, the United States today is the most thoroughly and deeply globalized of all the world's advanced economies. Yet this position also represents a major break with at least a century of American attitudes and practices. From the early years of the twentieth century to its last two decades, America was singularly self-sufficient economically. While Europe's economies were too small to produce everything each needed, and their proximity to each other allowed them to trade among themselves easily and relatively inexpensively, the United States faced oceans separating it from most other markets, and instead developed its own vast, national market across its states and regions. For nearly a century, America sustained its place as the world's largest economy by largely using its own natural resources and its own technological and human capacities to produce just about anything its businesses and people needed or wanted. Through the 1960s, for example, all U.S. trade—the total of the nation's imports plus its exports—equaled just 10 percent of its GDP.[20] In the same years, the value of everything traded by French or Italian firms equaled 28 percent of the GDP of those countries, and in Great Britain the figure was 42 percent.

One reason why the United States could be less engaged in international trade than other advanced countries is that Americans were (and are) richer and consume more. In the 1960s, their per capita income was 40 to 50 percent higher than the average for Europeans in the major countries, so they could generate sufficient demand at home to keep most of them employed.[21] Most Europeans weren't so fortunate. Their relatively low incomes—and the higher taxes they paid (and still do) to finance their more generous government supports—left too little domestic demand to support jobs for everyone. Then as now, their governments have had to rely on exports to tap demand from other places, principally the United States, to keep workers employed. The other reason why Europe has had to depend more on trade than America is more extensive government regulation, often tied to promoting exports from a handful of "strategic" sectors, which has made its economies both more specialized and less productive.

Europe, and eventually Japan (and China), too, found the United States a willing partner in their export strategies. From the early cold war, a succession of U.S. administrations saw access to America's economy as a prime geopolitical asset and, much like Britain in the mid-nineteenth century, offered to unilaterally cut tariffs for countries that allied with the United States against the Soviets. This open-door policy made classical economic sense, even if that wasn't much in the minds of its political sponsors. American consumers enjoyed a wider range of goods at lower prices than they would have behind the protectionist walls that prevailed in Europe and Japan for decades more, and American companies could concentrate on becoming dominant producers of the higher-end goods still out of reach of Europe and Japan's recovering economies.

Today, America is still only half as dependent on foreign trade as most other advanced countries. In 2006, imports and exports equaled nearly 28 percent of the GDP of the United States, or half the 56 percent share for the major European nations.[22]

Even so, the United States is the world's most global economy and, compared to Germany, France, Britain, or Japan, much better positioned for the next ten to fifteen years. This judgment differs sharply from what might be called the "Chicken Little" view of America's place in global trade, which sees the United States as fundamentally weak in trade and heading into decline. This view is fundamentally wrong, and insofar as U.S. officials buy into it, their responses could weaken the country's economic prospect. Consider the following economic contrast, one of the most striking and important in the world today: In 2004, more than 44 percent of all U.S. exports went to the developing nations driving much of the world's growth and integration, and those same nations provided more than 50 percent of U.S. imports.[23] By contrast, the nations in the EU-15 still hold most of the rest of the world at arm's length, selling less than 15 percent of their exports to developing markets and buying just about 20 percent of their imports from developing nations.[24] Instead, almost three-fourths of EU exports go to other EU countries, and more than two-thirds of EU imports come from other EU countries.

The United States imports and exports less than Europe as a share of GDP, not only because it doesn't have to, but also because its businesses have more extensive networks of foreign subsidiaries and affiliates to produce and sell in foreign markets. The huge U.S. trade deficit poses financial risks for the American (and global) economy—but it comes from the burgeoning demand of American consumers, not from the so-called faltering competitiveness that worries commentators who don't consult the data.

No major country will do well economically over the next decade without a solid economic presence in the world's fastest-growing and developing countries. Underneath America's more extended global network of imports and exports, U.S. companies also have a much larger presence in those nations and their economies. From 1995 to 2003, about 28 percent of all foreign direct investment from the United States went to developing nations—nearly double the developing-nation share of the foreign investments coming out of Germany, almost three times the share of foreign investment coming from France, and four times the share flowing out of Britain.[25] Among the advanced economies, only Japan focuses its foreign investment on developing markets as much as the United States.

This difference extends to China, the developing nation that will be pivotal for every advanced economy in coming years. In 2003 (the most recent international data), China held more than 3 percent of all U.S. foreign direct investments around the world and more than 11 percent of U.S. direct investments in all developing countries.[26] That may not yet be a huge amount, but it's ten to twenty times more than the direct investments in China from any European country, and it's been increasing quickly. Even measured as a share of their foreign direct investments, American companies commit two to four times as much to the Chinese market as do German, French, or British companies. As China's economy and those of other fast-developing nations expand and mature over the next decade, the western companies with the largest and firmest footholds there will be in the strongest positions to tap those growing markets and benefit from their progress.

Given America's modest trading position just forty years ago, its current full embrace of globalization is remarkable. It appears that America's large companies simply have adapted more successfully to the changes in global economic conditions. For example, in the 1970s, when energy prices abruptly soared and U.S. productivity slowed, many American managers began to look to Asia and Latin America for new markets and ways to cut other costs. When many of those markets began to develop and grow quickly in the 1980s and 1990s, many more American companies began to approach them as places where they could not only produce their products more cheaply but also sell what they made there. Europe's major economies and large corporations experienced the same changes, and they started out more experienced in international trade and more familiar with developing country markets, yet they've remained much less engaged in the world's fastest-growing and developing places.

There are many ways to understand why this is happening. Americans may be more comfortable operating globally, because the country's population and culture are more globally diverse and, at least in recent years, generally deal more calmly than many European countries with increasing racial and ethnic diversity. U.S. companies also probably have less choice about adapting quickly and thoroughly, since America's relatively weak regulation and trade protections make competition more intense. American companies certainly face fewer barriers to adapting to new conditions, because they don't have to deal with the kinds of regulations and laws in France or Germany that restrict a firm's ability to reorganize itself and fire or reassign workers as they chose. Or perhaps U.S. corporations are simply following the general path of U.S. military and geopolitical power. Whatever the precise explanations, the American economy has been changing in ways that create real and growing advantages for a period of globalization. That's fortunate for the United States, because no program or politician's promises will stop or even significantly slow globalization.

Who, precisely, benefits from these advantages is a complex matter. America's global corporations clearly benefit, judging by their record profits in recent years. These high corporate profits also help shareholders; who in this case extend to nearly half of all Americans, principally because more than 40 percent of U.S. stocks are held by pension plans and personal retirement accounts.[27] American workers have racked up impressive productivity gains in the same years—more so than most Europeans or Japanese—yet, as we have noted, those gains aren't translating into significant wage growth for many of them. It's not much consolation for stagnant wages, but all the cheap U.S. imports from developing countries help most Americans by stretching their wages further. Low-priced imports are an important reason why according to several studies, shopping at large discounters such as Wal-Mart saves average consumers several hundred dollars a year.[28] Moreover, the current problems with wages come not from globalization per se, but mainly from the collision of the competitive pressures it promotes and fast-rising health-care and energy costs, along with rising returns on capital—and these problems are as acute in Europe and Japan, which haven't adapted to globalization nearly as well.

Other measures also point to the underlying strength of the American economy, compared to the world's other large, advanced economies. Employing everyone who wants to work is a traditional test of an economy's health and strength; and even with the new problems emerging in job creation in the United States, this remains one of its strong suits. Over the last decade, the average share of the U.S. labor force looking for work but unable to find it, about 5 percent, has been about one-fifth lower than in Britain, two-fifths lower than in Germany, and half the share in France and Italy.[29] Only Japan has consistently lower unemployment than the United States. And when Americans lose their jobs, they usually find new ones much faster. Over the last ten years, about 80 percent of unemployed Americans found new jobs within six months, compared to 65 percent of out-of-work Swedes, 60 percent of jobless Britons, just 50 percent of jobless Japanese and Irish, and only about one-third of unemployed Germans and French. Americans with jobs also work more hours in a year than the people in other advanced countries (except Japan)—10 percent more than British workers, 15 percent more than the Italians and Irish, and 25 percent more than the Germans and French. Shorter hours may seem appealing, but the difference is one of the reasons why most Europeans earn less and their economies produce less per capita.

In these times, a country's ability to develop and use advanced information technologies is nearly as important for its economic health as its capacity to keep people employed. This, too, is an area of comparative American strength. Over the last decade, a number of nations have successfully made computers and the Internet integral parts of the way they conduct their business and personal lives. Yet, with the exception of the United States, all of them are small countries—Sweden, Finland, Denmark, Singapore, Hong Kong, South Korea, Bermuda, and Australia—that deliberately set about to do that. Despite America's vast size, deep economic inequalities, an economy with thousands of subsectors, a stupefying variety of businesses of every sort, and no national policy to support the spread of these technologies, in 2004 it had more than seventy-six PCs for every hundred inhabitants, and 63 percent of its population used the Internet.[30] No other large, diverse economy came close. Japan had nearly as many PCs per hundred people, but only 50 percent of them were online, while Britons were online at nearly the same ubiquity as Americans, but PCs were much less common. By both measures, Germany, France, and Italy all trail England and Japan, much less the United States.

Europe and Japan are catching up in these areas, but twenty years of lagging behind points to important differences that suggest that America's preeminence among advanced economies probably will expand further over the next ten to fifteen years. Part of America's continuing edge here reflects what some economists call a "compound, first-mover advantage," or the benefits that come from developing many new technologies first. America's research networks, entrepreneurial culture, and business environment don't create an advantage for producing some important global products—Japan, for example, is the world's strongest auto producer—but they are particularly conducive for developing and spreading new technologies. The early technological leadership of American inventors and IT companies created reservoirs of critical knowledge and business processes, as well as networks of important relationships, that extended these early leads to most of the industries' subsegments and market niches. So American inventors and companies came up with not only most of the initial rounds of IT and Internet innovation, but also much of the subsequent generations, because the United States has more of the intellectual, cultural, and organizational capital to do so cheaply.

Some of these developments were probably happenstance (economists call that "path dependence"). The Americans who first developed most of these technologies were no smarter than their counterparts in England, Germany, Japan, and India—but when IBM went looking for someone to develop a new PC operating system in the 1980s, all the clever young Europeans and Asians in the field were an ocean away from the informal networks spreading the word. Today, everyone in IT development, wherever they are, can learn about such opportunities through the Internet. But the job opportunities and potential rewards are still usually greater in the United States because its IT sector is much bigger—and its taxes on a big payday are generally lower. Those attractions also help explain why one-third of all U.S. doctoral degrees in the sciences and three-fifths of those in engineering are awarded to foreign or foreign-born students, why nearly half of all professionals with PhDs working at the U.S. National Institutes of Health also are foreign born,[31] why an estimated 40 percent of Silicon Valley startups in the 1990s were founded by Indian inventor-entrepreneurs, and why tens of thousands of budding European and Asian technologists now work in the United States. Developers in Europe and Asia will achieve breakthroughs in coming years that will startle everyone; but America's overall lead in IT innovation is also quite secure.

These American advantages are not restricted to IT. Social scientists usually analyze a country's R & D investments by how much of its GDP is devoted to it. But in the race to lead global innovation, what matters is how much is invested period, and how well a country's economy commercializes what comes out of it. Here, the United States also appears to have broad advantages. In 2003, for example, the U.S. spent nearly $300 billion on R & D, compared to $210 billion by all of Europe, barely $100 billion by Japan, and less than $80 billion by China—and the gap in 2003 was larger than it had been in 1990 or 1995.[32] That's one of the reasons why U.S. inventors and companies have early leads in many promising areas of biotechnology and nanotechnology, including genetically modified food (which most European governments atavistically resist), personalized medicines, filtration systems for highly polluted water, and advances in solar energy technologies. America's early leads in these areas may not ultimately matter, since no one can say which of these or other emerging technologies will have far-reaching economic value and effects. But if any of them strike gold, it's more likely to happen in the United States, with so much more annual R & D, so many more of the research universities and young companies that come up with technology breakthroughs, and private equity investors ready to place more than $20 billion a year in long-shot bets on infant technologies.[33]

If the United States is such a powerhouse in advanced technologies, why does it run large trade deficits in these areas? Does it mean that, as one public commenter on these matters, Clyde Prestowitz, warned recently, "America is well on its way to surrendering (technological) leadership"?[34] He and others conclude that in order to stay competitive in advanced technologies, America needs a new industrial policy that will provide more government support to produce and consume those technologies. But the data show not that the United States is losing its technological edge, but that its technology companies are fully globalized. That becomes more obvious when you know that half of the imports driving the high-tech trade deficit come from the foreign subsidiaries of U.S. technology companies. Moreover, the National Science Foundation reports that American companies over the last generation have actually and vastly increased their worldwide preeminence in the manufacture of high-technology products. Less than twenty years ago, Europe, Japan, and America each claimed a little more than 25 percent of the world market share in this area; by 2003, the U.S. share had reached almost 40 percent, while Europe fell to about 18 percent, and Japan had just about 10 percent.[35]

Much the same thing is happening in R & D, since much of the growth in those investments in places outside the United States and Europe comes from

American companies in those places. Intel, for example, has advanced research facilities in Israel, India, and Russia, as well as California; and IBM research labs are found in Switzerland, Israel, Japan, China, and India, as well as New York and San Jose.[36] Most serious R & D still takes place in advanced countries with large, sophisticated research establishments. The rapid growth of research operations by western multinationals in a few developing countries—in 2005, one-fifth of the Fortune 500 did some R & D in India—so far has mainly focused on adapting established products for fast-growing local and regional markets.

The United States seems likely to maintain these leads over the next ten to fifteen years, and globalization will increase their significance. For one, American hardware, software, and Internet companies will have a leg up in China and India as they go increasingly digital. In 2004, India had barely one PC for every one hundred Indians and just 3 percent of its population was online, while China had about four PCs for every one hundred Chinese and just a little over 7 percent of its people used the Internet. By 2020, China at least should be as digital and wired as most European countries today, and India also will make substantial (if smaller) strides. American companies will provide much of what will then be the latest generation of these technologies—even if much of it is produced by foreign subsidiaries and affiliates—as well as the IT services that will accompany their spread.

These technologies are also part of the essential infrastructure of globalization, from supply-chain management to local product development and marketing. As global production and demand for almost everything continue to increase over the next ten to fifteen years, the manufacturing, retail, and financial service companies that have integrated IT more broadly and successfully into their operations will be in the best position to meet much of that demand. Again, compared to the typical global firm in other large, advanced countries, America's global companies for now have a distinct advantage here, too.

Yet America's greatest advantage in advanced technologies lies not in their development and spread, but in how Americans use them. Despite what most people learned in introductory economics, a succession of American and European studies have found that how much a company or a country spends on IT makes little difference in how productive they become. Over the last decade, European businesses invested nearly as much in IT as U.S. firms, relative to the size of their economies. Yet the productivity of the industries that spent the most on these technologies increased by 3 to 4 percent a year in the United States, compared to no change at all or even slight declines in Europe.[37]

One reason is that American companies are managed differently, and the differences seem to affect how they use technology. For example, American businesses more often base their pay increases and promotions on performance, which wittingly or not creates powerful incentives for U.S. managers and workers to get more out of the IT they're provided. By contrast, large European and Japanese companies still typically base their employees' pay and promotions on tenure and union rules. Other rigidities in the European and Japanese economies make it harder for firms there to derive much advantage from their IT investments. For example, the labor laws and social conventions that sharply limit firms' freedom to fire or reassign most workers often prevent them from reorganizing their domestic operations to make their IT investments work for them. So a French or Italian company can invest in a state-of-the-art accounting or an automated phone system and find itself unable to reassign or let go any of its current accountants or operators.

A worker's education level also seems to affect how well he or she adapts when a company introduces new technologies. Here's a surprise for many Americans: The World Bank reports that U.S. adults average more than twelve years of formal education, compared to barely seven years for Italian adults, less than nine years for adults in France, and less than ten years for British, Japanese, and Germans adults.[38] Substantially more young Americans also enroll in some postsecondary education or training—69 percent, compared to between 45 and 58 percent in the major European countries and Japan.

Some of these differences will probably diminish over the next decade, although efforts in France a few years ago to modestly modify labor protections sparked riots by young people determined to keep their job security at any cost. While the share of young Europeans going on to college still trails the share of young Americans, it rose sharply in the last fifteen years. It's also indisputable that many large European and Japanese companies deeply involved in global markets are at least as productive as any American firm. U.S. business leaders certainly recall learning the basics of lean, high-quality manufacturing from Sony and Toyota in the 1980s and 1990s. Moreover, globalization is fast spreading the best practices of the most successful and IT-enabled American corporations. Wal-Mart, the world's most profitable and IT-intensive retailer, bought Britain's second-largest supermarket chain, Asda, in 2005, and Britain's number one retailer, Tesco, promptly adopted much of the Wal-Mart model.

Still, American multinationals as a group appear to be more productive and successful than European or Japanese multinationals. For example, researchers measuring the productivity of all kinds of plants in Britain found that those operated by U.S. multinationals were not only almost 40 percent more productive than the average domestic British plant, but 10 percent more productive than the plants of British multinationals in their own home market. The most important factor was that the U.S. multinationals got better results from each dollar of IT investment.

The fact is, the productivity gap between the United States and Europe and Japan has increased steadily for more than a decade, pointing to America's single most important economic advantage at a time for rapid globalization: Competition is more intense inside the U.S. economy, because there's less regulation and protection for both businesses and workers, especially for the service companies and workers that dominate advanced economies. Japan, and Europe's large countries, still maintain regulatory walls around much of their retail, wholesale, financial, business, and personal service sectors, so they're still dominated by millions of inefficient small companies with little incentive to change almost anything. America's more bare-knuckled competition at almost every level of its economy leaves its workers and companies less secure, especially in a time of galloping globalization and technological progress. It also forces those that want to prosper to change all the time, by using the latest technologies and business practices to improve something they make or do, or even come up with new products, processes, and ways of doing business. And American companies and workers accustomed to dealing with these intense competitive pressures will be better equipped over the next ten to fifteen years to deal with the increasingly intense competition in foreign markets created by globalization.

WHY NOT RUSSIA OR INDIA?

Globalization will surely accelerate modernization in scores of countries over the next generation. Korea could match Japan's singular achievement as the only country to move from low income to high income from 1870 to 1970—and do it in a fewer years. Poland and the Czech Republic could make as dramatic progress as Ireland did in the last twenty years. Our focus, however, is on nations whose economic impact will in some sense be global—a select circle of countries limited to the United States, the major European economies, Japan, and China.

Some analysts might make a case for Russia and India. Less than a generation ago, Russia was the center of a great world power vowing noisily to bury the West, with a weighty (if covertly weak) economy that reached well into Eastern and Central Europe, India, and parts of the Middle East. There's little doubt that Russia will be a force in global politics over the next ten to fifteen years, which we consider later. But with a GDP today barely larger than Mexico and smaller than Spain or Brazil—and per capita income that trails such countries as Libya, Poland, and Botswana—Russia today is a marginal economic force in the world and likely to remain so for many years. There are certainly some signs that conditions there are improving. For the last five years, its unemployment has been low, its growth has averaged 7 percent a year, and the ruble has been convertible since July 2006. One international expert, Nivedita Das Kundu of the Institute for Defense Studies and Analysis in New Delhi, forecasts that in another twenty years, Russia could be the world's eighth-biggest economy.[39] The IMF, citing low Russian wages that could help make Russian exports more competitive worldwide, also has concluded that its economy could grow at healthy rates for at least another decade—but only if Moscow is committed to serious structural reforms.

The IMF's caveat reflects several critical weaknesses that will most likely continue to marginalize Russia's global economic impact. In a time of rapid globalization, no country can make rapid progress without drawing on the best technologies and business methods from the rest of the world, principally through foreign direct investment—and Russia to date has not established the domestic conditions that attract those transfers. The OECD (Organisation for Economic Co-operation and Development) found that in 2002, for example, foreign direct investment in Russia amounted to fifteen dollars per person, or barely 13 percent of the per capita levels for its former satellite, the Czech Republic, and 7 percent of those for Hungary.[40] Nor have conditions improved much since then. The OECD report on Russia's future prospects for attracting foreign transfers is subtitled "Battling Against the Odds," and flows to all sectors except mining still remain low, especially following the Putin government's seizure of the Yukos oil empire and its continuing inability to curb corruption.[41] Every year since 2000, the flow of direct investment out of Russia has exceeded the flow coming in, which looks a lot like the capital flight seen in unpromising Latin American countries. And the most sophisticated recent analysis of Russia's prospects, produced by the neighboring Bank of Finland, concludes that without large, sustained increases in investment, both domestic and foreign, growth in the former superpower may soon approximate the rates in Latin America.[42]

Some Russian planners, tacitly acknowledging their country's problems with foreign investors, see high energy prices as the key to future national growth: According to recent forecasts by the Russian Economic Ministry, ten more years of oil prices averaging $75 a barrel will generate the resources required to substantially modernize infrastructure and expand industrial capacity, especially in the oil sector itself, and so keep Russian growth strong.[43] Yet, ominously, Russia's oil sector appears to be unraveling economically, despite the world's largest reserves outside the Middle East, Canada, and Venezuela. Russia's oil companies earn smaller profit margins than major producers almost anywhere else because the government keeps domestic prices low to subsidize Russian farmers and military. The Putin government also keeps oil sector employment high, and the combination of artificially low prices and high labor costs has squeezed investment and modernization throughout the industry. Drilling new wells and building new pipelines have virtually stopped, the industry's recovery techniques are outdated, and it still uses standard bits that last one-fifth as long as those used in Texas. And the large western oil companies that could provide new capital, technologies, and expertise largely avoid Russia because in such an uncertain environment they'll only come in as majority stakeholders, and the politicians and oligarchs that control the reserves won't agree to those terms.

Russia's low investment levels also undermine other sectors that were strong during the late Soviet regime. In 1990, Russia was the world's number-two steel producer, just behind Japan (the whole Soviet territory was the world's largest steel producer); and the three major Russian steel complexes at Magnitorgorsk, Sverstal, and Novolipetsk turned out their product, principally for the Red Army, nearly as efficiently as America's best steel companies.[44] But when demand for steel fell precipitously with the end of the Soviet empire, Moscow directed the major producers, as well as some thirty-three smaller and less-efficient plants, to keep operating even as their revenues (and government subsidies) plummeted. By 1998, productivity in Russia's steel sector had plummeted to 28 percent of U.S. levels.

The sharp deterioration of these major industries is critical to Russia's economic prospects over the next ten to fifteen years because its unique and terrible demographic conditions will demand much greater revenues and investment. Today, Russia has the lowest fertility rates and life expectancy and the highest rates of infant and youth mortality of any advanced or major developing society in the world. With per capita health-care spending below that of countries like Panama and Tunisia, more than half of Russian children reportedly suffer from a chronic disease; and the general health of Russians is worse today than that of most people outside sub-Saharan Africa and the Middle East.[45] Under current trends, by 2020, Russian life expectancy could fall below that in India or Bangladesh. The economic consequences of these developments are sobering. Evgeny Andreev, an expert at the Institute of Economic Forecasting of the

Russian Academy of Sciences, forecasts that his country's low fertility rates and high disease and mortality rates will reduce Russia's working-age population by 14 to 25 percent over the next fifteen years.[46] Nor will it end there: Andreev sees the number of Russian children falling by 20 to 25 percent over the same years.

Russia's falling fertility and high mortality rates produce the unique anomaly of a relatively poor nation with a rapidly aging population. Like many European countries, Russia will need a lot more revenue to support its growing elderly population just when the working-age population that has to produce that revenue is shrinking—and Russia will face this prospect in a much poorer state than anywhere in Europe. One nearly inevitable consequence is that millions of elderly Russians will retire into terrible poverty over the next ten to fifteen years.[47] The only way Russia can avoid this and materially improve conditions for everybody else would be to sharply increase the productivity of the workers it will have and the businesses that employ them. That would require much greater public investment in education, health, and infrastructure—the heart of the structural reforms urged by the IMF and others—along with much greater private investment by domestic and foreign companies.

Yet today, Russia devotes just 3.1 percent of its GDP to education, or less than one-third the share that advanced economies spend, and less than India or such countries as Kenya and the Congo.[48] Russia's current commitment to health care is no better, claiming a smaller share of the country's GDP at 5.4 percent than Eritrea, Cambodia, or Honduras. As Russia's economic ministry has recognized, the most realistic source for the revenues needed to substantially expand those investments is expensive oil. Yet even that won't help much unless Russia's deteriorating oil sector turns itself around soon. Here, too, the outlook is unpromising. President Putin made it clear in the Yukos affair that Russia's oil oligarchs will follow his script or risk everything; and Putin has shown no interest in changes that might make the companies more efficient and profitable, especially if it means higher prices for Russia's army and farmers, or ceding some control to western oil companies. Nor is his government even remotely disposed to the more difficult reforms required to attract substantial foreign investment in other areas. For now and the foreseeable future, the combination of dismal demographics and low investment will cast "a long shadow over Russian growth," as the chief economist for Deutsche Bank, Norbert Walter, put it recently.[49]

Claims for India's growing economic power are more common, and western commentators often pair it with China as the world's two great, emerging economies. Here, too, hard reality does not justify this pairing. Measured in dollars, India's GDP is smaller than Russia's (or Brazil's), so these claims are usually based on other measures. One such measure is purchasing power parities (PPP), which factors in differences in cost of living and elevates India's economy to number four in the world, behind America, China, and Japan. The second measure is India's recent success in certain advanced sectors, notably software programming and generic pharmaceuticals, which in certain respects are competitive with the global leaders from the knowledge-based economies of Europe and America. Close analysis shows, however, that while India is indisputably large, and its software and generic drugs are sold around the world, its basic conditions and development path will not make it a global economic power over the next ten to fifteen years.

India's extreme poverty is the first sign that it remains a long way from becoming an economic powerhouse for the rest of the world. Its per capita income, adjusted for each country's cost of living, ranks number 118 in the world, behind countries like Nicaragua, Armenia, Ecuador, and Guyana, and 53 percent lower than China. More telling for a country with global economic ambitions, India has one of the world's lowest ratios of trade to GDP. From 2000 to 2003, India accounted for barely seven-tenths of 1 percent of all world exports, roughly the same as Poland, 60 to 70 percent less than such small economies as Belgium and Taiwan, and less than 15 percent of China's total. One reason is that 60 percent of India's population still works in agriculture, compared to 14 percent in Russia and less than 50 percent even in China. Moreover, productivity in almost every sector in India remains far too low to make the country a force in global trade, regardless of how low its wages are. The productivity of India's vast agricultural workforce is about 1 percent that of American farmers.[50] Another 20 percent of India's workers hold jobs in very small businesses—street vendors, tailors, and so on—and their productivity is about 7 percent that of workers in the same sectors in advanced countries. The final 20 percent hold jobs in reasonably large businesses in manufacturing or services, and even their productivity is about 15 percent that of Americans in similar jobs.[51] India is still a backward economy decades away from global economic influence, akin perhaps to China twenty-five years ago.

Pundits who focus on India's huge labor pool earning wages about half of China's levels miss the fundamental differences between the two countries' strategies to modernize themselves. China has made itself one of the world's largest (and fastest-growing) global export production platforms by combining authoritarian politics, extreme openness to foreign investment, and expertise in most areas of manufacturing. By choice and political necessity,

India has tried to modernize by going alone, choosing a more balanced path tied to a nineteenth-century model of gradual, indigenous development. This course reflects in part its more democratic politics, in which countless political compromises and the efforts of every significant group to avoid the dislocations of rapid modernization slow or stall economic progress. Several generations from now, whether China or India will provide greater prosperity for its people is unknown. What is certain is that India's strategy will not remake its economy into a global power in the next ten to fifteen years.

The result is that India's markets are much less open to the rest of the world, foreign technologies and expertise play a very modest role, and growth depends on domestic consumption rather than investment and exports. In such economically pivotal areas as most heavy industry and retailing, India long ago barred most FDI (foreign direct investment). In 2004, when global companies invested $233 billion in developing countries, China received $60.6 billion, or 26 percent of it, while India attracted $5.3 billion, or 2.3 percent. Despite its rank as one of the world's largest economies in total GDP, India attracted less FDI than Poland and barely more than Malaysia. And 2004 was a record year for foreign investment in India. Taking a longer view, India can claim less than 2 percent of the total stock of foreign direct investment in all developing nations—a share one-third smaller than Portugal, whose economy is one-fourth India's size, and just one-quarter greater than Vietnam, with an economy one-fifteenth the size of India.[52]

Even in areas where India has allowed foreign investment, most foreign businesses site their modern plants and organizations in other countries, because India's dismal infrastructure and suffocating government regulation make it so hard to generate profits in the subcontinent.[53] For example, state governments get to approve layoffs in companies with more than one hundred employees. Indian law also stipulates that more than eight hundred major industrial products can be produced only by business so small that the value of their plant and equipment does not exceed $200,000—not enough to operate one modern production machine. Moreover, many businesses cannot build new facilities even if they want to, because more than 90 percent of land titles in India are subject to legal dispute. The anarchic land registration system also strangles the construction sector, since banks won't lend anyone the money to build a factory, office, or home on land that has no clear title. Where large factories do operate, the country's antiquated state railroad monopoly—a last-ditch employer for millions of poor Indians—poorly maintained road systems, and chaotic and government-run seaports and airports cannot ensure the reliable delivery of anything for export.

In principle, India could lift these restrictions and reform land titles, and its national and most state governments almost certainly will do so in time—although given the country's fractured and interest group–driven politics, it will take many years. Even that won't change what many observers consider the single greatest barrier to rapid progress in India, the country's desperately antiquated and inadequate energy systems. *The Economist* surveyed these arrangements in 2005 and reported, "power cuts are a way of life in India, at least in parts of the country lucky enough to regard them as an interruption rather than the norm. . . . Where electricity is available it is often only for a couple of hours a day, unusable for industry and of such poor quality that power surges routinely wreck equipment."[54]

The national and state governments still own most of the country's electrical power generation and distribution operations, but an electricity grid a half-century behind much of the rest of the world isn't merely the legacy of decades of inefficient public ownership. China had a government energy monopoly, too, but opened its energy sector to American and European imports of thermal and clean coal power–generating equipment, gas, hydropower, and wind turbines, large-capacity pump storage units, advanced nuclear power station equipment, gas desulphurization equipment, and middle- and high-voltage capacitors.[55] China also has regular power shortages, but its electricity generation has grown by more than 125 percent in the last decade, and China now produces more current than any other country except the United States—and more than four times as much as India.[56] One reason is that India's electricity monopolies still have neither the resources nor interest in importing modern plant and equipment. The western companies building new generating capacity in China stay away from India—the notable exception being an ill-fated joint venture with Enron that imploded with the company in 2001. Why would they invest there, when they can't get clear title to most land, their equipment imports are subject to high tariffs, and Indian law provides free power to farmers and other favored groups? All told, some 40 percent of the country's electricity output is either given away or stolen. These conditions not only destroy normal economic incentives to electrify the vast parts of the country that are still off the grid, they also force the country's industrial and commercial businesses to foot most of the nation's energy bills, with the predictable result that India's business sector is one of the world's least energy-intensive. With no serious steps to change these conditions, and national demand for power growing twice as fast as capacity, India's shortages and other energy problems will only grow worse over the next decade.

The issue for India is not whether it will be a global powerhouse in ten or fifteen years—that's not in the cards—but whether it can sustain sufficient growth and incremental progress to ultimately become a broadly prosperous and modern economy further down the line. India has clear economic strengths; but if it intends to keep up with countries such as Malaysia and Thailand, it will have to expand them in ways that take serious account of globalization.

Let's start with capital. With no near-term prospects of attracting large amounts of foreign direct investment, India will have to rely mainly on its own savings. The outlook here is mixed. India's gross domestic savings rate is more than 20 percent—less than half China's rate but still respectable by global standards. Government deficits claim a fairly modest share of those savings today, although that could change if the government were to become serious about improving the nation's roads, bridges, energy grid, and water and sanitation systems—as it must do to grow. The larger issue for savings and future business investment is that Indian households hold a relatively small share of their savings in the stocks, bonds, and bank accounts that fund business expansion. Instead, Indian households prefer to hold their savings in land and houses, cattle, and especially gold. Half of the personal assets of all Indian households are held in gold jewelry, making them the world's largest consumers of gold. The result, as the McKinsey Global Institute put it recently, is that "India simply has a lot less money circulating in its financial system than one would expect, given the size of its economy."[57] Until this changes, there will be little prospect of strong domestic investment to make up for the relatively meager flows of foreign investment, producing two negatives as compared to two of the most important economic positives in China's rapid progress.

Despite the relatively small holdings of India's banks, its banking system is more developed than China's, with private lending tied more closely to the real economic prospects of business borrowers. But special requirements set by the Reserve Bank of India mute that advantage as well, by directing domestic banks to allocate 45 percent of all business loans to small and medium-size enterprises. These rules may hearken back to Mahatma Gandhi's veneration for the self-sustaining village. But they preserve and protect the inefficient mom-and-pop enterprises that account for half of nonagricultural jobs in India—70 percent of apparel industry workers, for example, are tailors who work alone[58]—and needlessly limit financing for the larger native businesses that could help drive faster and more broad-based modernization.

India is changing, and in recent years its government has taken important steps to encourage more foreign investment, lower trade barriers, cut business taxes, and relax industrial licensing and currency controls. Much of the credit goes to the country's current leader, Prime Minister Manmohan Singh, the chief architect and driving force behind these reforms. But much like Japan's recent, reform-minded Prime Minister Junichiro Koizumi, Singh's effectiveness is limited by the staunch resistance of the National Congress Party that he leads. Jagdish Bhagwati, a noted Indian economist now at Columbia University, describes Singh as a man "walking in minefields defined by special interest groups."[59] For his own part, Singh recently said, one suspects wistfully, that, "it takes a lot of time for us to take basic decisions (in) many things, like getting land, getting water, getting electricity . . . We recognize that globalization offers us enormous opportunities in the race to leapfrog in development processes. It also obliges us to set in motion processes which would minimize its risks."[60]

India's auto sector provides a snapshot of the strengths and weaknesses of these modernization efforts, especially since the automobile industry helped drive American, British, and German industrial development in the first half of the twentieth century and more recently played a similar role for Sweden, Japan, and Korea. For decades, all but the richest Indians had just two choices in automobiles, both produced by state-owned companies. Then in 1983, New Delhi agreed to a single joint venture between Suzuki and a new Indian automaker, Murati. The tale is told that the impetus came from then-Prime Minister Indira Gandhi's son, Rajiv, who wanted to get into the auto business. Whether or not that's true, Suzuki's modern technologies and production methods catapulted Murati to an 80 percent market share by the mid-1990s. Yet, despite this success, no other foreign auto company was granted a right to produce cars in India until the late 1990s. In 1998, the government finally ended state licensing for auto production, and since 2000, American and Japanese automakers have set up modest production facilities there. But high tariffs on equipment imports, high taxes on foreign companies, and the sorry state of Indian infrastructure keep India out of the expansion plans of most global automakers. The result is that the fierce competition that characterizes automaking in advanced countries—and China—is still muted in India. Murati still dominates the national market with more than 60 percent of sales, while a 44 percent tariff on auto imports precludes competition from foreign imports. While China's two largest privately owned auto companies, Chery Automotive and Geely Automotive, today produce cars for about $2,000 each that already are selling in foreign markets, Murati remains an India-only enterprise.

A number of foreign automakers led by Toyota and DaimlerChrysler now source some of their components from Indian suppliers such as Bharat Forge, Rico Auto, Sundaram Fasteners, and Amtek Auto; and since the late 1990s, Indian auto parts exports have grown 25 percent a year. But at $2.8 billion in 2006, these exports still account for less than 1.5 percent of worldwide auto parts exports, far behind other low-wage auto parts centers such as Brazil and Mexico.[61] India's universities—much like Sweden's—stress engineering, and a number of European and Japanese automakers recently have set up design and engineering centers there.[62] But here, too, India is a very small player in an industry dominated by more developed economies.

The facet of India's economy that draws the most attention in Europe and the United States is the success of a number of Indian companies in knowledge-based industries that advanced countries consider their exclusive preserve, especially software, pharmaceuticals, and entertainment. These successful idea-based companies all draw on India's considerable pool of young, creative talent and skilled workers. Well-educated young workers are a critical resource for any country bent on modernization. But in India, they represent a very small slice of an economically destructive educational system that for decades has neglected basic education for almost everyone else. By the rankings of the United Nations, for example, India's national literacy rate is 146th out of 177 countries, well behind the Congo, Rwanda, and Laos. There are no positions in a modern economy for illiterate workers, and countries with large numbers of people unable to read or do basic math can modernize on an economy-wide basis only very slowly. It will be at least another generation before the average Indian has the basic capacities for modern economic life.

And while nearly six thousand Indian institutions of higher education graduate 2.5 million people a year, most of those graduates have far fewer skills than their American or European counterparts, or even new graduates in some other developing countries. A recent survey of human resource managers at U.S. and European multinationals reported that just 25 percent of Indian engineering graduates meet their standards, and 10 percent of those with degrees in the arts and sciences—in both cases, much smaller percentages than recent graduates from Central Europe.[63] While more young Indians receive post-secondary diplomas every year than French or Germans, and nearly as many as Americans, their average knowledge and skills remain substantially lower.

Still, every year, India's university system produces some fifty to one hundred thousand graduates who compare favorably with their counterparts elsewhere in the world, especially in the sciences and mathematics. They provide the highly skilled labor for the jewels of Indian modernization, the software and pharmaceuticals sectors. But the Indian companies that have made these sectors world-competitive do not owe their success to the larger dynamics of Indian development. Rather, their success has been path-dependent, owing as much to serendipitous conditions at their founding as to the neutral judgments of markets. (Many economists who study Microsoft attribute much of its market dominance to path dependence as well.)

The Indian drug industry's path began with the country's particular patent regime, which until recently protected the way a drug was produced but not its formula. With western pharmaceutical companies largely bypassing the Indian market—discouraged by the country's strict price controls on drugs and its readiness to award licenses for foreign-developed drugs to domestic companies—a handful of entrepreneurial enterprises led by Ranbaxy and Sun Pharma began developing alternative manufacturing processes for drugs already patented in other countries.[64] India was not the first country to create a drug industry by ignoring foreign patents. Switzerland did the same thing in the late nineteenth century so it could produce aspirin, and actually didn't introduce product patents for pharmaceuticals until 1978. But India is the only developing nation that used this approach to create a world-class pharmaceutical industry; and today, Indian generic versions of western drugs meet 70 percent of the country's own demand, and their producers export 60 percent of their production, mainly to other developing markets.

For all this global success, India's generic drug sector has had little impact on the rest of the economy. In terms of jobs, Indian pharmaceutical companies employ about 500,000 people out of a national labor force of nearly 400 million. The drug industry in India also generates few of the "spillovers"— which happens when technologies and business expertise developed in one sector are adopted by other lines of business—that can help drive broader development. The reason is that competition drives such spillovers, and most sectors in India are dominated by small enterprises protected from competition by strict regulation. Moreover, India's large pharmaceutical companies will likely face much harder times in the next ten to fifteen years. In January 2005, India's long free ride on the R & D of American and European drug makers ended when it officially accepted the patent rules of the World Intellectual Property Organization, a condition set by the United States for India's WTO membership. Since the old patent regime strongly discouraged Indian companies from developing new drugs, the country's world-class producers still invest a pitifully small fraction of their revenues in R & D. With the Singh government's decision to finally end restrictions on foreign direct investment in the sector, most large Indian drug makers will have to form joint ventures with U.S. and German generic producers to stay in the game.

India's software and IT service companies comprise the nation's leading sector in the global economy. While they don't account for more jobs than India's drug producers—barely one-tenth of 1 percent of all Indian employment—the sector's $28.5 billion revenues in 2005 were more than 3.5 percent of the GDP. That's a greater share for the sector than every advanced country except the United States and Ireland. Moreover, the industry's $15.2 billion in exports account for more than 20 percent of the value of all Indian goods and services sold outside the country. A recent McKinsey study estimates that before the end of the decade, Indian software and IT service companies will earn revenues of nearly $90 billion a year, or 7.5 percent of the nation's GDP, and their annual exports will top $50 billion, or a 35 percent world market share.

Much like the generic drug markers, Indian software owes part of its success to happenstance events. In 1986, when U.S. software makers were battling each other for the American market, a forward-thinking Indian national commission determined that the country's banks should adopt a standardized system based on a Unix platform. That one act triggered a fierce competition to design the platform; and in a few years, dozens of young programming outfits that are now household names in New Delhi and Bangalore—Infosys, Wipro, Softek, TCS, and others—were turning out cheap accounting packages to work with the new Unix platform. The budding industry received another boost in 1991 when the U.S. government directed American software companies to pay their foreign workers prevailing U.S. wages. Once advanced telecommunications went global in the mid- to late 1990s, Oracle, Microsoft, and others figured out that they could sidestep the prevailing wage requirement by outsourcing some programming to India, where thousands of young, English-speaking workers were skilled with code. Y2K provided more impetus for this outsourcing, and the strategy has produced some notable market successes, including Ramco's Marshall program, Banco 2000 from Infosys, and the EX program developed by TCS.

Indian software also has produced more spillovers than Indian pharmaceuticals. Most notably, the rapid growth of programming jobs has attracted hundreds of thousands of Indian students to computer sciences and engineering, providing a labor force for a fast-growing IT service industry. But other facts of Indian economic life hamper the sector's potential to help propel broader modernization. Rampant software piracy discourages budding Indian geniuses from developing the next Windows or Norton's Utilities; and for all its heft, Indian programming has developed few breakthroughs. Moreover, in other countries with strong IT industries, from Ireland and Israel to the United States, the big economic punch comes not from developing new IT technologies, but from their natural application and spread to other industries.

India's large industrial and commercial companies have gone digital to a greater extent than their counterparts in many other developing nations, especially in areas where the government has finally allowed significant foreign direct investment. But the country's huge and primitive agricultural sector has no use for IT; and its spread to other important sectors such as retail, apparel, construction, and personal services is stalled by the dominance of very small enterprises and labor regulations that prevent companies that might adopt new technologies from firing or reassigning employees. Instead, America has become a beneficiary of spillovers from Indian IT. As noted earlier, it's commonly said—although the source is elusive—that 40 percent of the founders and top executives of Silicon Valley startups in the late 1990s were Indian,[65] and a substantial brain drain of top Indian programmers and software entrepreneurs to America (and Britain) continues today.

India's other idea-based sector with global dimensions is its movie business. What westerners call "Bollywood" produces more movies and videos than any other country save the United States. Many of them are produced on the outskirts of Hyderabad, at the world's largest studio complex, the two thousand acres and more than five hundred soundstages and set locations that make up Ramoji Film City. At any time, some twenty foreign films and forty Indian films are in some stage of preproduction, production, or postproduction there.

The Indian movie industry today resembles the heyday of the contract hyperstar Hollywood system of the 1930s, dominated by larger-than-life personalities such as Aishwarya Rai, Miss World 1994, and Salman Khan, the bad boy of Indian celebrities. From 1997 to 2006, Ms. Rai appeared in thirty-six films, including six English-language features, and the glamorous actress is well on her way to becoming India's first global movie star. In recent years, she became the first Indian jury member at Cannes, the first Indian woman to be immortalized in Madame Tussaud's Wax Museum in London, the first Indian beauty to win a worldwide contract with L'Oreal cosmetics, and the only Indian actor named to *Time* magazine's "100 Most Influential People in the World." Mr. Khan cultivates a different and equally successful public image. While the Bollywood press machine has linked Mr. Kahn to Ms. Rai (and to a succession of lesser Indian actresses), his larger-than-life image is defined by regular run-ins with the law, including charges of hunting the endangered Blackbuck antelope (believed in Indian mythology to be the steed for the moon god Chandrama), drunk driving, including an ugly case in which he ran over and killed three people sleeping on a Mumbai sidewalk, and rumored connections to the Indian underworld. A dashing figure by any standard, Mr. Kahn has made fifty films in the last ten years for his tens of millions of devoted Indian fans.

In a business dominated everywhere else by Americans and Europeans, Bollywood owes its unusual success to three features of Indian economic life. First, India's GDP relies on old-fashioned consumption rather than investment and exports; and where consumer prices are very low—the average movie ticket in India costs twenty cents—the volume of domestic demand is enormous. Indians purchase more than 4 billion movie tickets every year, three-quarters of all movie admissions in Asia. In addition, the sad state of Indian infrastructure that bedevils most industrial enterprises is incidental to the movie business, especially one so centralized in Ramoji Film City. Finally, filmmaking in India, as in America and Europe, is a business tailor-made for talented, entrepreneurial personalities. Culture matters, and India's entrepreneurial ethic is also evident in the dominance of all the small enterprises in much of the economy. In the Indian film industry, the uncommon business leader that led the way is Cherukuri Rama Rao, India's version of a combination of Samuel Goldwyn, William Randolph Hearst, and Donald Trump. Ramoji Film City is his creation, and he's a genuine media baron who has set the standards in the Indian newspaper, television, and film industries for more than thirty years, and whose business empire now extends to hotels, banking, food products, advertising, and numerous other enterprises.

For all its success, Indian movies are a domestic sector with few exports and little global impact. While Rao's Ramoji operation produces features much more cheaply than American or European studios, other intangibles largely drive the global entertainment industry, from stars and writers to the ineffable capacity to tap into worldwide popular taste. Compared to the huge global entertainment conglomerates that operate worldwide and serve global audiences, the Indian film industry is a parochial small-fry and likely to remain so. Bollywood's revenues are still less than $2 billion a year, compared to worldwide revenues of $30 billion to $45 billion a year each for Time Warner, Disney, Vivendi Universal, and more than $50 billion a year for American-made movies. To make it in the global industry, Indian stars such as Ms. Rai or Chinese director Ang Lee end up working for Vivendi or Time Warner.

Indian software and generic pharmaceutical companies do have global reach, and as global platforms, they occupy a similar place in India's economy as manufacturing does in China. That comparison, however, helps explain why India has little prospect of becoming a global powerhouse in the next ten to fifteen years. Compared to Chinese manufacturing, India's modern, idea-based sectors cannot catalyze economy-wide progress because they're too small and have few ties to the rest of the economy. In China's case, the flood of foreign direct investment over the coming decade will give its manufacturers a growing advantage over their rivals in other developing countries. For India, lower costs in software and generic drug production will make them major players, but mainly for outsourcing by western IT and pharmaceutical companies. The U.S. Food and Drug Administration has approved more plants in India to produce drugs for the American market than any other country outside the U.S.[66] Otherwise, India's software and pharmaceutical companies will have to compete with the world's most advanced and R&D-intensive western enterprises; and there's little prospect in the next ten to fifteen years of Infosys displacing Microsoft or Oracle or Raxbury overtaking Pfizer or Astra-Zeneca.

While India will not be another China in the near future, it does have the potential for steady economic progress. For the last four years, New Delhi has moved to relax the Indian bureaucracy's stifling grip on new businesses by slowly dismantling the old raj system of licensing and production controls.[67] Recent reforms in telecom and electricity, and advanced equipment also should help productivity over the next decade.[68] Where Singh's liberalization policies have taken hold, the results are encouraging. In the last few years, for example, ABB, Honeywell, and Siemens have set up production facilities for electrical and electronic parts and products, LG Electronics of Korea has shifted some of its handset manufacturing there, and Degussa and Rohm & Haas manufacture chemicals. A decade ago, western multinationals couldn't get a foothold almost anywhere in India; today, their share of the Indian market has risen to 49 percent in breakfast cereals and washing machines, 51 percent in televisions, and 27 percent in autos.[69] And the prime minister and his reform-minded allies can drive home the examples of software and pharmaceuticals, which flourished once the government began to deregulate foreign direct investment and relax some labor restrictions.

Demographics also could favor rapid progress for India. The country's boomers are just now reaching adulthood, and by 2020 two-thirds of the population will be of working age. India could reap the same kind of demographic dividend in the next two decades as the Asian Tigers did in the 1970s and 1980s—if, like Taiwan and Korea, India devotes resources to educating its young people and rolls back regulations that still protect millions of mom-and-pop businesses from competition. One of India's leading economists, Jagdish Sheth, recently proposed an even more radical approach: Align his country's policies with the American model of privatization, incentives for innovation, a convertible currency tied to the U.S. dollar, markets open to foreign investment and goods, and massive infrastructure improvements.[70] But if India fails to put in place the "appropriate economic, social and political institution and policies," as the director of the International Institute for

Population Sciences in Mumbai, P. N. Mari Bhat, has put it, India's demographics will "only lead to higher levels of unemployment."[71]

It's a large undertaking. The IMF still rates India's economy as one of the most restrictive in the world. On its present course, Nobel laureate Amartya Sen wrote recently, India combines the worst aspects of capitalism and socialism—"it hasn't made the gains in health care and education that socialist nations do, and has failed to leverage the profit incentive of capitalism to produce a dynamic economy."[72] If that course doesn't change, India will still matter in the global economy, but mainly as a consumer market for everything produced in China, the other Asian Tigers, and the advanced nations. Today, India has some 1.2 million households with annual incomes of $10,000 or more (equivalent to $50,000, measured by purchasing power parity) consuming western-branded products.[73] More important is India's emerging middle class. Some 40 million households have annual incomes of $4,000 to $10,000—equivalent to $20,000 to $50,000 a year in buying power—and their numbers should reach 65 million households by 2010 and 80 to 90 million by 2020. They will comprise a huge constituency for lowering India's prohibitive tariffs on foreign goods. When that happens, as it almost inevitably will under WTO rules and future negotiations, India will become the world's number-three market for thousands of products—most of them produced somewhere else.

The extraordinary power of globalization is evident wherever we are and whatever we do. It enables a developing country to not only import advanced goods and equipment, but secure entire advanced industries and the latest management expertise through transfers and transplants, and then export what those industries produce to countless markets around the world. In so doing, globalization enables tens of millions of people in developing economies to work with modern technologies and business methods for the first time, and thus lift themselves out of poverty. For the advanced countries, globalization enables large corporations to locate each stage of the production of their goods, and soon many services too, wherever in the world it makes economic sense. These global companies can tap markets across the world whose governments and people are eager for the new, advanced products, processes, and technologies. The people in the advanced economies gain access to countless lower-priced goods and services, while their employers achieve productivity gains that eventually may produce the income growth for their workers that leads to upward mobility.

Technology plays a critical but still supporting role in all of this. Advanced telecommunications and information technologies are the backbone of every global corporate network, and these and other technologies are the transfers that matter most to the progress of developing countries. While anyone with the requisite skills can use these technologies, creating a theoretical possibility of what Thomas Friedman famously calls a world that's flat, everyone cannot use them with comparable effectiveness. To start, these technologies are useless outside modern business organizations. And the real world is no flatter for entire nations, because the capacity to take serious advantage of globalization differs so much from country to country.

At this time, China is deriving greater benefits from globalization than other developing nations, even though the relative human and natural resources of many others rival China's. But China's political commitment to globalization has been greater, especially in its extreme openness to foreign direct investment and its large public investments in education and infrastructure. Moreover, the Chinese people thus far have been willing to accept the wrenching social costs of their country's turbocharged modernization. China's extraordinary economic progress is apparent to everyone. The obstacles it will face in extending this progress for the next ten to fifteen years have been less well appreciated.

The United States similarly is deriving greater benefits from globalization than other large advanced nations. America begins with certain advantages in globalization, especially in the development and spread of new technologies and the relative education and skills of its workforce. Most of all, the country's more intense competition drives its companies to make better and often more extensive use of both the new technologies and the new opportunities in developing markets. Curiously, while America's current difficulties with stagnating wages and slow job creation have become apparent to most people, its strengths for sustaining its preeminent position are less well appreciated, especially in much of the U.S. and European media and in American political debate.

CHAPTER FIVE

THE NEW ECONOMICS OF DECLINE FOR EUROPE AND JAPAN

THE SLOW UNRAVELING of the economic strength of Japan and the big three economies of Europe is as important as China's rise and America's sustained success. In all four instances, once-strong economies and societies to date have been unable to accommodate the basic changes of globalization. Since a country's global influence and power are tied closely to its economic capacity and clout, these developments are the material of world history.

It's ironic, since England, Germany, and France, individually and collectively, led industrialization and dominated global commerce and production for more than two hundred years, and their wars and domestic politics reverberated worldwide as America's do today. As recently as 1940, Europe's big three accounted for nearly 20 percent of worldwide production, down only slightly from 1913 and a little more than the United States in the same year, while Japan was Asia's first modern military power and economic powerhouse. Even two decades later, when America and the Soviet Union fully dominated global politics, the defeated Germany and Japan became postwar models of economic recovery and progress. Their economic achievements seemed so awesome that they were widely dubbed "miracles," while postwar England and France created new worldwide standards for social welfare protections. On top of all that, two generations of European leaders moved gradually to make their continent mean something more than geography. The European Community and then the European Union adopted the remarkable ambition of creating a new global standard for transnational comity and civilization, and restoring its member nations to economic and political eminence. Japan, which seemed to take its military defeat harder than Germany or Italy, focused on becoming the world's richest society.

For all their achievements and expectations, for fifteen years Europe's major countries and Japan have been falling behind in most ways that matter economically and, for the rest of the world, politically, too. By 2006, Europe's big three accounted for just 10 percent of world GDP, less than half of what America produced that year, while Japan continued to struggle with the worst economic record in the advanced world. Most sobering of all, there's little reason to expect a real turnaround for any of them over the next ten years. And they cannot claim that the United States and China simply have unique native advantages for this period. Ireland and Korea show that any country can create a successful place for itself in globalization—as successful as the United States or China, if on a more limited scale.

From 1990 to 2005, the combined output of France, Germany, and the United Kingdom, adjusted for inflation, grew about 34 percent—barely 2 percent a year—and Japan grew at half that unremarkable rate. Over the same fifteen years, the American economy expanded 58 percent, which translates into more than 3.5 percent a year or 70 percent faster than Europe's major economies and 150 percent faster than Japan. The unsurprising outcome is that Americans are a lot better off economically than Europeans or Japanese, with income per person in Japan and Europe's major countries ranging from $29,200 to $32,700, compared to $42,000 in the United States. So an average American now earns 28 percent more than an average Briton, 36 percent more than an average German or Japanese, and 40 percent more than a typical French person; and those gaps have widened since 1990. Unless many things change, by 2020 a typical European and Japanese will live on about half what an average American will earn.

Ignore the chorus of American economic triumphalism—which is not much different from Germany's view of itself in the 1970s, Japan's in the 1980s, or Britain's through most of the nineteenth century—and it's apparent that the U.S. economy performed well over the last decade, but not extraordinarily so. The relatively eroding economic conditions of so many Europeans and Japanese come mainly from their own countries' substandard performances. Britain's record is better than France's and much better than Germany's, and Germany's performance has been stronger than Japan's. But all of their recent records are markedly worse than not only the United States, but also than their own performance in the preceding decades.

The pressing matter is not this recent past or the better decades that preceded it, but the very likely prospect that these economies will do as badly or worse over the next ten to fifteen years. Put together thirty years or more of substandard economic performance, and by 2020 they could be economies in genuine and serious decline.

WHAT'S GONE WRONG IN EUROPE

The usual explanations for Europe's economic difficulties begin with their welfare states. For example, in Germany and France—Italy, too— unemployment rates for nearly two decades have run 50 to 100 percent higher than America's. For most economists, the culprits are overly generous unemployment benefits that make jobless Europeans very choosy about going back to work, along with high minimum wages, sky-high payroll taxes and legal barriers to firing anybody that make employers reluctant to hire them. It's all true, but it's probably only a small part of their larger economic problems. Take a broader view and notice, for starters, that over the same twenty years, most new American jobs came from new businesses or fast-growing young ones. Now the low rates of new business creation in Japan and most of Europe—less than half as strong as America's, according to McKinsey studies—look at least as important as all the worker protections. There's no shortage of young or youngish people in all of these places with driving entrepreneurial ambition. But as OECD studies regularly complain, anyone setting up a new business in Germany, France, or England faces much greater bureaucratic hurdles and much greater difficulty finding bank or venture capital financing, and then has to compete with rivals that receive generous government subsidies and protections. McKinsey consultants have found, for example, that it takes two hundred days to register a sizable business property in France, versus two days in Sweden, and months or years to get approval to build a new factory in France, Germany, or Italy.[1]

That still doesn't solve the mystery, because the same barriers and impediments made less difference for growth and even unemployment in the 1970s and 1980s. Moreover, high unemployment hasn't been a factor in Japan. It turns out that high unemployment—allegedly the hallmark failure of Europe's welfare states—matters less, for example, than differences in how much of a country's potential labor force actually seeks or does work. Every country has its own way of measuring this, which economists call the "labor participation rate." So the U.S. Bureau of Labor Statistics recalculates everybody's data so they can be compared,[2] and it turns out that compared to Americans, a smaller share of working-age Europeans and Japanese are in their countries' labor force at all. Some 63 percent of working-age Britons work or want to, as do 60 percent of Japanese, 56 percent of French, 58 percent of Germans, and just less than 50 percent of Italians—that's compared to 66 percent of Americans.

The share of Europeans that could work and actually do or want to is depressed by some of the worker protection programs that dry up potential jobs, and by truly self-defeating policies encouraging early retirement. But a bigger factor is probably culture, since much of the differences come from the share of women working or looking for jobs. Nearly 60 percent of American women are in their country's labor force, and English women come close to that at 56 percent; but only half or less of all working-age Japanese, French, and German women have jobs or say they want them, and an astonishingly low 38 percent of Italian women. Experts on the left often say that this is a result of inadequate public child care; those on the right point to high taxes on second earners; and those in the middle sometimes blame short school days. All those factors apply as much or more to America as to Germany, France, Britain, or Italy. Economically, women are simply less the working equals of men in continental Europe and Japan than in America or England. In the United States and Britain, the big change in the numbers of women going to work occurred in the 1970s, when high inflation and small productivity gains squeezed family incomes and drove more wives to work. But families in Germany, France, Italy, and Japan felt the same squeeze without a similar response, which further suggests that the persistent differences in women working today substantially reflect culture.

For economies that create much of their value by applying human and technological capital, encouraging those with the most experience to retire early and tacitly discouraging women from working at all are very costly mistakes. If these attitudes persist, as they seem likely to, they could cut into European and Japanese growth even more in coming years, as globalization leaves advanced economies with little else to drive growth but their human and technological resources.

That's not all of it. What matters just as much is how hard and how well the people who have jobs actually work. Here, too, Europe's major economies operate differently. First, the average working German, French person, Italian, or Briton today works fewer hours than they used to and a lot fewer than the average American or Japanese. Americans and Japanese work on average about 1,800 hours a year, which comes to 37.5 hours a week (assuming you take off four weeks a year for vacations and holidays). The average Briton works 8 percent fewer hours than that; the average Italian 13 percent fewer hours; and the average German and French working person, about 25 percent fewer hours.[3] That comes to an average work week in Germany and France of less than 30 hours per week. As it happens, Americans and Japanese look like slackers, too, compared to the South Koreans, who work an average 50-hour week or one third more than Americans or Japanese and two-thirds more than the French or Germans. As we will see a little later, that's one reason why per capita incomes in Korea went from less than $600 in 1960 to about $18,000 today.

On top of all of this, people in Japan and many European countries (but not all) don't produce as much for every hour they do work. That's what "productivity" measures, and the variations are striking and important. The marquee case here is the Japanese: Their hours are as long as any in the advanced world, their unemployment rates are lower than almost anyone's, and more older people keep on working in Japan than most other places. But much of those gains are offset by the worst productivity rates of any major advanced economy (as well as the low rates of women working). Japan's workers produce goods and services worth on average $34.40 per hour, which is more than 40 percent less than the average value produced each hour by American workers. France is the opposite case: French workers produce as much for every hour they work as Americans—about $49 in value of GDP per hour—but suffer very high unemployment and work relatively few hours.[4] (Workers in Belgium, Norway, and Luxembourg produce even more per hour, but they're in niche economies with special circumstances.) Germans don't do much better than the French in the share of people working or the hours they work, and on top of that, the average value of what they produce in an hour is about 8 percent less. And Britons work longer hours and have higher labor participation rates than the French or Germans, but they produce 18 percent less per hour than Americans.

With these national variations, some combination of being more likely to be unemployed or out of the workforce, and for those with jobs, working fewer hours and producing less per hour, means it's no surprise that the average European and Japanese now earns so much less than the average American, and that their countries consistently grow less than the United States.

If Japan, Germany, and France—and Britain to a lesser degree—intend to avoid even harder times over the next decade and beyond, their best bet lies in goosing their productivity. At this point, attracting more women into the labor force or persuading older people to keep working would mainly mean higher unemployment for their young people and lower-skilled men. Moreover, whatever they do in those areas, their harsh demographic realities will inexorably worsen the problem of too few people working or working too few hours. Already in Japan and soon across most of Europe, the retirement of millions of boomers is beginning to shrink their workforces, and that will continue for at least the next generation.

If German or Japanese policy makers can figure out why, for example, the productivity of their manufacturers grew 60 percent less than their American counterparts from 1992 to 2006, or British manufacturers 90 percent less,[5] they might be able to affect the economic decline looming out there for them. There's no shortage of standard explanations. The global consultancy McKinsey and Company has made a cottage industry out of pinning productivity problems on government regulation in the countries of its major foreign clients. It's true that in Japan and much of Europe, zoning makes it much harder to build factories or shopping malls in the suburbs, or large-scale stores in many urban areas, and labor regulations prevent retailers from staying open in the evening and manufacturers from adjusting work shifts.[6] International surveys of global executives regularly report that France, Great Britain, and Germany, usually in that order, are major offenders when it comes to imposing huge regulatory burdens. It's also certainly the case that the thousands of small, family-owned grocery, clothing, and what-not shops in Tokyo, Marseilles, Bonn, and Liverpool stay in business mainly because various regulations protect them from competition from larger, more efficient retailers.

Yet big American cities have thousands of mom-and-pop stores, too, many run by recent immigrants; and most of these stores are wildly inefficient compared to Wal-Mart, or for that matter, Carrefour in France and Tesco in Britain. Anyway, Germany, France, and England all have trimmed their regulatory regimes over the fifteen years during which the productivity gap has widened.

Other often cited culprits don't explain all of this much better. Corporate tax rates, a favored target for conservatives on both sides of the Atlantic, have come down in Europe fairly sharply in recent years, so much so that the U.S. rate is now higher than almost anywhere else. And since 1990, slow-growing France, Germany, and Japan all devoted more of their GDP to business investment than the United States. Frustrating as it is, the economic problem facing Japan and most of Europe is deeper—and a lot harder to address—than anything that simple tax or regulatory changes could affect.

Here's the best guess: The guts of the problem now lies in the basic orientation of much of European and Japanese business, economic policy, and commercial culture, which is at odds with the role that globalization offers for advanced economies. The role for developing economies is clear: About a score of them have become serious production and assembly platforms for every good in global trade that can be standardized, from T-shirts and plastic caps to computers and automobiles. Most of this simply follows from all of the capital, technologies, managers, and, finally, entire modern business organizations that global corporations have transferred to the developing world over the last two decades. On top of that, the stupendous growth of Chinese exports is forcing other developing nations to stay ahead of the Sino export tsunami by accelerating the modernization of their own manufacturing.

These developments create new economic imperatives for advanced countries. Small, advanced economies can succeed by becoming preeminent in a few highly advanced niche areas for worldwide customers—offshore software development in Ireland and cellular technology in Finland, for example, and cellular electronics and advanced packaging in Sweden. Larger advanced economies have additional hurdles. Pressed by countries such as Korea and Brazil in advanced manufacturing and by the successful niche countries in high-end services and very-high-end production, they have to take advantage of globalization on a large scale—or ultimately face decline. As a start, their big businesses have to seize on new opportunities in developing countries by sourcing from around the world and selling around the world, so their imports, exports, and foreign direct investments come to span the global economy.

Equally important, their corporations, scientists, and quirky pioneers have to do what can only be done consistently in advanced places—innovate across their sectors and industries by coming up with or broadly adopting new technologies, materials, and production processes—new ways of financing, distributing, and marketing what everybody produces, and new approaches for organizing and conducting their businesses. No country or culture has a monopoly on innovation—Northern Europe is the home of cellular breakthroughs, for example, and Japan is the primary source of best practices in organizing large, modern production operations. But an innovation in one place can be everybody's boon, because globalization and global information technologies give nearly every business some access to the new things and new methods created down the street or ten thousands miles away.

In fact, research shows that a society's capacity to quickly and broadly adopt and adapt the innovations developed by others is at least as important economically as its capacity to come up with breakthroughs itself. Finally, globalization requires that businesses in advanced economies be able to use the most sophisticated management methods, information technologies, and Internet-based tools available to build and manage global networks that tie together everything. In most of these areas, Europe and Japan are flagging or failing outright, and have been for a long time.

Start with the basics: Europe's major economies aren't really global. The vast majority of their trade and investment is still focused on themselves, plus the United States, while they let the rest of the global economy pass them by. In 2005, more than 80 percent of all exports by France, Germany, and Britain stayed in Europe or went to America, with just 14 to 18 percent going to the entire developing world. Even more striking, only 17 to 21 percent of French, German, or British imports come from developing nations, including the Chinese export behemoth and their petroleum imports. By neither sourcing from nor selling into the developing world on a substantial state, Europe's big three sacrifice most of the efficiencies of globalization.

No one can afford to ignore so much of the world. Germany, long the world's largest exporter, saw its share of world exports in pharmaceuticals fall in the 1990s from 17 percent to barely 10 percent, in machinery from 20 to 16 percent, and even in autos, a decline from 18 to less than 17 percent.[7] German high-tech exports have recovered a bit in the last five years but those of others have fallen sharply. From 1995 to 2003, Britain's share of world exports in drugs fell from nearly 12 percent to just over 9 percent, and it has lost an even greater share of worldwide exports in computers and communications equipment.[8] The numbers are even grimmer for Japan, which saw its share of world exports in computers collapse from nearly 14 to barely 8 percent, and fall in communications equipment from nearly 17 to 10 percent.

Here's what the same measures can look like in a nation adapting to globalization: As noted once before, some 44 percent of American exports go to developing nations, and more than 50 percent of all its imports come from those developing nations—in both cases, between two and three times the share for Europe's Eurocentric big three. In this regard, Japan performs well. With all its economic problems, Japan sends 49 percent of its exports to developing nations, and they account for 63 percent of its imports. For Japan, this largely comes from oil and from its major manufacturers shifting much of their production to China and other Asian countries over the last decade, so they now export lots of intermediate goods to their plants across Asia and then import much of the final production. The same happens with America's global businesses—nearly half of all U.S. imports from China, for example, come from the Chinese subsidiaries of U.S. companies. The difference is that Japan's ties to developing economies are highly concentrated in Asia while America's foreign operations and networks span more of Latin America, Africa, and the transition economies of Eastern and Central Europe, as well as Asia.

The cost to Europe of largely ignoring the developing world—including, remarkably, many of their own former colonies—lies not only in denying

German, French, and British consumers access to many of the world's cheapest sources of standard products, and their businesses access to the lowest-priced materials and intermediate goods. Just as important, they're closing themselves off from the world's fastest growing markets at just the time when the businesses and peoples of many developing countries are finally able and eager to buy what more advanced countries produce.

Europe's economic preoccupation with itself cuts even deeper when it comes to where British, French, and German companies set up foreign operations or go into business with local firms. That's measured by where a country locates its foreign direct investment. In this area, France seems almost phobic about the developing world: In 2004, less than 5 percent of the total stock of French foreign direct investment was in developing countries, with China hosting less than 1 percent, while 94 percent was located in other European countries or the United States. Germany and Britain have spread their stock of FDI only a little more broadly—8 to 9 percent is in the developing world, and 1 to 2 percent in China—while 85 percent remains in Europe or America. It's especially remarkable, since countries like Poland, Hungary, and the Czech Republic offer German and French companies the same advantages that Mexico offers the U.S. and China now offers Japan—much lower wages and other costs, quite nearby.

Only the foreign direct investment patterns of America and Japan look truly global. Both hold 27 percent of their total stocks of FDI in developing economies, or *three to five times* the share of Europe's big three.

When the leading companies of another country source from the lowest-priced and fastest-developing places in the world, set up operations there, and learn how to operate and sell there, to two, three, or more times the extent of companies from your own country producing the same things, the costs are large and telling. This basic attitude towards the world beyond their own continent is one of the most powerful reasons why, at this particular time, Germany, France, and Britain have been growing so much more slowly, have developed much lower incomes, and are now generally less productive than the United States. Certainly, European companies in autos, mobile technologies, and energy-efficient production, to name three areas, are at least as globalized as their American rivals. One European airline recently moved its billing and collection operations to India, for example, and cut 40 million euros from its costs and collected another 64 million euros in previously unpaid invoices.[9] But in countless other areas that drive large economies, German, French, and British companies have been steadily losing market share in both third country markets and their own, to companies from not only America but also Scandinavia, Korea, Taiwan, and even Brazil. So, since just 1990, the global market share of European manufacturers shrank from 18.5 to just over 14 percent, while the global market held by American companies rose from 21 to 23 percent.

JAPAN'S PATH TO ECONOMIC DECLINE

Japan's weakness as a global player is different and quite unique for an economy that has been one of the world's most advanced for decades. Its global trade and investment patterns show that its companies are out there in the world. Its weakness is that it keeps the rest of the world out of its own economy almost as single-mindedly as India. It probably goes back to the 1950s and 1960s, when Japan built impenetrable protectionist walls to insulate itself from any competition from anywhere else. Behind all that protection, elaborate subsidies helped make the domestic auto and chemical sectors into world-competitive exporters (but not Japanese aviation and pharmaceuticals, which were just as subsidized). The serious misstep came in the 1970s, when the oil price shocks shook the economy and Japan's government redirected its subsidies and regulatory and other protections (as well as private bank credit) from strong export candidates to floundering sectors. This was the real beginning of Japan's current two-tier economy, split between strong, innovative exporting corporations with global reach in a handful of areas, and hundreds of weak, protected domestic industries, from food processing and textiles to retail and financial services.

Just as important, through the 1970s Japan continued to legally bar nearly any kind of foreign direct investment in the country. Japan's powerful finance ministry eased these bars a little in 1980, but not in ways that mattered, and a little more in 1992, again not enough for anybody else to notice. Many of these restrictions did finally end in 1998, but their legacy is enormous: Nearly a decade later, foreign direct investments in Japan as a share of its GDP are one-seventh the level in the United States, and one-twentieth and one-thirtieth the levels in Germany and Britain, respectively.[10] Japan is so far off the scale here that foreign-owned operations in Japan, relative to the economy's size, are one-fifth that of Turkey. (Barred from pitching a direct stake in the Japanese economy, foreigners have bought up stocks in Japanese companies and currently hold 22 percent of the Nikkei market.)

Moreover, a lot of serious restrictions remain in place. For example, Japanese law still bars foreign companies from carrying out most mergers and acquisitions of Japanese companies. This matters because everywhere else, M & As account for more than half of all FDI transactions—Wall-Mart or Japan's own Mitsubishi, for example, typically establish themselves in new foreign markets by taking over domestic incumbents and remaking them in their own image. But Japanese laws still effectively bar foreign companies from using their stock to buy a Japanese company, hostile takeovers are virtually impossible, and friendly bids that don't use foreign stock are routinely rejected by boards still dominated by "cross-shareholding" or interlocking directors. That's the common Japanese corporate practice of a group of companies holding each other's shares and using them to protect each other from all comers. So, even as Japan's economic stagnation and deflation have made lots of its companies cheap to buy, the world's second-largest economy accounts for barely 2 percent of global, cross-border M & As.

The record of foreign companies that have maneuvered their way into Japan's market should encourage others. The productivity of the Japanese subsidiaries of foreign manufacturers, for example, is 60 percent higher than their domestic Japanese counterparts, and the gap in services is 80 percent. That's not normal: In France and the United States, the average domestic company is nearly as productive as the average foreign affiliate or subsidiary. So those differences in Japan ought to give foreign companies huge advantages, but they don't because they still face Japan's extensive subsidies and regulatory protections for the less-productive native companies they compete with. So, in 2005 (the most recent year for which there are data), FDI into Japan fell to the lowest level since 1996, before most of the formal legal restrictions on FDI were lifted.

Japan's way of denying global economic reality is just as destructive as Europe's. While Germany, France, and Britain each largely limits itself to the relatively slow-growing markets and expensive labor and materials of other advanced countries, Japan walls off its domestic companies from contact with what global leaders in their industries or sectors are doing. The innovations redoing the companies and economy of the United States are largely lost on Japan. The combination of extensive regulatory protections for tens of thousands of inefficient small companies and the absence of competition from companies from other advanced countries destroys the need to develop their own technologies and best practices, and to adopt those of others. The result is that Japan is now the world's least productive advanced economy and, despite its global investments and trading patterns its worldwide market share in manufactured goods has contracted from 19 percent down to 13 percent since 1990, or even more than Europe's.

Some Japanese companies buck these trends, especially exporting companies that have to compete outside the country, and most obviously Toyota, now the world's number-two auto and truck maker and poised to overtake General Motors. (On the same day in February 2006 that GM closed a factory in Linden, New Jersey, and Toyota opened one in Tijuana, Mexico, GM chairman Richard Wagoner insisted, "We've been ahead for seventy-three years in a row, and the betting odds are we'll be ahead for the next seventy-three years."[11] He's either bluffing or dangerously uninformed.) Each of Europe's big three also has world-class exceptions—from Germany, companies such as the energy giant E.on, electronics maker Siemens, and insurance leader Allianz; from France, the energy giant Total, insurance leader Axa, and retailer Carrefour; and from Britain, British Petroleum, HSBC Holdings in financial services, and Vodafone in telecommunications.

Behind these marquee enterprises, most other companies in all of these countries have had terrible difficulties adopting, and adapting to, innovation, which globalization makes imperative for all of them. The biggest issue here isn't who commits the greatest relative resources to developing innovations. Since 1990, Japan has spent more of its GDP on research and development than the United States, and while America leads Europe's big three, much of the difference comes from America's R & D in defense areas.[12]

The best clue to what's really going on comes from studies of the source of America's productivity gains. Here, a consensus has emerged among most American economists and sociologists who've thought about it. First—no shock here—it's linked closely to the spread of information technologies. The surprise is that the extent of the actual benefits from those technologies comes not from how much a company or a country spends on them, but from the way they use them.

That's where Europe's and Japan's regulations and subsidies take a serious toll. All the subsidies for core industries and farmers make their returns healthy, almost regardless of what they do, while zoning and price control regulations discourage new competition. So, even though French, German, and British firms have invested as much in the same information technologies as American (or Swedish) companies, all of their subsidies and regulatory protections blunt the market pressures, which in America force them to change some of the ways they do business to make decent use of those investments.

European and Japanese managers are just as capable as Americans in figuring out what changes they need—and if they're not, there are plenty of international consultants ready to do it for them. But it's difficult and unpleasant for most people to make the changes required to adapt a business to new technologies—cut and reorganize departments, jobs, products, and processes, bring in people with new skills, get rid of those unable to work with all the new tools, and much more. So most people won't take the trouble and the grief to make such hard changes when subsidies and regulations will let them go on doing most everything much as they've always done.

If that's not enough, in Germany and France, firms that want to adapt sometimes find that it's against the law: There, all but the smallest employers cannot dismiss or reassign workers without cause, and reorganizing to make good use of new technologies doesn't constitute legal cause. A bank or retailer that wants to install a new inventory control system or a new automated phone operation can do it, but it can't let go the employees who ran and operated the old systems or, in many cases, even reassign them. That's one of the main reasons why the OECD concluded recently that outside Germany's auto and machinery industries, the German economy has benefited hardly at all from new information technologies.[13]

All the regulations and subsidies, along with high payroll taxes, also hold back productivity progress in Japan and much of Europe by making it harder to start new businesses. That matters a lot, because new enterprises are more prone to adopt, and adapt to, new technologies and business methods, as a way to make their mark in a marketplace dominated by well-known incumbents. And when they succeed, it creates pressures on those incumbents to follow their lead or strike out in their own direction. We cannot know for certain, but Barnes and Noble might still be a traditional bookstore if regulations and subsidies had prevented Amazon.com from taking hold.

Workers in France, Germany, Britain, and Japan also have more trouble adapting to new technologies and new business methods. One reason that's rarely acknowledged is that working-age Europeans and Japanese are significantly less educated than the more productive Americans. Japan and Europe's big three spend between 15 and 25 percent less each year on education, as a share of their GDP, than the United States—with some of those differences coming from American families' greater private expenditures, mainly to send their children to college. But the bigger factor is that for a long time, Europeans and Japanese have been attending school for fewer years than Americans—and there's ample research showing that workers with more education tend to be more flexible and adaptable to new technologies and new ways of working.

As we noted earlier, according to the OECD and the World Bank, Germans and Japanese over age twenty-five today have spent an average of less than ten years in school, which is more than the English at a little over nine years of formal education and the French with an average of just over eight years. They're all well behind Americans age twenty-five and over, who average more than twelve years in school, or 20 percent more than Germans or Japanese, one-third more than Britons, and an astounding 50 percent more than the French.[14] The only countries that come close to America in this respect are Sweden . . . and Korea. Moreover, the differences extend to higher education, with the people who get it presumably being most effective of all with new technologies. Some 23 to 24 percent of French and German adults, and 28 percent of English adults have some education past high school—compared to 38 percent of Americans.[15] In this respect, only the Japanese, at 37 percent, come close to the Americans.

These findings shed new light on reports that American students score lower than students from many other countries in standardized math, science, and reading tests. In fact, American students score well in all subjects at fourth- and eighth-grade levels, compared to students from other advanced countries.[16] And American high schoolers score higher in reading skills than older students from all other G-7 countries except Britain. However, the results also show that American math and science education fails many older students, especially those from lower-income families. At ages fifteen and sixteen, students from three other G-7 countries outperform Americans in math, and students from four G-7 countries outperform Americans in science. But that's not the whole story, because a larger share of American kids are still in school at ages fifteen and sixteen than kids in other advanced countries, removing from the sample relatively more non-American young people who would score low.

Europeans and Japanese appreciate as well as anyone else that in a time when technologies are changing rapidly and competition from distant, low-cost countries is increasing mightily, more education can make most people more productive and competitive. But in much the same way that subsidies and regulation insulate many European and Japanese businesses—perversely so—the elaborate formal and informal protections for workers in these places shield them from the hard edge of the realities inexorably narrowing their own prospects.

WHAT LIES AHEAD FOR EUROPE AND JAPAN

Like our own time, the 1970s and 1980s were challenging for advanced economies. Everyone largely did what the times dictated. They responded to persistent inflation by raising interest rates and soldiering through the sharp slowdowns that followed. They dealt with nasty oil price shocks by making their economies a little more energy efficient through new auto-mileage standards and gasoline taxes. And they responded to an unexpected drop in productivity growth by cutting business taxes to boost capital investment.

Voters complained and leaders fell, but the major countries mostly controlled their own economic course and managed to adjust reasonably well without seriously disturbing the way their societies worked.

Globalization is less accommodating; and this time Japan and much of Europe have less control over their economic fates. Already, they've fallen behind its adjustment curve so quickly that their options for the next decade are limited. Moreover, adapting successfully to economic realities will require major and unpopular changes in some of their basic social arrangements.

That's not likely, so the outlook for Europe's big economies looks pretty dismal. Starting with the basics, if they persist in ignoring most of the developing world in their trade and investments—and there's no evidence of a learning curve here—they will increasingly relegate themselves to selling into and buying from the world's slowest-growing markets—their own—along with the United States. Problems with their domestic economies will make shifts in their direct investment habits costly. In America, every dollar sent abroad in foreign direct investment produces $1.14 in gains at home, from the cost savings for outsourcing businesses and the new jobs that the displaced workers find and fill.[17] But France and Germany are so weak at creating jobs that for every euro sent abroad, their economies earn just 0.86 euros and 0.74 euros respectively. Moreover, American, Japanese, Korean, and the most foresighted European companies already have grabbed the best investment opportunities in China and many other developing nations—which is just the economic law of diminishing returns in action.

The outlook looks grim for Europe's manufacturing workers, who are perilously dependent on exporting what they make to the United States. That's a perilous position, because German, French, and British manufacturers produce largely the same things as American companies—and their American rivals are more globalized than they are. So their U.S. competitors draw more on cheaper foreign labor, materials, production, as well as more of the best ideas and technologies from anywhere in the world. Europe's global market share in manufacturing has been shrinking for years. Over the next decade, it is likely to contract even more in not only third-country markets where everybody competes, but also the mammoth U.S. market and, ultimately, in Europe's home markets as well.

Turning to globalization's other demands on advanced economies—a healthy capacity to innovate and to adopt and adapt innovations from other places—the next decade will be very difficult for Japan and Europe's big three, if they don't get serious about making better use of information technologies. The catch is that it requires chipping away at lots of worker protections at a time when years of slow growth have made workers even more attached to them. Similarly, helping European firms create a lot more new jobs—so those who lose theirs in business reorganizations can still contribute to the economy—will mean lower minimum wages, perhaps lower payroll taxes, and certainly less generous unemployment systems. Again, all this would have to happen at a time when the incomes of most European workers are stagnating and Europe's governments need all the revenues they can get.

The recent election of a Nicholas Sarkozy, a conservative reformer, as French president has been widely seen as the most encouraging news in years for France's economic prospects. Sarkozy is far from a free-trade, free-market liberal on the American or British model. A few years ago as budget minister, he led a government rescue of the French power and railroad giant Alstrom and opposed the purchase of the huge steel maker Arcelor by Mittal Steel from India. As a presidential candidate, he regularly called for "economic patriotism" and maintaining state supports for large industries. Nevertheless, since taking office in May 2007, Sarkozy has proposed changes that would constitute a genuine break with the past, including large cuts in the state bureaucracy through attrition, significant or symbolically powerful pension reforms, new incentives for people to work longer hours, a minimum-wage reform, potentially important tax changes, and critical modifications in workers' legal protections against being fired.

Sarkozy's program would be rough on many French workers, but it would better align France's economy with the needs of globalization. However, the early results suggest that the president's promises may exceed his political reach. He managed to cut the top income-tax rate and inheritance taxes in short order; but broader, more important proposals to spur job creation by cutting payroll taxes (while raising VAT charges) and to stimulate home construction by making mortgage interest tax-deductible haven't moved at all. His plan to effectively lengthen the thirty-five-hour workweek by making overtime tax-exempt also has stalled. Moreover, his initial effort to build a grand coalition with France's powerful unions—for far-reaching plans to reduce some state pensions, raise the retirement age, slow future increases in France's current minimum wage of more than $12 per hour, and make it easier for employers to dismiss workers—failed as well. Some of this might change, but these early returns suggest that like Japan under Junichiro Koizumi, France is more prepared for a reformer than for real reforms.

Creating jobs and attracting investment by cutting payroll taxes, as President Sarkozy proposes, seems well beyond the reach of all the major European countries—especially as the health-care and pension costs they help finance are set to soar with the retirement of Europe's older boomers, while the workforces that finance it all begin to shrink. The unhappy fact is that

Europe has put off its adjustment to globalization for so long that now it's set to bump up against the additional pressures of adjusting to their demographic changes. Even if Germany, France, Italy, and Japan had economies attuned well to globalization, their dismal demographics would still put greater downward pressure on their standards of living than any force since the end of World War II.

European governments have studied the issue of looming pension costs for thirty years. In the one case of Margaret Thatcher's Britain, a country did something about it. In fact, Thatcher's reforms cut so deeply that today 20 percent of elderly Britons live in poverty. Without reversing course and expanding state pensions—which is increasingly likely—by 2020 nearly one-third of British retirees will live in poverty. Not so in France and Germany, which still have such generous pension systems that they're plainly unsustainable even for the next fifteen years. France took a stab at reform in 2003. The prospect sparked public protests by two million people in a hundred cities, even though the changes were so modest that the system still faces funding shortfalls by 2020 equal to 4 percent or more of the country's GDP. Sarkozy's proposals to modestly and gradually increase the current retirement age of sixty and end special provisions for even earlier retirement in certain professions and by workers in state-owned companies already have drawn attacks from the unions whose support is probably indispensable. Germany also enacted some modest pension changes in 2001, and the latest estimates of their effects see total costs rising by 1 percent of GDP every three years from now on.

The basic quandary for European reformers over the next decade is that globalization depresses the productivity gains and competitiveness of those who don't adapt well to it, which in turn slowly and surely reduces people's living standards. As European living standards stall or slowly erode, people's attachment will only intensify to the distinctive guarantees against firing or being reassigned, the generous unemployment benefits, and the high minimum wages that make Europe Europe. And across the vast service economies of Germany, France, Britain, and Italy, the lumbering, domestic market leaders will cling just as tightly to their own subsidies and regulatory protections.

The European Union is supposed to ease some of this quandary by creating a single market as large and stable as America's, within which everything can pass without tariffs or quotas, capital can flow without regulation, and any person can move without restriction. The EU's dilemma is that its member states don't want such a free market union. For example, in 2000, the members adopted the "Lisbon Strategy" to make the EU "the most competitive and dynamic, knowledge-based economy in the world." Seven years later, almost none of the recommendations have gone anywhere. The strategy called for more R & D outside the four industries that have claimed two-thirds of it in Europe, and nothing has changed. It urged patent reforms to bring down those costs, and nothing has happened. It called for full privatization of Europe's telecom and electricity state companies, and that mostly hasn't happened. It urged member countries to encourage new business creation by ending subsidies for monolithic "European champions," and here, too, little of substance has changed. Finally, Lisbon called for lower taxes; and while corporate taxes have come down in many countries, the overall tax burden hasn't declined at all.

For all of the EU's pronouncements, its core countries seem to prefer a strategy closer to Fortress Europe—an EU that resists global pressures—than one open to the global economy. So Germany has called on the EU's ten new members to *raise* their taxes to German levels "in the name of fairness"; and Brussels actually ordered Poland's government to raise more taxes. In the same vein, Sweden was ordered to increase its farm subsidies, and the Czech Republic has had to adopt stricter labor market regulations. And recently, Germany and France stopped moves to lower barriers to sales of services from companies in one EU country to another, out of their well-grounded fears that low-cost Eastern European businesses would overwhelm their own, well-protected and inefficient service companies.

It's hard not to suspect that the EU no longer believes in itself—and based on the rejection of its constitution by the French and Dutch in 2006, neither do many Europeans. EU skeptics have been on the political ascendancy in Britain, Poland, France, Sweden, and Holland; and the Italian, French, and Spanish governments all have taken steps to insulate their domestic companies from takeovers by others *inside* the EU. And in France, where the EU was born, a "Don't Tread on Me" attitude towards the Union was a point of rare agreement in the 2007 presidential campaign. The socialist Marie-Ségolène Royal called for new industrial policies, a "social treaty" that would impose France's worker protections across the EU, and tariff walls around Europe, while Sarkozy blamed the euro for undermining French competitiveness and spoke of neutering the European Central Bank. On top of these problems, the Union new faces new and explosive social pressures stoked by bringing in Central and Eastern European members. The EU's enlargement not only has opened the labor markets in Germany, France, and Italy to millions of workers prepared to work for lower wages. It is also increasing Muslim immigration and ugly responses to it in some of Europe's most tolerant countries, developments that eventually could lead to reinstating national borders inside the EU.

Most recent forecasts suggest that at a minimum, being part of the EU won't help Europe's major economies over the next decade. According to the OECD, France should expect to grow by about 1.5 percent a year from 2010 to 2020—a miserable rate—and Germany's outlook is no better. One of Germany's leading think tanks, the Institut fur Weitwirtschaft, is even more pessimistic, forecasting gains of just about 1 percent a year. The experts at Deutsche Bank take a rosier view and figure that with decent doses of reform, Germany and France could grow, respectively, by 1.9 percent and 2.3 percent annually. But the bank's experts also warn that without reform and strong immigration, this growth path could fall to as little as one-half of 1 percent a year for a decade, which would be social dynamite. Amidst all this gloom, the OECD's chief economist, Jean Philippe Cotis, offers the typical European consolation: "in the end . . . our own way of life is worth so much more than a vulgar GDP per capita statistic."

One bright spot is Britain's outlook. After two decades of Thatcher and Blair, British government is smaller, its taxes and unemployment rates are lower, its regulations are less stringent, its subsidies are less generous, and its labor participation rates are higher than Germany, France, or Italy. Even if a remarkable one-quarter of adult Britons lack literacy skills, young Britons on average are as educated as young Germans and more so than the young in France. Its businesses also invest more per worker than German companies, although still less than French or American firms. The experts' outlook for Britain over the next decade averages about a half-percentage point higher than Germany, and at least as good as the outlook for France.

Not so in Japan, where resistance to reform is almost a public virtue. To western eyes, Japan offers a baffling case of a democratic society that accepted more than a decade of economic stagnation with little protest and no more than cosmetic changes in its politics and policies. Many Japanese blame their problems on a singular event that allegedly no one could foresee or be held responsible for—the collapse of the "bubble economy" in the early 1990s, when real estate and stock prices crashed and thousands of business loans went sour. The bubble and its bursting were no acts of nature, but the clear and direct results of certain tax and monetary policies, subsidies, regulations, and business arrangements well known to nearly everybody. Moreover, even if Japan's one-party politics and social norms make it difficult for Japanese voters to hold their political and economic elites responsible for anything, the burst bubble should have run its course in a few years. The United States has shrugged off stock market crashes and serious banking problems in recent times. The shock to Japan's system was also certainly less severe than the one dealt to Korea in the financial crisis of 1997–1998. But there, as we will see,

Korean reformers took charge and within a few years Korea once again was one of the world's fastest-growing economies.

In Japan, decisive economic reforms seem out of reach—at a minimum, they haven't happened; and without them the economy has stagnated and slipped into recession five times in fifteen years. Through it all, a succession of prime ministers have done little to reform public and private economic arrangements as outdated as communism. Even when the charismatic reformer Junichiro Koizumi managed to come to power, his own party stopped all but the most modest changes; and Koizumi was succeeded by party functionaries that would have fit right in during the do-nothing 1990s. Most subsidies, protections, and regulations have changed little; the cross-shareholding arrangements that help keep thousands of inefficient firms afloat continue, and the country's financial system remains underdeveloped, with almost no IPO or commercial paper capacities as alternatives to a weakened banking system. Outside the small circle of Japanese exporting giants, led these days by the hypersuccessful Toyota, economic productivity continues to languish.

There also has been little progress towards creating an environment for foreign direct investment, Japan's best shot at introducing the bracing competitive forces that globalization insists on. And the nation's economic blind spot about the role of women in the economy remains strong. Koizumi's former chief advisor on this issue, Mariko Bando, describes her nation as "a developing country in terms of gender equality,"[18] and she doesn't exaggerate: Women earn barely half of what Japanese men do for the same hours; and despite the world's fastest-shrinking labor force, less than one-third of Japanese companies say they would consider recruiting women for management or technical positions.[19] Public pensions are the one area where Japan has embraced real reform—although it's easier to raise the payroll tax in a country where, according to the OECD, nearly 40 percent of the people manage to avoid or evade at least part of it.[20]

The way to get out of a hole when you won't or can't climb out is to wait for someone to rescue you. In modern Japan, that typically has meant commercializing a new gadget or equipment that the world will want—copying machines, color televisions, VCRs, and the Walkman—or better still, achieving a real technological breakthrough. Mikio Wakatsuki, head of the Board of Counselors of the Japan Research Institute, has said that with so much of his country's traditional manufacturing going offshore, its future lies in robotics, biotechnology, and nanotechnology. Honda, Sony, and Mitsubishi all have major R & D programs in robotics, and there's considerable interest in solar cell technologies. But for nearly twenty years, as *The Economist* recently put it, "Japanese firms have not counted for much in software, the Internet, biotechnology, and other high-growth industries," and there are no Japanese equivalents to Microsoft, Google, Amazon, or even Apple.

The cultural, social, and political forces contributing to Japan's ongoing economic weakness are stronger than even a few serious technological breakthroughs. Ryosei Kokubun, a Keio University expert on Japan's place in the world, has found that most Japanese still see globalization as just a threat and not an opportunity.[21] That attitude reinforces Japan's political and social attachments to all the arrangements that globalization dictates must change. With a labor force contracting 1 percent a year and desultory efforts at best to raise productivity, Japan now looks forward to another ten or fifteen years of growth averaging 1 percent a year or a little more, punctuated by occasional spurts of stronger gains (as recently) and periodic slowdowns. On this likely path, the incomes of millions of average Japanese will stagnate or edge downward, as low rates of return and fast-falling savings rates also erode the real wealth of Japanese households. On this course—and it takes a vivid imagination to see how Japan can get itself on a better one—by 2020 Japan will be a national economy in unequivocal decline and a diminishing factor in the world economy.

While Germany, France, and Japan all are failing to adapt successfully to globalization—and to some degree, Britain, too—Ireland and Korea are using globalization to promote themselves to the top tier of world economies. Neither has the size or breadth required for global effects, but when European countries such as Greece, Hungary, or Portugal look for a new economic model these days, or Asian countries like Thailand, Malaysia, and Indonesia do so, they are more likely to think about Ireland and Korea than the much larger and, of late, much-less-successful modern economies in their regions.

THE CELTIC TIGER

Korea is the greatest small economy success story of the last half century, while Ireland is the biggest success of the last generation. The Irish success already draws delegations from Central and Eastern European governments and research institutes, looking for lessons. It's obvious why—the new Celtic Tiger has been the world's fastest growing developed economy over the last fifteen years. From 1990 to 1995, the country grew nearly 5 percent a year and then accelerated to almost 10 percent annual growth over 1995–2000. Even from 2000 to 2005, Ireland's real growth averaged nearly 6 percent a year, about twice as fast as the United States, more than three times the rate for the Euro zone countries, and nearly four times as fast as Germany.[22] And over the last twenty years, Ireland's real GDP tripled, producing by 2003 (the last year for this comparative data) the world's highest GDP per employee.[23] It has also escaped European-style unemployment: Since 1993, even as the country's labor force grew by more than 60 percent, its jobless rate fell from typically high European levels of almost 15 percent to the lowest in North America or Europe.[24] *The Economist* sums up these recent achievements:

> Fifteen years ago Ireland was deemed an economic failure . . . Yet within a few years, it had become . . . a rare example of a developed country with a growth record to match East Asia's, as well as enviably low unemployment and inflation, a low tax burden and a tiny public debt.[25]

Nearly everyone's living standards in Ireland have improved sharply. From 2000 to 2005, the average real income in Ireland grew nearly 20 percent, several times faster than Americans and twenty times or more faster than the French and Germans. By 2005, the average Irish had a considerably higher income, at $40,150, than the average Japanese, Briton, French person, or German, and nearly as much as the average Swede or American.[26] There is one unusual feature that qualifies some of these achievements. The Irish economy depends enormously on foreign investments, and much of their earnings, which are included in Irish national income, flow back to parent companies in the United States, the United Kingdom, and continental Europe. If we adjust Ireland's national income for these large outflows of profits and dividends (technically, by using "Gross National Product," or GNP, instead of the standard GDP), the average Irish person still earns nearly as much or more as the average person almost anywhere else in Europe. But adjusting as well for differences in the cost of living secures Ireland's current edge over the rest of Europe. And by any measure, Ireland today is one of the world's ten richest countries.

Ireland owes this rapid progress to a strategy that combines the paths of the Asian Tigers in the 1970s and 1980s and some of China's policies in the last decade. Like the Asian Tigers, decades of large social investments produced economic dividends when Ireland's boomer demographics hit home. While Ireland's boomers were growing up, Dublin substantially increased support for higher education and created a business-friendly environment for companies that would later employ the demographic bulge, by spending liberally on public infrastructure and offering tax relief and special incentives for foreign investors. This new approach by a country that for decades had shielded itself from progress with traditional protectionism and meager public investments began in an unusual way—with a government study that broke decisively with the government's then-current policies. In 1959, the Secretary of Finance, Thomas Ken Whitaker, directed a policy review which to nearly everyone's surprise, proposed deep cuts in tariffs, a shift in public priorities from European-style worker supports to new investments in industry, education, and roads, and sweeping tax and regulatory changes to attract foreign companies. One recent analysis didn't exaggerate greatly when it called Whitaker "the architect of modern Ireland."[27]

More surprising, over the next decade Dublin followed his advice, ending restrictions on foreign ownership of Irish companies, repealing corporate taxes on profits from exports, unilaterally lowering import tariffs, negotiating a free trade agreement with the U.K., and declaring its intention to join the European Economic Community (EEC), the EU's predecessor. Ireland wasn't admitted to the EEC until 1973, when France finally stopped vetoing it as part of Paris's policy of blocking Great Britain's entry. Yet even in the 1960s, Irish growth jumped to more than 4 percent a year, as its ample numbers of low-wage, English-speaking workers—along with the new tax breaks— began to attract U.S. and European companies. By 1970, more than three hundred and fifty foreign corporations had set up facilities there,[28] so when Ireland finally made it into the EEC, it promptly became a low-cost place for American and British industrial firms to sell into Europe. And while Ireland's neutrality in World War II had disqualified it for the Marshall Plan, its relative poverty within the EEC entitled it for substantial "convergence" payments that helped fund the new infrastructure and education spending.

This was a good beginning, but the real force of Ireland's new economic strategy took longer to gather. Through the 1960s, it remained heavily dependent on basic agriculture, much of it exported to Britain, and unemployment stayed high even with substantial emigration. Drawing on a 1965 OECD analysis of Ireland's prospects that stressed education, the government in 1967 ended fees for secondary education and offered all Irish youth free transportation to school. Educational levels began to rise; and by the mid-1970s, employment, population, and national income all began to grow at the strongest sustained rates since Irish independence, a half-century earlier.

While its large public investments linked Ireland to the East Asia Tigers' policies, its turn away from protectionism and appetite for foreign transfers anticipated later Chinese policy. From the 1970s, Ireland set out to become a production platform for the European market, focused on high-technology products that would need highly skilled and highly paid labor, especially in computers, pharmaceuticals, and medical equipment. By 1975, the two U.S.

mainframe makers Amdahl and Wang, the medical device producer Baxter Travenol (now Baxter International), and drug makers Merck Sharp and Warner-Lambert all had set up new facilities in Ireland, and more than four hundred and fifty foreign-owned operations accounted for two-thirds of the country's industrial output.[29] As these foreign transfers rose, the country's expanded commitments to public education produced their skilled labor. From 1965 to 1980, Ireland doubled the number of students in secondary or post-secondary education, and in the following twenty years, it redoubled the number of students in colleges, universities, or other postsecondary institutions, a sixfold increase in thirty-five years.

These investments set up Ireland for what the leading scholars of demographics and economics, Harvard's David Bloom and David Canning, dub the demographic dividend—when a boomer-size generation enters the labor force with more education and the economy creates the additional high-skilled jobs for them to fill.[30] For sustained progress, Bloom and Canning's research found that a baby bust generation has to follow the boomers, so the share of the nation's population that's actively working ratchets up. That drop in Irish fertility rates began in the 1980s, when contraception was finally legalized and the country's economic opportunities began to draw more Irish women into the workforce and away from having large families. Bloom and Canning calculate that demographics account for one-third of the increase in the growth rates of Ireland—slightly less than in East Asia.[31] Europe's large economies and Japan also reaped the same kind of initial benefits—in the 1970s and 1980s pundits called them "miracles"—but didn't sustain them with additional reforms that could take account of globalization.

Ireland's progress nearly derailed in the late 1970s and early 1980s, until once again it followed the experts' advice. When the oil shocks of the 1970s began to eat into the rising incomes of Irish workers, Dublin reverted to big budget deficits and easy money policies; and by the early 1980s, inflation and unemployment were back and growth was sputtering. In 1985, the national deficit reached nearly 10 percent of GNP, and the IMF warned that Ireland was headed for a fiscal meltdown and currency crisis. More up close and personal than the IMF's warnings, Irish academics and journalists noted that many of the country's best and brightest were leaving for better prospects, including virtually entire classes of graduates from Trinity College, the Harvard or Oxford of Ireland. Overall, some 200,000 Irish emigrated to other countries in the 1980s, almost 6 percent of the total population.

To its credit, Ireland's government took the warnings seriously and began to quickly bring down its deficits. In 1986, funds for health and education fell by 6 percent, defense lost 7 percent, roads and housing took cuts of 11 percent, and agricultural spending fell 18 percent. The following year, a new Fianna Fail government put in place even sharper cuts and worked out a new social agreement among large employers, unions, and farmers to accept very limited wage increases in return for modest cuts in direct taxes. In three years' time, Irish government spending as a share of GNP had fallen by one-quarter, from 55 to 41 percent.

Ireland's real good fortune, however, was that a strategy intended to make it part of greater Europe positioned its economy quite well when globalization left much of the rest of Europe off balance. For years, the government did everything it could to attract American and British firms in the IT hardware and software businesses. By the 1990s, much like Washington under Bill Clinton, Dublin moved from budget deficits to small surpluses, privatized telecommunications, invested 5 percent of GDP a year in upgrading its ports, train systems, and roads—twice the EU average—and increased funding for public health and water projects and, always, education. Public spending on education has grown 150 percent since 1985, especially once the government abolished college and university tuition fees for Irish young people in 1995. Although Ireland still spends less as a share of GDP on education than many other places in Europe, a higher percentage of its young people attend school than in Germany, Italy, or Japan. Moreover, 37 percent of Irish men and women ages twenty-five to thirty-four have some postsecondary education, compared to 27 percent across the EU (in the United States it's 40 percent).[32]

So, when the American IT boom gathered force in the mid-1990s, Ireland was well-positioned to participate. By 2004, Ireland's $230 billion stock of direct foreign investment was as large as Italy's, an economy eight times Ireland's size;[33] and today the stock of foreign investment, relative to Irish GDP, is ten times that of Germany, five times that of France, and four times that of the United Kingdom. Little Ireland also has about one-third the FDI of China, the world's leading new destination for foreign investment and an economy nearly twelve times its size.

Ireland's commitment to education and foreign direct investment have created a special form of globalization synergy, as the foreign transplants expose thousands of the country's highly skilled workers to leading-edge technologies and operating practices, which enable them to further improve their skills. Those direct transfers are generally focused on information technologies, pharmaceuticals, and medical equipment. Today, seven of the world's top ten IT companies—IBM, Intel, HP, Dell, Oracle, Lotus, and Microsoft—have Irish production facilities. Thirteen of the world's fifteen major pharmaceutical companies also have major operations there, including Wyeth; Schering-Plough; Merck, Sharp and Dohme; Pfizer; Novartis, Allergan; and

GlaxoSmithKline; and just outside Dublin, the world's largest biotechnology campus was recently completed. Finally, fifteen of the world's top twenty-five medical equipment makers are in Ireland, including Boston Scientific; Becton Dickinson; Bausch and Lombe; Abbott; Johnson & Johnson; and Stryker. One result is that advanced technologies play a much larger role in manufacturing in Ireland than anywhere else in Europe outside of Finland—more than 20 percent, compared to an EU average of barely 8 percent, and not far behind America's 26 percent. And foreign companies in these three high-technology areas employ 85,000 Irish, or one-third of the country's manufacturing workers.[34]

Nearly 85 percent of all this manufacturing is exported, so that where a few decades ago Ireland's principal exports were foodstuff and its young people, today it's mostly the high-tech products of its 1,200 or so foreign-owned companies. Intel, Dell, and Microsoft together, for example, account for 20 percent of all Irish exports; and on a per capita basis, Ireland's exports are now seven times those of the United States, six times those of Japan, and four times those of Great Britain.

Remarkably, Irish modernization has also spread to some high-end services, by the same path. The educated workers, tax incentives, and telecommunications investments that attracted the advanced technology transplants also have drawn major FDI in financial services; and in the last decade or so, with little notice from the *Wall Street Journal* and the *London Times*, Ireland has become a significant location for global finance. More than one-third of the FDI in Ireland is now concentrated at the Dublin International Financial Services Centre, and more than half of the world's twenty largest insurance companies and fifty largest banks are now located there. Strange as it may seem to traders at the Paris Bourse or the Bank of China in Hong Kong, Ireland is a genuine hub in global capital markets, with the total value of the international funds managed there now topping more than $500 billion, or twice the country's GDP.

American and British conservatives like to insist that Ireland's transformation was propelled by a 12.5 percent corporate tax rate and zero tax on profits from exports. While they exaggerate the role of low tax rates, lower taxes are better than higher taxes for any economy, so long as they cover the public spending the country wants. Ireland's total tax burden at 28 percent of GDP is far below the 40 percent average for the EU 15; and as a share of labor costs, Irish taxes are about ten percentage points lower than any other advanced European economy.[35] The convergence payments from the EU also played a role, paying for a good part of infrastructure and educational spending in the 1980s and early 1990s. But low taxes would have had little effect if Ireland hadn't opened its economy to foreign competition and foreign investment; and EU subsidies wouldn't have mattered if Dublin hadn't used them wisely.

Moreover, these factors make little difference today, since corporate taxes have come down in many countries and Ireland hasn't qualified for EU subsidies for some time. What matters now is that the technologies and business methods of many of the world's most advanced companies are firmly embedded in Irish manufacturing and business services. Moreover, these huge investments are part of the world-spanning networks of those global corporations, linking Ireland's economy to the most successful aspects of globalization. The result, in the judgment of the global business consultancy AT Kearney, is, "the world's most globalized economy" with the best business environment in Europe.

Whether Ireland will sustain its advantages over the next ten to fifteen years is an open question. On the plus side, the strategy is still producing the strongest productivity gains in Europe, and the country doesn't face the shrinking labor force that haunts the rest of the continent and Japan. Not only are Irish birthrates higher than most of the rest of Europe; but its prosperity has become a powerful magnet for immigrants. Like the United States, immigration accounts for more than half of the total growth in the country's population since 1995,[36] and foreign nationals comprise nearly 10 percent of all Irish residents. And unlike the waves of poor Muslim immigrants meeting fierce resistance in France, Germany, Denmark, and the Netherlands, Ireland's immigrants are western, generally educated, and eager to assimilate. Half of them come from the U.K., and most of the rest are either returning Irish who had emigrated in the 1980s and 1990s or new immigrants from Poland, Lithuania, and other new EU accession countries.

Ireland also will avoid much of the public pension crisis looming for most advanced countries. Now and for the next generation, the share of the Irish population age sixty-five and over will be the smallest in the EU. Moreover, as a recently poor country, Ireland's public pensions have never been generous. And despite its extraordinary recent prosperity, Dublin actually pared back future benefits in 1999 and, alone among advanced economies, puts aside 1 percent of its GDP every year in a National Pensions Reserve Fund.

But globalization will present a new and difficult challenge for Ireland over the next decade. The country built its prosperity on waves of direct investment by offering a cheap, educated workforce, tax benefits, and duty-free access to the rest of the EU. This strategy still works: The annual return on American investments in Ireland averaged over 20 percent from 1995 to 2002, or twice as high as investments in other EU countries.[37] Moreover, in 2005, American companies still invested more in Ireland than in China. But a generation of strong growth and productivity gains has erased Ireland's wage advantage; and with the EU's recent expansion, new members such as Hungary, the Czech Republic, and Poland can compete for Ireland's gateway status to the European market.

The test for Ireland over the next decade—one China also will face in a different environment—is to wean itself from its dependence on foreign investment and develop more domestic companies that can compete in advanced manufacturing and services. In the last decade, a handful of Irish companies have become players in worldwide markets, notably: CRH, a global cement and building materials corporation and Ireland's only company in the Fortune Global 500; the Kerry Group, a food and flavoring manufacturer headquartered in Tralee, Ireland; and the international hotel chain Jurys Doyle. But the nation's vaunted productivity and export gains still come mostly from foreign-owned operations. To address the issue, the Irish government in 2000 created the Science Foundation Ireland and the Technology Foresight Fund to promote and support domestic R & D in areas such as DNA chips used to decipher complex genes, genetically modified agriculture, parallel processing systems, wearable sensors, and nanoscale assembly. Even so, Irish public spending on R & D, as a share of GDP, is still the lowest of any advanced EU country.

Over the next decade, Ireland's explosive growth will almost certainly settle down to near-U.S. levels. Of greater consequence is the evidence, recently compiled in a Microsoft-funded study conducted by one of the country's economic gurus, Dr. Paul Tansey, that the country's productivity progress "masks a dual economy" in which native companies still lag far behind.[38] Ireland's big challenge for the next decade will be to apply its technologies and operating methods to the domestic service companies where the majority of Irish workers still work. If it cannot manage that, the Celtic miracle will come down to earth for most Irish.

KOREA: MAKING IT ON ITS OWN

Ireland's record is impressive; but the Republic of Korea's development from one of the world's poorest agricultural societies to a modern industrial economy in less than half a century is one of the more extraordinary in economic history. It's also an important political story, since most of the progress was directed by a succession of brutal military dictators. As one native commentator has put it, Korea has been a "developmental dictatorship"[39] that created from virtually nothing world-class export industries and the best-educated workforce in the developing world. These national assets have carried the economy through a wrenching financial crisis and prepared it reasonably well for the demands of globalization. And unlike Ireland or China, Korea achieved this with almost no foreign direct investment from global corporations, although it did get substantial financial aid from the United States.

When the Korean War ended, the South was one of the poorest places on earth. Through the Japanese occupation from 1910 to 1945, Korea's railroads, ports, hydroelectric dams, factories, and mines were all developed in the north, around the Yalu River; and by 1945, the factories and plants of northern Korea accounted for one-quarter of Japan's industrial base.[40] The southern provinces that became the Republic of Korea had to settle for being Japan's breadbasket; and what little modern infrastructure and factories the south had were decimated when the civil war was fought mainly in the south.

In 1960, seven years after the armistice, South Korea's GDP of $2.3 billion and per capita income of $79 were among the world's lowest.[41] Measured in 2006 dollars, the 1960 income of an average Korean was $534, which today would rank 148th in the world behind Kenya. By 2005, the income of the average South Korean was $16,800, a thirtyfold real increase since 1960 and ranked thirty-sixth in the world; and the country's GDP reached nearly $800 billion, number ten in the world and a fiftyfold real increase since 1960.[42] As for the once-prosperous North, its GDP in 2005 was $30 billion, less than 4 percent of the South's, and its per capita income of $1,700 was barely 10 percent of its southern cousins.

Despite the presence of tens of thousands of Americans in Korea throughout this period, this extraordinary progress owed more to Chinese and Japanese approaches, along with fortuitous developments that the country's elites used to advantage. The kinship to China was mainly political. In the crucial years from the 1953 armistice to the late 1980s, Korea's leaders decreed all of the terms of economic development, like the orders of a military campaign. In fact, in the 1970s, President-General Park Chung Lee reissued Sun Tzu's "The Art of War" in a Korean version that cast economic competition as a war won by achieving dominant market share. The nation's sprawling chaebol cartel conglomerates that arose in those years were the artillery, created by government design not to maintain wealthy dynasties—most of those families were wiped out during the colonization and the civil war[43]—but because it was easier for the government to control a handful of large companies than thousands of smaller enterprises.

From 1953 to 1987, every Korean government began and ended in uprisings and military coups.[44] Yet a singular feature of modern Korea was its dictators' consistent commitment to economic changes that improved the lives of most Koreans. Syngman Rhee, who ruled from 1948 to 1960, closed the national assembly, purged or executed most of his political opponents and potential rivals, and banned independent political activity. He also instituted universal primary and secondary schooling and invested heavily in the country's basic infrastructure. When Park Chung-Lee seized control from Rhee, Park also purged his opponents, kept the assembly powerless, and created a new internal police force and personal political party that dominated the country's public life into the 1990s. But he also drove rapid industrialization, starting with a unique and ingenious plan: Park arrested hundreds of business leaders for corruption, imposed huge fines, and then told them they would pay their fines by investing their assets in a number of labor-intensive light industries and "donating" large blocs of shares to the government.[45] To protect his new portfolio, Park created a system of loans, tax breaks, and bureaucratic favors for the businesses, which became the chaebols that propelled the country's industrialization.

Park's plan won America's blessing when President Kennedy's national security advisor, Walter Rostow, persuaded him that Korea could become Asia's best showcase for western capitalism. More important, he convinced JFK to maintain U.S. aid for Korea's defense, infrastructure, and education, which from 1953 to 1965 totaled some $12 billion—nearly as much as the entire Marshall Plan.[46] And from 1972, when Park declared himself president for life, to his assassination in 1979 (on orders from his close political ally Kim Jong-Pil), Park ensured that the chaebols had the capital they needed to carry out his new plans for developing the heavy and high-tech industries that later would become world-competitive.

The 1980s followed the same patterns. Generals Chun Doo Hwan and Roh Tae Woo seized power after Park's death, promptly declared martial law, and directed the arrests and executions of thousands of political activists, banned political activity by everyone else, and closed the universities. Their brutality didn't end there: When the first serious recession in decades led to large-scale protests in early 1981 in the ancient city of Kwangju, Chun's soldiers killed thousands in what became known as the Kwangju massacre. The general then named himself president and purged eight thousand officials and business leaders—and like his predecessors, turned to economic reforms that strengthened the industrializing chaebols and expanded domestic businesses for Korea's emerging, new consumer class.

The shift to democracy began in June 1987, when Chun tried to pass on the presidency to General Roh and massive protests erupted again. This time, Chun decided to hold the country's first real elections and General Roh squeaked into office. New President Roh, acting like Boris Yeltsin to Chun's

Mikhail Gorbachev, persuaded his opponent and one of Chun's most prominent critics, Kim Young Sam, to join a coalition government; and then in
1992, Kim defeated fellow dissident Kim Dae-Jung to become his country's
first nonmilitary president.

The basic strategy of Korea's long line of dictatorial economic reformers—
jump-starting selected industries by giving favored companies such as Samsung and Hyundai huge loans from state-run banks, special tax breaks,
government contracts, and other bureaucratic favors—came from Japan. But
Korea's rulers rejected Japan's other key policy of protecting young industries
from foreign competition and using its own people as captive customers. This
approach, which economists call "import substitution," was followed not
only by Japan, but also by the United States in the second half of the nineteenth century and India in the second half of the twentieth century. But Korea's dictators shrewdly recognized that their country was too poor in the
1960s and 1970s to generate enough economic demand to keep their protected industries growing. So, from the 1960s—and virtually alone among
poor developing nations in the postwar era—Korea opened its economy to
imports of raw materials, parts and machinery, so its budding companies
could use them to produce exports.

There was nothing free market about this "Exports First" program. Park
and his successors set specific export targets for each company and directly
tied its loans, tax breaks, and other government preferences to meeting its
target. They also made it illegal for Koreans to purchase many of the products
intended for export, including phonographs, color televisions, portable tele-
phones, and mink coats. Only when the domestic market was finally growing
strongly in the mid-1970s did Park introduce a parallel track of import protections for the chaebols' new ventures in heavy manufacturing. The new
Korean producers of steel, heavy machinery, autos, industrial electronics,
shipbuilding, metals, and petrochemicals also soon received 60 percent of all
Korean bank loans and more than 75 percent of all manufacturing investment.

These policies worked, and the companies became the second generation of
Korean exporters. By the end of the 1970s, one-quarter of the country's fast-
growing GDP was being exported; and in the 1980s, the average jumped to
one-third. (Most of those exports went to the United States or to Japan, which
had normalized relations with Korea in 1965.[47]) This flood of exports slowed
a bit in the 1990s, when the government ended restrictions on Koreans buying their own country's products; but since 2000, they have jumped again to
nearly 40 percent of Korea's GDP. This latest flood of exports is driven not by
government loans and state favors, but by global demand for many Korean

products and by the expertise and relationships in the world's most demanding markets that companies such as LG Electronics, Samsung, and Hyundai built up over decades of meeting their state export targets.

So, unlike Ireland and China, Korea didn't import a modern industrial base by inviting the world's most advanced companies to set up business there. Quite the opposite—Rhee and Park barred foreign direct investment in most industries, and until the 1980s foreign businesses couldn't even own land or buildings in Korea. Instead, they channeled their subjects' high private savings into state banks that loaned it to the growing chaebols, and then directed the chaebols to purchase or license foreign technologies from abroad. Along with loans from western and Japanese banks attracted by the country's rapid growth, this strategy made Korea one of the world's highest-investing countries. Even in the 1960s, when the country was poor and backward, investments in fixed business assets claimed almost 18 percent of GDP— roughly America's gross investment rate over the last forty years. By the 1970s, Korean business investment jumped to 27 percent of GDP and kept on rising to nearly 30 percent of GDP in the 1980s and 36 percent in the 1990s.[48]

Add it up, and since 1970, Korea has devoted more of its GDP to business investments than any other significant economy—about ten percentage points more per year than Mexico or Brazil, eight percentage points more than the countries of the European Union, and two percentage points more per year than China or Thailand.[49] Even in the decade from 1995 to 2004, which includes the financial crisis, Korea's gross investment rate nearly matched China's.

So, when events distant or nearby created new opportunities—to supply U.S. forces in Vietnam, for example, or participate in the Middle East building boom of the 1970s and 1980s—Korean companies, virtually alone in the developing world, had the investments required to take advantage of them. And the decades of high investment, mostly channeled to the chaebols, produced an unusually large number of successful global companies: Fourteen of the world's five hundred largest corporations today are Korean, which is more than Spain, Sweden, Italy, or India, and twice as many as Singapore, Thailand, and Taiwan combined.[50]

The other critical element in Korea's singular progress is summed up in one startling fact: Since the 1960s, the Korean government and its people have devoted more of their GDP to education—an average of 10 percent a year—than any other developing country and almost all advanced nations. The farsighted vision of a succession of dictators played the largest role; but much of the credit also goes to Korean parents who've paid for half of all that spending. Today, about 40 percent of the country's high school students and 80 percent of its college students attend private schools, the highest percentages in the world.[51]

The end result is one of the world's most highly educated workforces. Some 28 percent of Korea's male workers have college training, a larger share than France, Germany, or the United Kingdom (Korea also leads Germany in the share of female workers who attended college, but lags behind France and the U.K.).[52] The numbers are still rising. In 2005, more than four out of five Korean high school graduates went on to college, the highest share in the world.[53]

Globalization has made these investments pay off big, because as it forced Korea out of the lower-value industries now dominated by China and other low-wage nations, Korea's leading companies could draw on an extraordinarily well-educated workforce to move into markets dominated by the world's most advanced economies. The extraordinary education of the Korean people is one of the critical reasons why today Samsung outcompetes Sony, Toshiba, and Philips; and why Hyundai Motors, even after its setbacks in the late 1990s, is larger than Volvo, Renault, or Mazda and more profitable than Ford, Volkswagen, or Fiat.

Higher education on such a democratic scale also makes Korea the world's most equal developing society. By an economist's standard measure, Korea today is the eleventh-most-equal society on earth—well ahead of France, Italy, Great Britain, and the United States, and behind only Scandinavia, Japan, Germany, Austria, Luxembourg, the Netherlands, and Bosnia-Herzegovina.[54] And the steady economic progress of average Koreans almost certainly helped keep the country stable through its long years of dictatorships, military coups, and financial setbacks. Perhaps this will be a lesson not lost on Beijing's leadership.

Korea also offers an important example for France, Germany, and Japan about how to adapt when economic conditions change. Korea faced serious economic problems twice in the last generation, and both times government and business elites swallowed hard and chose painful reforms. In 1980 the worldwide oil shock, the worst harvest in a generation, mass unrest following Park's assassination, and reestablishment of military rule all brought the Korean economy to a full stop. The government pinned the blame on the huge and powerful chaebols, which in truth had grown so big, inefficient, and debt-heavy from their constant diet of state loans, subsidies, tax breaks, and export directives, that the top ten accounted for nearly half of the country's GDP.[55] More unusual, General Chun, fresh from the Kwangju massacre, carried out a series of sweeping and economically sensible reforms. He not only broke up some of the chaebols into smaller business units and began phasing out many of their direct subsidies, he forced them to focus more on the domestic market by cutting their export subsidies and repealing the laws restricting what Korean consumers could purchase. Even more far-reaching, he decreed that a majority of local bank loans and 35 percent of loans by the large national banks would go to small and medium-size companies. Within a decade, smaller enterprises accounted for nearly half of all Korean manufacturing, value-added, and two-thirds of manufacturing jobs.

Then in the late 1980s, as IT advances were remaking the American economy, Chun's successor, General Roh, made another fateful policy shift by rolling back restrictions on foreign direct investment—well before India or China—while encouraging Korean companies to develop their own advanced technologies. A few chaebols were already successful innovators—notably Samsung, which developed the first 64-bit DRAM chip in 1983 and a few years later pioneered a strategy IBM would follow later of offering extensive IT services to the customers for its chips, electronics, and other advanced products. Now, with new government support, Hyundai began to develop prototypes of cars fueled by hydrogen, electricity, and alcohol, all well before its western rivals. By 2000, Korea was investing more of its GDP in R & D—2.4 percent—than all but eight other countries.[56] Today, geeks and consumers around the world use such recent Korean advances as precision laser welding equipment, technologies that let cell phone users receive broadcasts traveling at high speeds, alternative fuels for automobiles, fuel cells that can run laptops for a month, and new techniques to culture and transplant skin for burn victims.

The capacity to adapt to new conditions also is clear in Korea's response to the financial crisis of 1997–1998, which upended the livelihoods or economic security of millions of Koreans. In one year, the country's GDP contracted by nearly 7 percent, the currency lost half of its international value, fifteen of the thirty largest chaebols and ten national banks went under, and unemployment tripled from less than 3 percent to nearly 9 percent.[57] A $57 billion bailout from the IMF stabilized the currency and the financial system, so foreign investors would return. But America's Alan Greenspan and Robert Rubin set strict conditions for the bailout: Jettison industrial planning and the mutual back-scratching of large companies, big banks, and government; and adopt the U.S. model of open trade, deregulated markets, and financial transparency. The American approach to globalization and the criticism of the chaebols happened to resonate with longtime dissident and newly elected President Kim Dae-Jung, who seized on the IMF conditions to announce his "Big Deal Industrial Program" for breaking up the largest conglomerates and forcing them to compete at home with western companies.[58]

Half of the major chaebols disappeared in two years—a record China's leaders must envy as they work to unwind their state-owned enterprises—and the Korean economy rebounded with 9.5 percent growth in 1999 and 8.5 percent gains in 2000. Foreign direct investors flooded in to buy hundreds of Korean businesses at fire-sale prices depressed by the depreciated currency; and by 2005, foreign companies owned or controlled more than 40 percent of the companies listed on the Seoul Stock Exchange and eight of the fourteen national banks.[59]

Under the Kim-Greenspan/Rubin program, the chaebols that survived had to remake themselves. Samsung sold off ten large business lines to foreign companies, including construction equipment (sold to Volvo AB) and the auto division (sold to Renault), and then spun off Samsung Electronics and invested even more in product development. It worked well: Today Samsung is the world's largest producer of DRAM chips, flash memory, optical storage devices, and LCD monitors; its aerospace division is developing space stations, its construction affiliate is building the world's tallest building in Dubai, and the company's thirteen thousand researchers are among the top-ten recipients of U.S. patents. Hyundai, the largest chaebol before the financial crisis, followed a similar strategy, selling off peripheral businesses and reorganizing everything else into five business groups with global markets. But Daewoo used the business panic to take over fourteen faltering firms as additions to the company's 275 subsidiaries, and vastly increased its own debt. The buying spree failed: By late 1999, Daewoo was bankrupt and its founder and chairman, Kim Woo-Jung, fled the country under indictment. Almost all that remains today of Daewoo is its consumer electronics business, whose products now sell at Target stores under the Trutech brand.

Judging by broad measures, Korea's economy should be in sound shape for the near future, but it's not a sure thing. Growth since 2000 has averaged more than 4.5 percent a year; and productivity has risen 4.3 percent a year, well ahead of Ireland and the United States, and two or three times the rate of Germany, France, Japan, and the United Kingdom.[60] The chaebols' eclipse and Kim's reforms also remade the country's labor markets along American lines—more flexible for businesses, with much less job security for millions of Koreans. Unemployment has remained less than 4 percent, although some 45 percent of working Koreans are now "irregular" employees who work without long-term contracts and often only part-time.[61] But over the next ten to fifteen years, Korean workers will also have to compete for the best jobs with a new influx of foreign workers from the United States, Japan, Europe, and other Asian developing countries. Young people are being hit particularly hard by these changes, with their jobless rate running four times the national average. Unsurprisingly, thousands now emigrate to America and Europe, including a new brain drain by many trained in computer science, electronics, and research and development.

Korea will need to keep its most educated workers at home, along with all the foreign workers it can attract, because it also faces demographic stresses that could slow its progress. Korea followed the baby-boom and baby-bust pattern of other developing societies, but with special features that intensified its baby bust. To start, its enormous investments in higher education and its prodigious growth created tens of thousands of attractive job opportunities for women, putting stronger downward pressure on birthrates than in many other places. On top of that, the Park and Chun governments introduced free contraception in 1962 and, from 1983 onward, cancelled insurance coverage for maternal care for pregnant women with three or more children. The upshot was a virtual collapse in Korean fertility rates, which fell from more than 4.5 children per woman in 1970 to 2.8 in 1980, 1.6 in 1990, and less than 1.2 today.[62]

The consequences could cramp Korea's growth down the road. UN projections suggest that working-age Koreans will top out at around 33 million from 2010 to 2015 and thereafter contract by about 1 percent a year. On this path, Korea's potential workforce will fall to 31.5 million by 2020 and less than 28 million in 2030.[63] At the same time, the number of elderly Koreans will soar from about 6.6 million in 2005 to 11.4 million in 2020, a jump of more than 70 percent—and keep on increasing to more than 15 million by 2030. These shifts will also produce serious problems for Korea's older population. The country only introduced state pensions in 1988, usually linked to jobs at a chaebol. So much like China, where the dismantling of the agricultural communes and state-owned enterprises unraveled its rudimentary safety net, Korean pensions (and health-care coverage) have eroded badly as first the government and the financial crisis drove most of the chaebols out of business. Even so, the OECD forecasts that by 2020 the contracting labor force and fast-growing numbers of elderly will make the country's public pension system "unsustainable."[64] As the numbers of elderly and near-elderly rise sharply, Korea almost certainly will face demands to expand pension coverage—in a society where even under military rule, tens of thousands of people regularly took to the streets to protest.

As the labor pool contracts and the elderly population expands, the only way to keeping growing—and generate the revenues for broader pension coverage—will be to increase the share of Koreans who work or enable those who do work to be more productive. The first approach could be more promising in Korea than in much of Europe. For example, there have been moves to end the practice by many large Korean companies of forcing employees to retire at age fifty. The largest pool of potential workers, however, are women: Today, only 56 percent of working-age Korean women hold jobs, much less than the 69 percent average for all industrializing nations.[65] As Korea's labor markets grow tighter in coming years, businesses that have adapted to other pressures may be expected to revise their attitudes towards women.

It's harder to see how Korea could improve its current rates of higher education in order to raise productivity. However, its businesses could expand the scope of their R & D beyond the IT and auto industries that currently claim nearly 80 percent of it. The biggest opportunities for higher productivity, however, lie in service companies—banking, retail, business services, health care, and so on—which are barely half as productive as Korean manufacturing.[66] The recent records of Ireland and China show that one direct way to lift the productivity of lagging sectors is to increase direct investments by much more productive foreign companies. Since the financial crisis, the Korean government has tried to do that by creating special "zones" where foreign firms receive the kinds of subsidies that used to go to the chaebols—exemptions from income and corporate taxes for seven years, permanent exemptions from customs duties and excise taxes and, in high-tech manufacturing or services, free government leases and medical care and education for employees. So far it hasn't worked. From 2000 to 2005, FDI to Korea averaged less than 1 percent of GDP, one of the lowest rates in the world.

There's good reason to expect that FDI to Korea will increase substantially over the next decade. Many foreign companies have stayed away because Korea's labor costs are now double, or nearly so, those in Taiwan, Singapore, and Hong Kong,[67] and the country is well known for its militant labor unions. But with most of the chaebols gone or well pared down, the large numbers of irregular workers have created a new nonunion labor pool. In time, perceptions should catch up with reality, and foreign direct investment should increase. More FDI, especially in services, would not only create jobs, perhaps for women and older people not working now, but also would expose domestic service companies to the best practices of the world's most productive service businesses and the competition to force some of them to adopt those practices.

Korea also can offer foreign companies the world's most intensely wired developing economy. In 2005, 80 percent of households had access to home computers, substantially more than the United States, Britain, France, or Germany, and an order of magnitude ahead of other developing nations.[68] In the same year, two-thirds of Koreans used the Internet and more than 26 percent of its households subscribed to broadband, again more than the United States and the large European countries, and leagues ahead of other developing countries.[69] For western companies, Korea provides a workforce and customer base that's already computer literate and Internet savvy, and the infrastructure to plug into their global networks.

Finally, while Korea's heavy dependence on exports—nearly 40 percent of GDP is exported each year—aligns its economy with strong forces in globalization, it also makes its national growth hostage to its major-country customers, especially the United States. A recession in the United States, especially one that spreads to the rest of the world, would quickly stall out the Korean economy. But downturns come and go, and the most promising sign for Korea's future is its consistent capacity to take what the global economy dishes out and make progress anyway.

When individuals, companies, or countries experience unusual success, they become role models to those aspiring to the same result. In America, China, Ireland, and Korea, the world has four economic role models today. The number-one economy, the United States, offers a big-country model of the benefits of relatively unregulated markets, strong public investment in education and infrastructure, entrepreneurism, domestically driven development, and aggressive globalization. The American model also rests on basic democratic politics, limited social benefits and protections, and a tolerance for relatively high inequality. It's likely to be the model for India and, perhaps in time, Germany, France, Brazil, and Mexico. China offers different guidelines for economic success—wages and other resources priced to attract massive foreign direct investment, a mixture of highly regulated and lightly regulated markets, low consumption and very high personal savings, strong public investment in infrastructure, educational investments targeted to only part of the population, and severely limited social benefits and protections. Most important for other countries, China's approach depends on authoritarian politics, as well as, again, a tolerance for high inequality. It's a model that could well appeal to African countries, the more authoritarian Asian countries, and perhaps, in time, Russia.

The world now has two other models. Ireland is a small country model for the benefits of tax and subsidy policies designed to attract large-scale foreign direct investment, substantial investments in education and infrastructure, and relatively limited social benefits and protections, all built on middle-class democracy and producing relative equality. Because Ireland's success is tied to its economic access to the European Union, it is becoming a model for transitional economies in Central and Eastern Europe. Finally, Korea offers smaller economies a different approach, based on massive public and private investments in education and health care, large initial subsidies for core industries accompanied by low barriers to imports and foreign competition, substantial social protections which, like the industrial subsidies, phase down, and support for entrepreneurism. Unlike Ireland, it relied initially on authoritarian politics, with a prospect of a transition to democracy. Korea could well become a model for some African and Asian countries with farseeing authoritarian leaders.

In the end, many countries won't follow any of these approaches; and they are not likely to develop quickly or prosper greatly in this period. That is the likely course of much of the Middle East and parts of Latin America. Yet, what is perhaps most striking in all of this is the historic development that the major economies of Europe and Japan no longer provide a viable economic model for other countries.

CHAPTER SIX

THE NEW GEOPOLITICS OF THE SOLE SUPERPOWER: THE PLAYERS

AMERICA'S GLOBAL MILITARY POWER is so commonplace that it's easy to overlook how historically unique it is. What's so unusual and world changing is not the extent of America's military, political, and economic capacities, but the absence of countries that come anywhere close. This creates what one prominent theorist on these matters, Stephen Walt, calls an "imbalance of power in America's favor (that) is historically unprecedented . . . in virtually every important dimension of power."[1] In the seventeenth and eighteenth centuries, France was a formidable military and cultural power, and so were Britain and Spain—and Britain was richer. In the second half of the nineteenth century, when Britain had the world's largest and most important economy, it wasn't so large that the United States couldn't overtake it by the century's end. And in this respect at least, the twentieth century was entirely typical, until its final decade. Britain, Germany, the United States, and Japan had competitive economic, military, political, and cultural clout for most of the first half of the twentieth century, and for forty-five years after World War II, America's reach was not greatly more daunting than the Soviet Union's.

All that has changed. America's historically anomalous position as a sole superpower with no near peer ended the balance-of-power geopolitics that organized much of world affairs for more than a thousand years, and will fundamentally shape a new geopolitics for at least the next generation. With globalization increasingly affecting how nations interact, America's singular world position depends substantially on its economy. It is clear that America's economic power goes well beyond being the largest economy and, among the major ones, the most productive and fastest growing. The geopolitical power that the United States derives from its economy is based, first, on how much and profoundly it affects other economies. As we've noted before, America's businesses and consumers buy unprecedented amounts of goods and services from the rest of the world—$2.2 trillion worth in 2006, or more than the entire economies of all but five other countries, and more than 6 percent of everything produced in the world outside the United States. This matters geopolitically because the jobs and incomes of tens of millions of people in other countries now depend directly on the United States growing enough to keep this going. The result is that powerful groups in other countries now identify their interests with America's, which in turn can limit the ability of their governments to take steps adverse to the United States. For example, Brazil in 2005 and Thailand two years later both announced plans to nullify the patents of American-made HIV drugs—an illegal but popular move at home that would have cut the costs of their public health programs and, in the bargain, enabled them to produce generic versions to export to other countries. They both backed down when their own business elites opposed them (and they both happily accepted sweet new discounts from the rattled drugmakers).

The United States also derives geopolitical power from its singular capacity to develop new technologies and other valuable intellectual property in large volumes, especially in the software and Internet areas that drive so much economic change and the processes of globalization itself. Other countries now lead in producing and improving the basic manufactures that American companies dominated a few generations ago—steel, consumer electronics, automobiles, and much more. And Northern Europe is home to a good number of the world's most innovative IT firms, especially in wireless technologies. But so far, there are no Japanese, German, British, French, or Chinese peers to Microsoft, Google, Amazon, Oracle, Dell, and Amgen—or in business methods, Wal-Mart and Goldman Sachs.

This capacity enhances America's global position not because it increases the profits that U.S. companies earn on their foreign sales; much more far-reaching, it subtly aligns the economic paths of other countries with the United States and, whether or not they like it, makes them a little more like America. The Internet's software infrastructure developed in a typically American way—by entrepreneurs working in areas largely untouched by government regulation—into a radically open and decentralized system. As a result, the Internet has become essentially "American" wherever it is, and not just because U.S. companies dominate its development and content. Much more than that, it creates American-style opportunities wherever it reaches, disseminating information without restriction and spurring the development of new services and products by newly formed companies operating in new ways. And that can produce geopolitical benefits for America, as tens or hundreds of thousands of newly successful people around the world associate their success with its continuing technological achievements.

Imagine if China were the world's leading innovative economy, its language was the international standard, and the Internet had come out of its government labs. As it is, China makes substantial efforts to restrict its people's Internet access to content it deems politically unacceptable, perhaps recalling how East Germans led the unraveling of the Soviet empire after years of direct access to uncensored news from West German television. A Chinese-designed World Wide Web almost certainly would be a closed system in which information could be centrally managed. And if it had happened that way, it could have stayed that way, at least for a time. Global markets are filled with instances of what economists call "path-dependent" outcomes—think of the VCR or, some would say, Microsoft Windows—in which a product's initial critical advantage created a path that became hard to alter, even when alternatives emerged that were more efficient or effective.

The heart of the geopolitical clout that America derives from its economic preeminence, however, is that so much of the world now embraces its basic approach to organizing their economies and doing business. In less than a generation, the alternative models that much of the world had followed for decades have been discredited and largely discarded. This reaches past the epochal collapse of Soviet collectivism and China's startling conversion to capitalism. The appeal of the more mixed models of a private economy with heavy government direction also has waned, after Asia's bumper economies melted down or stagnated and Europe's entered a decade of disappointing growth. Some leaders in Europe and Japan may deny it, but for the first time there are no credible, grand alternatives anywhere in the world to America's basic take on the limits of the government's role in the economy and how businesses should run.

This doesn't make the American economic approach "right" in an objective sense, like the composition of the atom. Rather, it's very broadly preferred right now, because it's in sync with the current demands of globalization. For a long time, the European and Japanese approaches—not to mention what passed for an economic model in the Soviet Union, Eastern Europe, and China—produced more equality and economic security for individuals than American-style capitalism. But for nearly a generation, globalization has crippled the capacity of those approaches to generate strong, sustained growth—and greater equality and security are less appealing when people also face a prospect of growing poorer. This simply is a time when growth in both advanced and developing economies depends on governments not only stepping up to invest in education, health, and modern infrastructure, but also stepping back from protections and regulations that slow or muck up the era's massive transfers of technologies, capital, and expertise. And almost every country also now supports the international institutions that enforce American-style rules of globalization, especially the World Trade Organization and World Intellectual Property Organization, but also the older International Monetary Fund, World Bank, and Paris Club (which deals with sovereign debt default issues among countries).

The core business of geopolitics is national security; and the other critical geopolitical fact about U.S. economic dominance is that it will indefinitely finance America's position as the world's sole military superpower. In 2006, the United States spent about $570 billion on defense, or roughly as much as the rest of the world combined. This asymmetry in military spending is also historically unprecedented. In fact, the U.S. military spends more on research and development into new defense systems—some $73 billion in 2007—than the entire defense budgets of every other country except China.

All this spending has bought the United States its remarkable military dominance as the first military superpower in more than a millennium with no near peer in sight. Other nations have armies, air forces, and navies capable of protecting their borders from just about anyone else; and with some considerable lead time, a few of them could send their soldiers, sailors, and pilots to fight in other countries in their own regions. But the United States alone has a blue-water navy capable of operating in and across the world's vast oceans, and a blue-sky air force that with little notice can project forces anywhere from U.S. bases and those around the world. When they arrive, they wield technologies at least two generations more advanced than anyone else's, including China, Britain, and Russia. These forces can prevent others from using their own militaries beyond their own borders and, as Saddam Hussein learned, no government can survive for long against their serious assault.

Geography reinforces America's awesome military advantages. The United States is the only country with thousands of miles of ocean separating it from anyone else with an army, navy, or air force to speak of. And the oceans and the air space above them and above every other country are part of America's military territory, since only its navy and air force can roam them freely. Most wars begin in conflicts that in some way arise out of geographical proximity; and even today, the proximity of Russia, China, and India, for example, will make it harder for them to work together to balance America's military advantages.

America's disproportionate military power is a direct result of generations of U.S. policy, but with the tacit agreement of much of the rest of the world. The United States created a peerless military by devoting the last fifty years to developing fearsome technologies and systems, and spending whatever it took to build and maintain them. Those commitments likely would have made the United States the mightiest military power regardless of what anyone else did. But the lack of a near peer capable of providing a measure of geopolitical balance is not simply a result of the Soviet Union's fatal defects. It also took the extraordinary decision by countries that had militarily dominated much of the world for centuries—notably, Germany, France, Britain, and Japan—to stop competing militarily for fifty years. Instead, Europe's major countries and Japan outsourced their security to the United States, leaving it to America to deter or fight wherever the Soviets or Chinese seemed too aggressive. These once-great powers became providers of largely token forces, serving for mainly symbolic purposes, in what historian Paul Kennedy has called a hollow, or "Potemkin," alliance: "The U.S. does 98 percent of the fighting, the British 2 percent, and the Japanese steam around Mauritius." It was an arrangement that suited everyone, since it also fit nicely with the shared American and Soviet conviction that their time to direct the world's affairs had arrived.

When the Soviet empire imploded, America made no effort to preserve some form of bipolar system—and why should it have?—and moved quickly to absorb the Warsaw Pact into the NATO alliance. Moreover, without a Soviet near peer to balance the United States, Europe and Japan continued to downgrade their own military capacities—as, without a Soviet threat, America continued to upgrade its defenses and fight new wars. Today, the United States has several hundred thousand battle-ready troops in more than one hundred countries—that's not counting Iraq—and provides military training for more than one hundred and thirty nations. Experts say that the entire European Union couldn't put sixty thousand well-equipped troops in the field in sixty days and keep them there for a year.

Europe's foreign- and defense-policy elites have raised alarms, so far with no results. The Venusberg group of high-level European experts, brought together nearly a decade ago by the Bertelsmann Foundation to examine EU security policy, issued its most recent and most sobering report in July 2007. They acknowledged that "no European state can be described as a truly Great Power in today's world" and concluded, "it is now or never for an effective EU foreign and security policy . . . (and) time that Europe ended its obsession with internal structure." Their conclusion was stark and accurate: "Fail and Europeans will lose any ability they now have to shape developments."[2]

It could have been different, and it would have been, if Germany, France, Britain, and Japan had approached the cold war as partner powers of the United States, with America as the managing partner. Even though they didn't do so, Europe's big three and Japan could have devoted more of their resources and attention to the West's competition with the Soviets. Taking that path, though, would have cramped the development of the modern European welfare states, since the costs of major defense buildups would have come from domestic programs. It can be fairly said that the choice by Europe's major countries to let the United States assume responsibility for how the West confronted communism allowed them to build and preserve the very networks of benefits, protections, and regulations that now are so at odds with the demands of globalization and the aging of their populations.

America's military power is not unlimited. China and Russia—and perhaps North Korea soon—can get a nuclear missile to Washington, New York, or Los Angeles, which limits America's military options, at least at the extreme. (Britain could, too, but that doesn't count.) More important, the Iraq War shows the world both how easily its sole superpower can defeat another government militarily, and how hard it is for that superpower to occupy a large foreign society. And as Al Qaeda demonstrated on 9/11, America's overwhelming military superiority inspires new forms of attack that leverage an aspect of its strength against it.

THE NEW GEOPOLITICS MEETS GLOBALIZATION

This new geopolitics intersects with globalization in three portentous ways. First, but for globalization, China would be a modest geopolitical force for several more generations. Instead, China's economic ascendancy through globalization will create an expansive international role for the country over the next ten to fifteen years. Second, globalization has helped make energy and terrorism central issues for geopolitics for at least the next decade. We will look at these matters later. First, however, we will examine a third geopolitical issue that is even more fundamental than China, energy, or terrorism: Does globalization preclude direct military conflicts between major countries and recast geopolitics into a series of small clashes between the countries engaged in globalization and the groups and states outside it?

This is not the first time that people have thought or hoped that countries extensively connected through trade and commerce will never go to war with each other. It's at least as old as the eighteenth-century Enlightenment, when most educated Europeans believed that their era's progress in science and reason would inexorably produce peace through the instruments of trade and commerce. They were obviously wrong. Science and reason only made the Napoleonic wars and the U.S.-British War of 1812 more lethal. The great advances in globalization in the second half of the nineteenth century also didn't usher in an era of peace. Rather, the major powers' intensive competitions in international markets combined with their nationalism and led eventually to the horrors of World War I. In the end, nothing in history is more common than major countries striving to make themselves richer; and in most instances those strivings are closely associated with a drive for military power. And when two or more countries strive for more national military power in the same region, they almost always have come to blows.

The idea that, this time, globalization will produce permanent peace between the major countries is pressed most avidly and cleverly by a Harvard-trained Pentagon analyst named Thomas Barnett. More than anyone else, Barnett has developed the view that the geopolitics of our time will be defined by a great divide between those engaged in economic globalization and those outside the club—and everyone in the club will figure out how to get along.[3] This isn't merely an update on the Enlightenment. Barnett figures that the spread of nuclear weapons will preclude even conventional wars between the major powers, while leaders in places cut off from the global economy will find they can only achieve their goals by violent means.

At a minimum, the optimistic parts of this view require a lot of faith. War between nuclear powers is clearly not unthinkable. The United States and the Soviets came perilously close to nuclear exchanges on more than one occasion, as have Pakistan and India. Moreover, the most senior military and foreign policy advisors to the heads of governments in every nuclear power have been thinking hard about it for two generations, and they're doing so right now in Washington, Moscow, London, Paris, Beijing, Delhi, Karachi, Tel Aviv, Pyongyang, and soon Tehran. And nuclear-armed countries that trade extensively with each other can still have important conflicts of interest over both economic and noneconomic matters—think of energy and then of territorial ambitions. So military planners in Washington, Beijing, Moscow, and other capitals have developed conflict scenarios and plans which, if possible, keep a confrontation with another nuclear power nonnuclear—and if they cannot, what they'll do. And while some will say that India is part of globalization but Pakistan is not, a conflict between these two powers would have little to do with that distinction.

The extensive interactions of this era's globalization also have not tempered the nationalistic tendencies closely associated with confrontations and wars. One of the reasons that China and the United States embrace globalization today is that it offers a way to become more consequential—more powerful—in the world. Modesty is not a trait of countries that consider themselves great and exceptional. China and America today—and, at other times, Europe's major countries and Japan—see themselves not as the pieces of a large global system, but as nations in history claiming larger places for themselves. And that's classic nationalism. Even as Chinese corporations become part of the supply and distribution networks that American businesses created, their societies show few signs of converging social values. Stated plainly, China is not enduring the wrenching social and economic disruptions that go along with rapid modernization because it wants to be America or Sweden; it wants to be what it's always been, China, only more so.

Even so, Barnett and his admirers are right that the wars of the next ten to fifteen years are most likely to involve internal conflicts in poor countries (outside the club) or retaliatory or preemptive attacks on renegade groups and states by the sole superpower, along with whatever allies it can muster. But the salient features of the geopolitical landscape will not be the great divides of globalization, critical as they are, but the traditional phenomenon of poor countries coming apart and the entirely new one of one superpower assuming responsibility for global security.

There's another serious and troubling reason why globalization cannot preclude a major war, even in the next ten to fifteen years. Gradually but now finally, the world has entered a new era of nuclear proliferation, in which weak countries well "outside the orbit" have gained access to nuclear weapons. For decades, a country had to be large and prosperous to build nuclear weapons. That era ended decisively with the nuclear arming of Pakistan and North Korea.

The most important—and threatening—development in proliferation has been the largely uncontrolled spread of nuclear power technology, and the ability of even weak states to eventually leverage that technology to develop nuclear weapons. Pakistan, North Korea, and in all likelihood Iran, soon enough, provide sobering cases in point. Moreover, across the Middle East, Saudi Arabia has expressed interest in building nuclear-power capabilities, Egypt and Algeria already have nuclear research reactors, and other Arab nations (including Jordan) hope to follow them within a decade.

The risk that these nations will use their reactors to produce plutonium can be contained only so long as their generally autocratic governments let the International Association of Atomic Energy maintain strict controls and oversight of their nuclear-power activities. North Korea and Pakistan demonstrate with sobering finality that those controls are not reliable barriers to proliferation once states decide they want to acquire nuclear weapons. The spread of nuclear power is not new; Eisenhower's Atoms for Peace initiative helped launch that spread decades ago.[4] What is new is the accelerating pace with which unstable countries both poor and rich are acquiring nuclear skills and technology, and the technological progress that has transformed the once-complex and customized components of nuclear arms into standard industrial commodities.

The next ten years will see more states go nuclear. Most likely, the trigger will be what political scientists call a "cascade scenario," where, for example, Iran's testing of a nuclear weapon triggers the rapid development and testing of weapons by Egypt and Saudi Arabia, both of which have been on the nuclear arms fence for some time. This scenario has happened once already, when China's bomb moved India to develop its own, and India's testing led Pakistan to produce one for itself. Taiwan's capacity to do the same probably helps check some of China's aggressiveness towards the island—and vice versa—but it takes only one to start a cascade.

To avoid regional doomsdays over the next ten to fifteen years, the new nuclear-armed states will have to replicate the deterrence relationship that the United States and the Soviets constructed over decades, and with a lot of good luck. Clearly, the most dangerous area for nuclear proliferation over the next decade is the Middle East, where everybody has major interests. It's not likely that the United States, Russia, or China can stop Egypt or Saudi Arabia from going nuclear if Iran does and doesn't behave—or stop Israel from doing whatever it believes it has to do. Weak states also may be more likely than stronger nations to use nukes against each other, because an adversary with only a few nuclear weapons is a more tempting target for a first strike.[5] Avoiding the catastrophic global consequences of a nuclear blowup in the Middle East will almost certainly be a major issue for America and China, as well as Russia and even Europe, in the next decade.

THE GEOPOLITICS OF THE U.S. SUPERPOWER

For the last fifteen years, one country, the United States, has exercised power and responsibility across virtually every part of the world. The most important single factor in the geopolitics of the next fifteen years is that regardless of the ultimate outcome in Iraq, this will not change. Just as very few people with great wealth live as if they have to worry about their next paychecks, superpowers almost always act based on what they are, which is very powerful.

And without a near peer to balance, block, or seriously challenge America's ability to take whatever steps it chooses in every region, U.S. interests and conduct will continue to be global in scope from now to 2020 and beyond.

In this respect, the presidencies of Bill Clinton and George W. Bush have much in common. While Clinton used U.S. military forces much more briefly and much less intensively than Bush (although also on more separate occasions), his geopolitical approach was as global and activist as Bush's.[6] Under Clinton, the United States sent aircraft carriers into the Taiwan Strait and later convinced Beijing to end nuclear assistance to Iran and Pakistan, while using access to its market to get China to accept the WTO's American-made rules on domestic competition and foreign investment. America in the 1990s also got North Korea to freeze its plutonium production, used NATO and the European Union to end the fighting and change regimes in the Balkans, brought Russia into the G-8 and the Warsaw Pact into NATO, denuclearized the Ukraine, Belarus, and Kazakhstan, contained Saddam Hussein's ambitions, used new trade arrangements to help end one-party rule in Mexico, restored a democratic process in Haiti, protected western capital markets from financial meltdowns in Asia and Mexico, and played the broker's role successfully in Ireland, India-Pakistan, and Turkey-Greece, and less successfully in the Mideast. The Clinton administration also targeted and tried to kill Osama bin Laden.

The traditional foreign policy realism of the administrations of Clinton and his predecessor, George H.W. Bush, led to different geopolitical strategies than the doctrines of George W. Bush and his close advisors. But those differences in worldviews did not affect the fact or inescapable logic of America exercising its power across the world.

While no other country can stop the United States from pursuing the policies it chooses, those policies can fail or be so wrongheaded that even in success they damage its own interests (and those of others). America's Iraq policy almost certainly will be judged a failure, and it also may well have been fatally wrongheaded. At a minimum, it has established that the superpower can overturn the leadership of any other country, but not necessarily occupy it successfully. Those results will affect the strategic judgments of not only future U.S. presidents and their advisors, but also other countries and groups. It also may well lead to a proliferation of the technologies and unconventional strategies used by insurgents in Iraq to counter the superpower's traditional military superiority. Since America's vaunted "network-centric warfare," it turns out, is useless against IEDs, its forces will confront them again wherever they fight in the future. The same fate probably awaits Russian forces in Chechnya. Yet America's unhappy venture in Iraq will not reduce U.S. activism over the next decade, even in the Middle East. The fact is, U.S. activism in that region—as in Latin America, Asia, and Europe—is built into the global interests and geopolitics of a sole superpower.

An open-ended, activist global role is just what top U.S. military and intelligence strategists see in America's future, based on the most recent *Quadrennial Defense Review Report* from the Pentagon and national intelligence estimates. This role doesn't come from an obvious, existential threat to the United States. In fact, in vivid contrast to the cold war years, only Russia on occasion and a handful of smaller countries seem willing to challenge the United States even rhetorically—North Korea, Iran, Syria, Cuba, and Venezuela—and all of them are also ready to do business with the United States in most areas. Nevertheless, the experts who explain the world to U.S. presidents all see a compelling case for U.S. global activism, based on the "capabilities" of those hostile to American interests and the capabilities of the United States in thwarting them. Its global scope starts with the worldwide reach of terrorist groups—said today to operate in eighty countries—capable of inflicting harm on U.S. interests and assets around the world. The Pentagon states this clearly: "The United States is . . . engaged in what will be a long war . . . global in scope . . . against violent extremists who use terrorism as their weapon of choice, and who seek to destroy our free way of life."[7]

The U.S. interests that concern these strategists, and the consequent goals of U.S. defense policy for the next decade, go far beyond "defeating terrorist networks" and "defending the U.S. homeland." Saudi Arabia has a capability to interrupt U.S. oil supplies, for example, and China could use its military buildup and economic clout to intimidate its neighbors into closing down their U.S. bases. Moving along, North Korea has the capability to sell advanced weapon and missile technologies to other rogue states, and Russia could withhold natural gas and electricity from much of Europe. Approach the world the way that America's senior defense and intelligence strategists do, and the United States has almost limitless options for exercising its superpowers to "shape the strategic choices of other countries," including military force if it comes to that. To prepare for the use of force, the Pentagon also is reorganizing its military power, "moving away from a static defense in obsolete cold war garrisons, and placing emphasis on the ability to surge quickly to trouble spots across the globe . . . and conduct multiple, overlapping wars."[8]

The most recent *Quadrennial Defense Review* organizes the universe of threats using a topology that also requires an expansive exercise of American power for the foreseeable future. First, there are traditional threats from states using "recognized military capabilities and forces in well-understood forms of military competition and conflict." While no other country today can take on America in this way, the Review looks ahead to a generation from now, when China might be a real military competitor, and calls for a new generation of strategic missiles and defense systems. The prospect of future traditional conflicts is also behind the Pentagon's current plans for new, advanced technologies for conventional warfare—should the superpower decide to take on another hostile state (the Iraq model) or if U.S. forces are nearby when a traditional war breaks out between two other countries (and they're stationed near almost everybody). The planners may be thinking about African oil states for the first possibility, and India-Pakistan, or some of the independent states from the old Soviet Union.

A second class of threats is what U.S. military planners call "irregular challenges from those employing unconventional methods to counter the traditional advantages of stronger opponents." That covers terrorism outside the United States, as well as an Iraqi-style insurgency, and its status as the number-two threat signals the U.S. military's determination to build the "capability" to win next time. For these strategists, there's little question that there will be a next time—if only to address their third class of threats, "catastrophic challenges (that) involve the acquisition, possession, and use of WMDs." The U.S. failure in Iraq, for all its political damage to George W. Bush and his party's brand, will not change the saliency of a WMD threat for future U.S. presidents. Few people in U.S. national politics—and perhaps not many in Europe, either—would object to a future preemptive strike that could successfully prevent a hostile country from arming itself with WMDs. The practical problem is coming up with the incentives or operations to prevent countries from going nuclear, much less biological or chemical. So future U.S. actions to counter these threats, of necessity, will focus on preventing others from actually using WMDs. This explains why the Pentagon is now emphasizing its pressing need for not only "advanced detection technologies for chemical, biological, and nuclear weapons," but also "new or improved capacities for persistent surveillance and intelligence, global mobility, (and) rapid strikes."

If anyone still doubts the global approach of U.S. military strategists, the fourth class of threats settles it: "disruptive challenges . . . from adversaries who develop and use breakthroughs in technologies to negate current U.S. advantages in key operational domains." These threats involve ultra-high-tech versions of a 9/11 type attack that require a nation's resources to carry out. The thinking here is that while it would takes China or Russia generations and trillions of dollars to match America's various military capacities, both countries and a few others could develop breakthrough technologies that could disable some critical superpower capacity. Perhaps scientists in

Shanghai or St. Petersburg could develop, for example, nanotechnologies that could disable tanks or airplanes; software viruses that could disrupt U.S. financial systems, or more ominously, cyber warfare programming that could penetrate the security of U.S. hydroelectric dams and nuclear power utilities, and direct them to open their gates or initiate meltdowns. (The latter scenario, which planners take seriously, became a plotline in the popular counter-terrorist, sci-fi TV soap opera 24.) The Pentagon also worries about space and directed-energy weapons that could take out its military systems.

Countering such threats will claim an increasing share of the U.S. military's huge annual R & D budget. Over the next decade, as defense R & D passes the $100 billion a year mark, much of it will go to develop better capacities for—and the Pentagon ticks them off—persistent surveillance and intelligence, global mobility, rapid strikes, sustained unconventional warfare, counterterrorism and counterinsurgency campaigns, medical countermeasures against genetically engineered bioterror agents, advanced detection technologies for chemical, biological, and nuclear weapons, more advanced nuclear deterrents, more unmanned surveillance, and the development of the next generation of long-range strike systems.

Beyond the R & D, the proposals that the Pentagon and White House are willing to describe publicly to address these threats include a major ramp-up of Special Operation forces in every service, thirty-five hundred more personnel for PSYOPS, or psychological operations, the advanced procurement of the Predator and Global Hawk unmanned aircraft, a new generation of conventional ballistic missiles for the Trident submarines, and a new generation of land-based missiles that can wipe out an enemy's military without using nuclear explosives.

While the principal stated focus of this planning and procurement is terrorism, U.S. military strategists also are focused on future threats from China, based on reports of its investments in "high-end, asymmetric military capabilities, emphasizing electronic and cyber-warfare, counter-space operations, ballistic and cruise missiles, advanced integrated air defense systems, next generation torpedoes, advanced submarines, strategic nuclear strikes from modern, sophisticated land and sea-based systems, and theater unmanned aerial vehicles." There was a frenzy of Pentagon concern in early 2007 when the Chinese destroyed one of their own weather satellites, since an effective antisatellite weapon could blind America's space fleet of more than one hundred military surveillance and communications satellites.[9]

There was even greater concern in June 2007, after successful cyber attacks on the computer systems serving the office of Defense Secretary Robert Gates were traced to the People's Liberation Army.[10] Similar cyber intrusions on

German government systems also were traced to Chinese government facilities. The United States and China—and most likely Russia, Israel, and others—regularly try to probe each other other's military networks. The problem here was that China succeeded in penetrating central Pentagon systems, suggesting a new capability to disrupt American military networks at critical moments. According to reports, "Hackers from numerous locations in China spent several months probing the Pentagon system before overcoming its defenses," and the Pentagon was forced to take down its network for more than a week during the attacks.[11]

For a global superpower with interests everywhere, the threats are even broader than terrorism, cyber attacks and the Chinese military in another generation. With increasing frequency, new perceived threats will arise, as matters that the United States never cared much about are seen as new security issues. The Bush administration's intense attention to AIDS in Africa, for example, has been driven largely by concerns that the epidemic's corrosive social effects could turn some Muslim African nations into terrorist havens or threaten flows of African petroleum. That's also the reason why the United States, the only country that divides the world into military "commands," recently created a new Africa Military Command. It takes its place alongside the Pacific Command (Asia), the European Command, the Central Command (the Middle East), the Southern Command (Latin America), and the Northern Command (created in 2002 to coordinate the defense of the homeland).

This tendency to reconceive nonmilitary matters as new security threats could well expand over the next decade, starting with climate change. Scientists and former Vice President Gore settled the domestic political argument over the risks of climate change in 2006; and by 2007, several former generals and admirals entered the fray. U.S. national security, they argue, will require strong measures on global warming, because even modest climate changes might strengthen the hand of terrorists by bringing on catastrophic storms, famines, epidemics, and water shortages. The former commander of U.S. Naval Forces in Europe, Admiral T. Joseph Lopez (ret.) makes this case: "Climate change will provide the conditions that will extend the war on terror . . . in regions of the world that are already fertile ground for extremism."[12]

THE SOLE SUPERPOWER AND THE REST OF THE WORLD

The geopolitics of the next decade will be shaped by both the steps America takes to remain the dominant power in Asia, Europe, Latin America, and Africa, and the responses of other governments. Foreign policy realists, who almost certainly will advise the next U.S. president, have long observed that countries become very uncomfortable when one nation is much stronger than everyone else. As Harvard's Professor Walt puts it, "states respond to unbalanced power by balancing against the dominant country." It's easy to appreciate why—a powerful country that no one can check can break a lot of other countries' crockery, even when it doesn't mean to. The United States didn't intend to destabilize Cambodia and pave the way for the Khmer Rouge when it pursued the Viet Cong there in 1970—any more than the Bush administration expected its invasion of Iraq to ignite a sectarian civil war and empower Iran. But that's what happened. Differences in values also can drive nations and societies to find ways to balance a superpower, especially when a superpower's alien values are forcibly introduced into places they're not welcome. Many Saudis hate the customs that Americans bring with them, and extensive Saudi financial support for Muslim fundamentalism is one of the results.

The Pentagon agrees with the realists: "Our leading position in world affairs will continue to breed unease, a degree of resentment, and resistance." Tellingly, it's also what many foreign officials say publicly. A few years ago, for example, Joschka Fischer, then Germany's foreign minister, warned that, "a world order in which the national interests of the strongest power is the criterion for military action simply cannot work," and his French counterpart at the time, Herbert Vedrine, offered that "the entire foreign policy of France . . . is aimed at making the world of tomorrow composed of several poles, not just one."

As a practical matter, while China and Russia show signs of following these axioms, Europe or most of Asia have not. Apart from public complaints, Germany, France, Britain, and Japan have made little serious effort to introduce more balance in geopolitics, nor are they likely to do so in coming years. Even as European countries and Japan worry about the consequences of unchecked American policies, they can remain detached, because those policies have posed no direct threat to their interests. In any case, after devoting little money and attention to the great conflicts of recent past decades, Europe and Japan are in no position now to exert a counterbalance in global affairs. The European Union didn't exercise leadership even in bloody conflicts in its own backyard, waiting for the United States to take the lead in Kosovo and Bosnia.

In truth, they don't seem to aspire to do so; and no major candidate in the most recent national elections in France, Germany, or Italy proposed that Europe play a serious global role. The only exceptions have been Europe's contribution to the U.S.-led overthrow of the Taliban in Afghanistan and Britain's role in the U.S.-created Iraq War—and Gordon Brown and his successors are unlikely to emulate Tony Blair in this regard. And if European leaders decide to reclaim a serious role in world affairs sometime in the next decade, they will need heroic persuasive abilities. Public opinion surveys show that most Europeans don't see global responsibilities as a tradition worth reviving. Those attitudes are only likely to harden as Germany, France, Italy, and most of the rest of Europe become preoccupied with the costs of supporting their rapidly growing elderly populations, especially if globalization continues to dampen their growth.

One reason that the United States had no serious conflicts with Europe or Japan under the first President Bush and President Clinton is that when those presidents exercised American power, when and where they chose to, they did it a within a UN or NATO framework and presented the policies as prudent responses to aggression. The lesson of the first Gulf War, Kosovo, and Bosnia is that the United States can project its military might as it chooses, so long as it acts as if there were a balance-of-power relationship with its allies. Outside the military realm, the United States also has successfully coopted other nations' impulses to balance by allowing international institutions to disagree with America on matters of marginal importance. This is most clear at the WTO, where the European Union and developing countries periodically win cases against the United States, even compelling the U.S. Congress to modify certain domestic subsidies that affect trade.

Russia presents a more complicated problem and could be more of a factor in the geopolitics of the next ten to fifteen years because it has the will to affect events. But it will not be a real power in geopolitics, because it no longer has the military, economic, and political resources to move any major country, much less the superpower. Russia's present role was largely settled when Boris Yeltsin came to power more than fifteen years ago, and the disintegration of the Soviet Union left him no choice but to slash defense spending by nearly 90 percent and cut military personnel by half. By 1999, it took tens of thousands of Russian troops more than a month to repel a ragtag band of less than two thousand Islamic militants back into Chechnya—longer than it took U.S. forces to defeat Saddam Hussein's army in either 1990 or 2003. In addition, Russia's once-fearsome strategic systems are now outdated and sufficiently deteriorated to be dangerously unreliable; and one U.S. expert dismisses Russia's current conventional forces as "impoverished, incompetent, sullen, and sluggish."[13] While that overstates their problems, the Russian army, navy, and air force are now so poorly trained and badly equipped that they could no more fight a real war today than could Italian or Japanese forces. And where the old Soviet Union's geopolitical clout could call on worldwide alliances and a worldview that appealed to many developing nations, less than a generation later NATO has absorbed the Warsaw Pact and the Soviet ideology is forever discredited.

Russia is also now a minor player in the global economy, and with no realistic prospects of rebuilding the economic means to restore its military might and geopolitical power. After five years of solid growth fueled by high world oil prices, Russia's $987 billion GDP in 2006 was still less than Brazil or Canada, and barely 7.5 percent of America's GDP. Even adjusting for costs of living across countries, the Russian economy is smaller than Italy and barely larger than Brazil. Today, Russia couldn't afford to rebuild its military strength even if it chose to.

Nor will Russia be able to do so anytime in the foreseeable future, since the country's likely economic trajectory points down. The Putin government's rollback of legal rights and its arbitrary property seizures have discouraged the foreign direct investment that the country desperately needs to modernize its critical energy sector and build new industries. In 2005, the entire stock of foreign direct investment across the Russian Federation totaled $132 billion—less than Singapore, Mexico, Brazil, or Belgium, and one-quarter that of China. Nor can Russia reassure foreign investors about its future internal stability. Moscow not only faces bloody challenges in Chechnya and other Muslim areas in the North Caucasus currently flirting with independence.[14] In the next few years, it also could be drawn into a series of conflicts in other parts of the former Soviet Union, including the dispute over Transnistria in Moldova (where Russia still has two thousand troops), a continuing effort by Nagorno-Karabakh to break away from Azerbaijan, and the secessionist movements in Georgia by the Abkhazian and South Ossetia regions.

Even if Moscow had the money and the purpose to rebuild its forces and alliances, it still would face the world's fastest-deteriorating demographics: Relatively soon, Russia simply may not have the manpower and womanpower of a state that matters in geopolitics. The life expectancy of Russian men has been deteriorating for more than a generation, and the trend has accelerated in the last decade. Today, it stands at less than sixty years, which means that men in Egypt, Guatemala, and Vietnam, on average, can expect to live longer than the men in Russia. Russian birthrates also are very low, with live births lagging deaths by 1 million a year since 2000. But the larger issue in the high death rates is the Russian government's real abandonment of health care. In the most recent year for which the World Bank has data, 2003, Russia spent less than $187 per person on health care—Costa Rica spent 50 percent more per person, Poland twice as much, and the United States, at $5,705 per

person, spent an astonishing thirty times more than Russia.[15] And beyond
the death rates, a recent Russian health census found that more than half of
Russian children now suffer from a chronic disease of some kind.[16]

According to projections by the Russian Academy of Sciences, by 2020 the
country will have 50 million fewer people than Nigeria or Bangladesh—barely
more than Mexico—and its life expectancy could lag behind India and
Bangladesh.[17] Between now and then, the working-age population is expected
to fall by almost 15 percent, and the number of children is forecast to decline
about 20 percent.[18] As for military ambitions, the combination of depressed
birthrates and high death rates for males age fifteen to twenty-four (three times
U.S. levels) will cut the pool of young men of draft age by nearly half. Given
these conditions, the issue is not whether Russia will recover its geopolitical
clout, but the likelihood that those conditions will lead to large-scale unrest.

Despite its marginal military, economic, and human capacities, Russia has
two cards it can play in the geopolitics of the next decade. First, with the
world's longest border, any conflicts that involve North Korea, China, Azerbaijan and, across the Caspian Sea, Iran will necessarily also involve Russia.
Second and more important, Russia is a genuine force in the geopolitics of
energy. Today, it is the world's largest producer of natural gas, the second-
largest producer of oil after Saudi Arabia, and number seven in coal production. Those positions will not diminish much over the next ten to fifteen
years, because Russia also has the world's largest natural gas reserves, the
world's second-largest coal reserves, and the eighth-largest oil reserves. In
fact, Russia could become a larger player in global energy over the next decade if it were to reinvest its oil revenues in modernizing its production and
take serious steps to encourage western oil companies to expand their foreign
direct investments there (a tough sale after Putin's seizure of the Yukos companies).

Russia today supplies more than one-third of Germany's oil and two-fifths
of its natural gas; and by 2020, according to forecasts by the International
Energy Agency, it will be Europe's main oil and gas supplier. One reason is a
major new pipeline linking the Black Sea port of Burgas, Bulgaria, with the
Greek port of Alexandroupolis, so Russian oil can be transported directly to
the European Union. The Russian electricity grid also supplies power to parts
of Europe, especially its former Eastern and Central European satellites. In
addition, Russia recently chose Japan (over China) to finance 2,600 miles of
oil pipeline from Siberia to the Pacific, which will make Russia one of Japan's
major oil suppliers.

Russia's oil and gas weapons will give it genuine influence, but little real
geopolitical power. It can use its energy supplies to wring political concessions

from countries that were once part of the Soviet Union, especially those in Russia's current confederation. It also will be able to use long-term contracts to secure economic and commercial concessions from European and Asian governments. But it doesn't have the strategy or the means to use its energy resources to balance or even affect U.S. policies, even when it involves countries on its borders such as Iran. It may be more distasteful for Russia than for Europe and Japan, but it, too, will have to look the other way as the United States pursues its global interests in the regions it once dominated.

CHINA AND THE SOLE SUPERPOWER

China and the United States may be seen as potential partners or potential competitors in geopolitics. However, given the gulf between their political systems, China's increasing economic clout in Asia, and its current military buildup, there's little prospect that the next ten to fifteen years will see the United States approach China as a genuine ally, as it does Europe and Japan. For its part, China seems increasingly uneasy with America's domination of geopolitics, especially in Asia, and these concerns will likely increase over the next ten to fifteen years. Its practical ability to balance U.S. global policies will also increase over that period, but even in its own region, China's capacity to check U.S. actions will be limited.

China's foreign policy is formally governed by the "Five Principles of Peaceful Coexistence" set down by Zhou Enlai in 1953 and written into China's constitution thirty years later. They can be seen as embodying the international posture of a country still stung by foreign occupation and still feeling relatively weak, expressed in the first three principles of "mutual respect for every nation's sovereignty and territorial integrity," "mutual non-aggression," and "non-interference in each other's internal affairs." The "Five Principles" also suggest a country committed to becoming sufficiently powerful to ensure that no other nation dominates it again, especially in the last two tenets, calling for "relations of equality and mutual benefit" and "peaceful coexistence."

Within this framework, China's foreign policies are unusual in the extent to which they are shaped by the government's overriding domestic priority, of modernizing and growing as fast as possible. Rapid modernization is not merely a necessary condition for defending the country against potential enemies. Beyond that, China's modernization strategy depends on maintaining amicable relations with the countries that provide its huge direct investments, as much of its growth depends on its relationships with the countries that buy its exports.

China's leaders often seem fixated on growth and modernization, and the foreign policies that help sustain them, because they apparently believe that's what will ultimately maintain the Chinese people's allegiance to their authority. It's not surprising, given how much the Chinese government has asked of its people for more than a generation. First, the Cultural Revolution wrenched millions of educated Chinese from their professions and homes. Then, the sharp U-turn from communism to capitalism jettisoned much of the national ideology, including a series of radical agricultural changes that cost more than 100 million peasant families their homes and livelihoods. That has been followed by the government closing down thousands of state-owned enterprises, a policy that every year costs another 10 to 20 million Chinese workers their livelihoods. And the dissolution of the agricultural communes and most of the state-owned enterprises has cost most Chinese any public retirement or health-care security.

For leaders concerned about how long people will put up with government policies that demand so much sacrifice, delivering the world's most rapid growth and modernization year after year can convey that better times do lie ahead.

Their concerns are fueled by the sharply rising incidence of public protests across the country. One Chinese expert, Sun Liping, estimates that every day, 120 to 250 demonstrations of more than a hundred people take place in urban areas, plus another 90 to 160 incidents a day in country villages. The country's senior police official, Zhou Yongkang, has acknowledged that nearly 4 million people took part in 74,000 significant public protests in 2005, rising in recent years "like a violent wind." And all these figures understate what's really happening, since local officials often try to cover up disturbances.

Reporters from *The Economist* magazine scanned Chinese press reports for a typical few weeks a few years back, and found, among the major incidents, a protest march by 5,000 retirees from a state-owned textile mill demanding an increase in their paltry monthly pensions, which brought much of Bengbu city in eastern China to a standstill; mass demonstrations for higher pay by 3,000 workers from a CD factory in the Shenzhen Special Economic Zone which disrupted the city's main highways for four hours; and a battle between 500 riot police and some 5,000 workers from a Japanese-owned printing facility in nearby Dongguan over substandard food.[19] More recent examples from March 2007 include 20,000 people protesting an increase in bus fares and battling 1,000 police with guns and electric cattle prods in Hunan province, and a few days later, thousands of villagers in Guangxi, near the Vietnam border, smashing cars and starting fires to protest fines of up to $1,300 for families breaking the one-child laws.

Some Chinese writers such as Shuguang Wang, a professor at the Chinese University of Hong Kong, publicly link the protests to the government's poor performance in ferreting out the corruption that many Chinese blame for their income problems, in fighting crime which has risen fivefold since 1980, and in improving dangerous working conditions that result in more than 100,000 fatal industrial accidents a year.[20] Judging by Public Security Minister Zhou's statements, the national government will live with the economic protests—but not with anything that suggests political dissidents are using them to pressure the Party. So most local officials tolerate the thousands of low-decibel, small-scale marches and sit-ins by peasants and workers with petitions and posters, especially if they seem spontaneous and disorganized. Demonstrations organized by people with political agendas or which disrupt a city's business draw harsher responses, and they're increasingly common.

When *The Economist* asked editorially, "Does China face serious instability?" its answer was, "In the longer term there are reasons to worry."[21] That conclusion may be overstated, but the concerns about social stability felt by the country's self-appointed leaders do impart a sense of urgency to foreign policies that may affect the country's growth and modernization. That connection alone has made a stable relationship with the United States "the central thread in China's foreign policy strategy," as President Hu has put it. It's no mystery: America is the critical market for Chinese growth, buying 30 percent of its exports (Wal-Mart alone imports more than $15 billion a year) and, through its huge imports from other countries, indirectly supporting global demand for Chinese products. The United States is also indispensable for China's modernization, providing much of the country's foreign direct investment both directly and indirectly through its role in other economies. China's leaders understand perfectly well that the deal they have tacitly struck with America goes beyond trade and investment. Since at least the late 1990s, when they acceded to WTO's entry terms, they have accepted a long transformation based largely on American-style capitalism. For now, they're also willing to live with America's role as sole superpower, even in Asia, because they cannot afford to challenge it and because it provides the global stability they need to maintain the flow of foreign technology, expertise, and energy.

America's central role in the path that China has set for itself helps to explain why its leaders become so upset when the U.S. Congress or, worse, the White House, periodically rattles a trade saber. In Beijing, American calls to cut back trade with China sound like a threat to the country's growth and, for more cynical and distrustful officials, part of a superpower plot to undermine the country's stability. To discourage those calls, the leadership recently intensified its efforts to increase China's imports from the United States, already up 300 percent in the last decade.[22] And over the next decade, China is likely to make most economic concessions that the United States truly demands, so long as they don't impede its growth. Unless U.S.-Sino relations rupture over other matters, China will continue to finance the U.S. current account deficit, continue to expand its imports from the United States, and open new areas to American businesses, especially in financial, business, and health-care services. What it will not do is let pressures from Washington dictate the value of its currency, because that would slow its exports to other countries and concede an element of its own sovereignty.

National sovereignty and power are the other reasons why the U.S.-Sino relationship will remain the central issue in Chinese foreign policy for the next ten to fifteen years. China's immediate concerns over sovereignty involve Taiwan, Tibet, and Xinjiang, which its leaders see as purely domestic issues that hardliners in the United States could still use to divide "greater" China. The secessionist unrest by ethnic Uighur in Xinjiang, possibly goaded by Al Qaeda in nearby Turkistan, is not widely followed in the West; but it has worried Beijing much more than Tibet. As for Tibet, it has become an issue that some American celebrities believe justifies U.S. action of some sort, but that view has never been held by U.S. policymakers. Taiwan is different. The worst period in U.S.-Sino relations in recent times came in 1996, when China tested a new missile that seemed designed to threaten Taiwan and the Clinton administration responded by sending U.S. aircraft carriers into the Taiwan Straits. The incident left a large impression in Beijing. The director of the China Center at the Brookings Institution, Jeffrey Bader, wrote recently, "China's military planning is overwhelmingly directed at one target—the use of force in the Taiwan Strait to prevent formal Taiwan independence."[23]

Assuming that Taiwan, America, and China all behave reasonably in the future, the larger sovereignty issue for China is how to live with a sole superpower whose global responsibilities make every region its area of influence. Britain, Japan, Germany, France, and Russia have to live with this as well, but their geopolitical roles have been winding down for some time. China's power is growing, and many Western observers now see the outlines of a Chinese strategy to build up the economic and military resources required to balance or constrain America's ability to pursue its interests, at least in Asia.

One sign is that China has been taking a more active role in the regional aspects of a number of global issues—helping to persuade North Korea to scale back its nuclear program, providing reconstruction assistance for Afghanistan, and voting with the United States to refer the Iranian nuclear program to the United Nations. Some prominent Chinese foreign policy experts also have begun to write forcefully about their country's regional ambitions. Wang Jisi, director of American studies at the Chinese Academy of Social Science, says that China now sees itself as a continental power that has to deal with a wide range of *zhoubian guojia,* or neighboring places, including India, Mongolia, Pakistan, Russia, and Japan, as well as Central Asia. Jisi also has written that China's growing military power will not threaten any of its "regional partners," but rather provide insurance against a U.S.-led coalition to contain China.

The specter of China as a balancing power in Asia is based mainly on its ongoing military buildup. China tells the world little about its military spending, and western experts cannot agree on even its order of magnitude. A few years ago, for example, Beijing said it spent $22.4 billion on the military—a figure that explicitly excluded military research and development, costs for the People's Armed Police, weapons purchased from abroad, and a few other areas. The American CIA and the Institute for Strategic Studies in London both concluded that the real figure was about three times the official one, which would make China the world's second-largest defense spender, but a long way from number one. The Rand Corporation, however, estimates Chinese defense spending at five to eight times the officially reported levels, which would come to perhaps half of U.S. defense spending. Whatever the correct numbers, even the official data show Chinese military spending growing by about 12 percent a year since the late 1990s.

More is known about the size of China's forces. The country maintains at least 2,250,000 active duty personnel, plus another 1,500,000 in the People's Armed Police. Five years ago, it had more than 70 known submarines, at least 750 navy planes and 2,000 combat aircraft—fleets that are meager stacked up against America, but larger than France, Germany, Britain, Japan, and India combined.[24] China also has a respectable ballistic missile force. Again, five years ago, experts counted more than 30 missiles with the range to reach Europe or America, 110 that could strike anywhere in Asia, and significant numbers of submarine-launched and short-range missiles for a place like Taiwan. Since then, China has expanded and upgraded these forces. U.S. defense planners believe, for example, that China recently developed and built five new "Jin class" subs armed with new JL-2 nuclear missiles that have a range of more than 5,000 miles, as well as a new mobile, land-based ICBM system that could target the entire United States.[25]

At a minimum, these capabilities, along with significant numbers of amphibious assault ships, fighter jets, and Russian-made submarines with missiles that can threaten an aircraft carrier, could pose not only an immediate threat to Taiwan but also, in the view of a former U.S. military attaché in Beijing, "deter or delay our ability to intervene."[26] If that becomes the common international judgment, it would begin to revise the terms of U.S. dominance in Asia. The recent improvements in China's nuclear forces have also moved some western analysts to ask if China's official policy of maintaining "only enough nuclear capacity to retaliate" has been superseded by a "more flexible use of nuclear weapons," as a former senior Asia advisor to George W. Bush, Michael Green, has put it.[27] Chinese officials dismiss the concern, pointing to America's growing ability to not only monitor and target China's existing deterrent but also to develop new ballistic missile defenses.

This is beginning to look like an arms race; and whoever started it, the outcome may well be that "what was gray before is now clear—China now can effectively fight a nuclear war," as a U.S. Naval War College China expert, Lyle Goldstein, wrote recently. He may be right—although, with an estimated 400 nuclear warheads, compared to 10,000 deployed by the United States, Chinese leaders know that any attempt to fight one with America would be the end of China. Compared to every other country in the region, however, China's nuclear and conventional forces will soon loom very large. An independent task force formed to evaluate China's military prowess, headed by former U.S. Defense Secretary Harold Brown, came up with what's probably a realistic and sobering assessment. It concluded that in addition to maximizing its options on Taiwan, its military buildup is intended to make China the leading power in Asia, whether that takes a decade or generations. In the meantime, as Harvard University's Professor of China in World Affairs, Alastair Iain Johnston, puts it, "China is willing to see the United States as a global hegemony, so long as it doesn't act like one in Asia."

Given China's size and growing economic importance, its regional ambitions are inevitable—much as is America's global agenda, given its worldwide military and economic dominance. The problem for the next decade is that China's long-term goal conflicts with America's continuing aim to be the principal power in Asia (as well as Europe, the Middle East, Latin America, and Africa). On this core issue, China's leaders reportedly discount the regular U.S. statements that America's interests are best served by a stable, prosperous China, believing instead that as foreign policy realists recover their influence in the United States, America will move to "contain" China.

This helps explain China's continuing hostility to Japan, which it sees as allied to the United States in containing China, as well as China's 2001 friendship treaty with Russia and its repeated efforts to cooperate with the

European Union. In the last few years, China has settled longtime border disputes with several Central Asian countries as well as with Russia, and opened serious border talks with India. It also is working to make its economic relations with other East Asian countries as important to them as their ties to the United States. In the last decade, its two-way trade has grown by nearly 350 percent with Singapore, 700 percent with Korea, 835 percent with Thailand, and more than 1,000 percent with India, Malaysia, and the Philippines.[28] And every one of those trade partners runs a surplus with China.

The U.S.-Sino competition in Asia is here to stay. While China's military today probably could not take and hold, for example, the islands disputed with Japan or territory disputed with India, by 2020 its military power will dwarf everyone else in Asia including Russia. Between now and then, America's Asia strategy will work to delay China's regional influence as much and for as long as possible. That's the strategic meaning of the pending U.S. free trade agreement with Korea, a model likely to be replicated with other Asian countries in coming years. And to shorten China's reach, America has been busy building a network of military bases across Asia and the Pacific—in Japan, Guam, Australia, South Korea, Singapore, Malaysia, Afghanistan, Pakistan, Kyrgyzstan, Tajikistan, and Diego Garcia in the Indian Ocean. (*Foreign Policy* Magazine calls Diego Garcia and the Manas Air Force base in Kyrgyzstan two of America's most important global bases.) All of this forward placement of U.S. military assets can also serve a "dual use" by tempering some of Russia's and Iran's ambitions.

China is unlikely to find a genuine ally against the United States in any of Asia's other major countries—especially Japan, Korea, and Russia. For reasons of proximity and oil, the Kremlin worries more about China challenging Russia than about the consequences of America challenging China. Korea is tied more closely to the United States economically, militarily, and politically than any other country, with the possible exceptions of Japan and Britain. And Sino-Japanese relations have been in a deep freeze despite Japan's extensive involvement in China's economy. Whenever a Japanese prime minister visits the Yasukuni shrine to the country's war dead—including convicted war criminals from Japan's occupation of China—Beijing permits massive anti-Japan public demonstrations and even has looked the other way at attacks on Japanese-owned businesses. Of larger moment, China and Japan compete directly over access to oil and natural gas from Siberia and drilling in the East China Sea; and China consistently blocks Japan's bid for a permanent seat on the UN Security Council.

At the same time, Japan is China's largest investor, and China is Japan's largest trading partner, which alone will make peaceful relations a natural priority for both governments over the next decade. That's certainly the view in Japanese business circles. As Motoya Okada, head of the large Japanese retailer chain Aeon said not long ago, "There was one time when people talked about the China threat, but I believe that the Chinese people will make us rich."[29]

It's also clear that as China has built up its military, Japan's strategic and defense ties to the United States have strengthened. For several years, Japan's National Defense Program guidelines have stressed both the centrality of the U.S. alliance and new procurement plans that seem driven by China's military buildup, including new transport aircraft, fighters, and destroyers "to respond to an invasion of offshore islands," new systems to detect and track incoming ballistic missiles (the latter is also aimed at North Korea),[30] and a new agreement to acquire ballistic missile defense systems from the United States.[31] Japan's ties to the United States are strong enough that despite wide public criticism of America in Iraq, Tokyo provided eleven hundred troops to the U.S. war effort from late 2003 to mid-2006.

China's leaders, however, can probably be confident that Japan will remain nonnuclear for some time. To start, it's unimaginable so long as Japan remains well within the extended defense perimeter of the United States, since the superpower wouldn't countenance it. While the Japanese public is less opposed to rearmament than it once was, strategists such as Matake Kamiya of the National Defense Academy of Japan argue that a Japanese bomb would sow distrust of Japan across much of Asia and could push South Korea and Taiwan to follow suit.[32] Even a prominent Japanese analyst who criticizes America's "global, unilateral" approach to the world and believes Japan should consider going nuclear, Nakanishi Terumasa of Kyoto University, agrees that China's military buildup enhances America's value as Japan's natural ally.[33]

The big unknown in China's approach to geopolitics and the United States is its own political evolution over the next decade. The Dean of Harvard's John F. Kennedy School of Government, Joseph Nye, has been an optimist, suggesting that China's dependence on foreign markets and its development of a middle class could begin to roll back authoritarianism, better aligning its interests with America's. In theory, widespread social unrest could lead to political reforms; but instability could just as likely produce a crackdown, especially on economic freedoms. For the foreseeable future, China's leaders are the only plausible source of meaningful political change. The current "fourth generation" of leaders will be in power until 2012, and they almost certainly will remain committed to avoiding the "Russian mistake" of rapid political reforms that ultimately weaken the Party's authority.

An insider's account of China's current leadership became available in late 2002, with the western publication of some of the internal investigations of candidates for the Standing Committee, called *Disidai* (fourth generation).[34] The most striking portrait is that of Premier Wen Jiabao, who rose through the ranks as the right-hand man first of Zhao Ziyang, the Party secretary general purged after siding with the students during Tiananmen, and then of Premier Zhu Rongj of the third generation. Wen is portrayed as favoring some far-reaching economic reforms and questioning some of China's massive public works projects, including plans to build three thousand-mile networks of canals and aqueducts to carry water from southern China to the northern provinces. But on political reforms, Wen is as conservative as Hu Jintao. The security chief, Luo Gan, who rose to the top as the key aide to the security head during Tiananmen, Li Peng, also is portrayed as favoring reform in some areas—including an end to police torture, fewer restrictions on people moving from place to place, and a single benefit system for urban and rural Chinese. But he is also a hard-liner on law and order and political dissent, presiding over a system that executed an estimated sixty thousand people from 1998 to 2001.

No one can say how the fifth generation will approach geopolitics and America's role in Asia, especially since no one knows who they will be. Every sovereign country has the ability to say no to the United States—Germany and France famously did so over the war in Iraq. But saying no to a sole superpower often doesn't affect what the superpower does, even in a country's own neighborhood. In another generation, China may have the economic and military resources to say no to the United States and back it up convincingly, especially in Asia. That's not likely to happen in the next ten years.

THE EXTENT AND LIMITS OF THE SOLE SUPERPOWER

A little later we will examine the likely shape of two critical geopolitical issues for the next decade, energy and terrorism. Here, we conclude our discussion of the basic topography of geopolitics by examining the sole superpower's largest current global action, the U.S. military overthrow of the government of Saddam Hussein and subsequent occupation of Iraq.

No one knows how the Iraq saga will unfold over the next year or the next decade; and as you read this, conditions may be quite different from when it was written in mid-2007, and in ways that no one could predict. However, the way that events in Iraq already have unfolded provides insight into critical features of the next decade's geopolitics.

Beginning with the obvious, this sole superpower can wage a major war with great economic and political implications for every other important country by its own decision and virtually on its own. The United States took this course over the objections of four other geopolitical players—France, Germany, Russia, and China—and with the endorsement of just two, Britain and Japan, and material support from only one. With the end of balance-of-power geopolitics, no country is in a position to alter war-and-peace decisions by the United States. America did secure from the United Nations what it interpreted as formal consent for its military action, while declaring its right—that is, its power—to proceed without anyone else's approval. The only support that the United States truly needs to undertake major military action is domestic political support, and securing that was widely seen as the Bush administration's main purpose in going to the United Nations.

Because a superpower can make these decisions without consulting other countries, it also does not have to appeal to traditional casus belli—as in clear aggression (World War II), a humanitarian catastrophe (Serbia), or an ally under attack (Vietnam and the first Gulf War). Instead, the United States offered a new, superpower doctrine of preemptive war and regime change. Weapons of mass destruction in the hands of an "irrational" enemy may well justify preemptive war in almost anyone's view—although the doctrine also would fit the Soviet Union in the 1950s. America never seriously considered a unilateral, preemptive war on the USSR because it also requires military superiority so great that it can be carried out at relatively small cost and without drawing in others. The degree to which a right to wage preemptive war has become part of the worldview of some of those who might lead the United States was evident during a televised debate by Republican presidential hopefuls in June 2007, when several said they would consider a preemptive nuclear strike to end Iran's WMD program.[35]

Even as the Iraq War plays out so badly that no future U.S. administration will try to replicate it, its failure does not change the superpower capacity to wage war at its own discretion, nor eliminate potential threats that could lead to future preemptory attacks. Nor have the countries that opposed the Iraq War from its start imposed political costs that might force the United States to secure their support for the next war. When a future U.S. president asks if another preemptive war would serve some vital American interest, the decisive issues will still be the prospects of success and the extent of domestic political support, not whether other countries will support the policy.

As likely as not, that decision will again involve the Middle East. Even as the United States approaches its interests in global terms and engages actively in every region, those interests are greatest today in Asia and the Middle East. Asia will necessarily receive a great deal of the superpower's attention, because that's where China is, and China is both vital to the U.S. and world economies and the only major country with the capacity and inclination to challenge America in any meaningful way. The Middle East will be the other major, permanent focus of American geopolitics over the next ten to fifteen years, because that's where much of the world's oil is, and every major economy still depends on its steady supplies. Unfortunately, most of the oil lies in the sands of nations ruled by fragile and insecure authoritarian regimes, all in close proximity to the continuing Israeli-Palestinian conflict that itself is a source of regional tensions and instability. As a sole superpower, the United States has tacitly become the guarantor of the region's stability and the sovereignty of many countries there, including Israel for the last fifty years. None of that will change in the next decade. Finally, the Middle East is also the primary seat of Islamic fundamentalism and support for terrorist groups that will continue to menace American interests for the foreseeable future.

Again, the failures of America's Iraq campaign will not reduce the stakes for the United States in the Middle East, and ultimately could raise them. Regardless of who Americans elect president, the United States over the next ten to fifteen years will be actively and aggressively engaged in the region and its conflicts—including the real possibility of another significant military campaign. Even if the United States finds it will not have permanent bases in Iraq from which to operate, it still will retain bases in Qatar, Kuwait, Bahrain, and Turkey, as well as its secret bases in other places in the region.

The Iraq saga also shows that the real constraint on America as a superpower comes from its own people: They can force the U.S. government to do what no other state can require of it, namely to change course. A significant minority of Americans opposed the Iraq policy from the start. It started to change only after a significant majority turned against it, not for what it had intended to accomplish but for its failure to do so. Americans will forgive many things in their government and presidents, but not failure.

The Iraq War also provides an important gauge of the American superpower's geopolitical capacities. Now and for the foreseeable future, the United States can defeat the military of any other country, quickly and convincingly, but by itself cannot bring about the democratic regime changes it seeks. It can wipe out a foreign government's legions, but not alter a country's social dynamics. The United States has already begun what will be a protracted debate over whether its Iraq policy fundamentally reached too far or merely was poorly planned and badly implemented. How the foreign policy and defense professionals resolve this question—and whether the American public agrees—could vitally influence how the United States manages geopolitics for as long as it remains the sole superpower.

So far, one point on which most experts seem to agree is that the United States is unlikely *on its own* to succeed in stabilizing a deeply divided society once it topples its government. James Dobbins, who directs the International Security and Defense Policy Center at the RAND Corporation, makes this case plainly. He argues that the right model for Iraq was Kosovo and Bosnia, where the United States led a NATO coalition and with the cooperation of neighboring countries, successfully pacified ethnic conflicts as bitter as Iraq's.[36] But Kosovo and Bosnia are daunting examples, because based on what worked there, Iraq would require an international force of a half-million with the support of neighboring Iran, Russia, Turkey, and Syria.

More important, the Report of the Iraq Study Group takes a similar view. The group embodies the worldview of Washington elders, being comprised of two former secretaries of state, a former attorney general, a former defense secretary, a former White House chief of staff, two former senators, a former chairman of the House Foreign Affairs Committee, and a prominent Washington fixer. Their key recommendation:

> The United States should immediately launch a new diplomatic offensive to build an international consensus for stability in Iraq and the region. This diplomatic effort should include every country that has an interest in avoiding a chaotic Iraq, including all of Iraq's neighbors. Iraq's neighbors and key states in and outside the region should form a support group to reinforce security and national reconciliation within Iraq, neither of which Iraq [and the United States] can achieve on its own.

In this case, plainly the insurgency will not end without cooperation from Iran and Syria, since they supply and support the Shia militias. More generally, wherever a state is in turmoil, its neighbors will involve themselves (if they can), since the nearby turmoil inevitably affects them. And so, six months after dismissing the study group's analysis, the White House in mid-2007 moved to open talks with Iran and intensify its contacts on Iraq with Russia and Turkey.

There is no sensible way to predict the ultimate outcome in Iraq and its implications for the stability of the Middle East and everything else which that entails. The lessons drawn from the experience by future U.S. presidents and Congresses, and by other governments, are also uncertain. On top of that, events and personalities unknown today may tip the balance in future decisions of war and peace. What can be known is the enormous capacities and real limits of the United States, the centrality of the Middle East in its geopolitics, and the likely disposition of future American administrations to organize global support for the policies that the United States will continue to set.

CHAPTER SEVEN

THE COMING CRISES IN HEALTH CARE, ENERGY, AND THE GLOBAL ENVIRONMENT

SOMETIMES, SERIOUS PROBLEMS that will shake and shape a nation's life are predictable, because what drives them is both deep-rooted and apparent. Fast-rising health-care and energy costs are two issues that almost inevitably will create enormous social and economic stresses in every major country over the next ten to fifteen years. How each society manages these stresses could have momentous effects not only for the health of hundreds of millions of people, but also for the overall growth of their economies. Combined with the intense competition coming from globalization, these accelerating costs may continue to stunt job and wage gains in advanced countries, and so profoundly affect the prospects of middle-class families around the world.

The health-care crises facing most countries all involve the combination of fast-rising numbers of older people and the equally fast-rising costs of the medical technologies used to treat them. The costs of treating younger people are rising, too. But over the next decade and beyond, boomers will turn sixty and then seventy, as their older brothers and sisters age into their seventies and eighties, and some of their parents survive into their eighties and nineties. Those are the ages when large numbers will develop conditions that cost $15,000 or $150,000 to treat, instead of $200 or $1,000.

The timing of this crisis—now—reflects not only the singular demographics, but also the effects of globalization. The potential markets for new health-care devices, equipment, and drugs all have gone global, driving more (and more costly) R & D, and down the road, more and more costly new treatments. When Gilead Sciences outside San Francisco or South Korea's LG Life Sciences considers investing $1.2 billion in a promising line of biotech research—that's the average cost today of developing a new biological treatment—it estimates future revenues from potential patients in not only America and Europe, but much of Asia, Latin America, and anywhere else people or governments may be capable of paying for the treatment. Many of these newer markets are expanding very rapidly, and not only because of their demographics. In addition, globalization is making more of the world prosperous; and everywhere as people become more affluent, they value small improvements in their health care more highly. And the intellectual property protections that help drive medical progress also have gone global in the last five to ten years, insulating developers from most direct competition and often keeping prices at least fairly high in most places.

No country today is prepared to pay the bills that will come due in 2020, and before, from the combination of so many expensive technologies and treatments, and so many more people who could use them. In some places, this problem will become a political and social crisis as people seek treatments that neither they nor their governments can afford. Health care is unlike almost anything else in the marketplace, because while people can substitute a used Ford, or public transit, for a new BMW that they can't pay for, or a banana for raspberries with their morning cereal, there are no replacements for chemotherapy or dialysis. So, even when prices are very high, people are very reluctant to cut back on their own health care—especially when their government ostensibly guarantees everyone access, as it does in every developed country except the United States. As the ranks of the elderly and the array of expensive new technologies continue to grow, that guarantee will crack the financing arrangements for health care almost everywhere. And as those demanding the care face off against those who finance it, both will take their cases and their frustrations to the political arena.

Each country has its own special circumstances. In Japan and much of Europe, as the demand for costly treatments marches up, the number of working people around to finance it with taxes will be shrinking—and slow growth and globalization will probably be holding down their wages. This crunch will be especially difficult in France and Germany, where health-care costs already are very high and the total tax burdens already are 45 percent to 50 percent. Japan and Britain spend considerably less on health care, and their taxes are lower. But Japan has the world's most elderly-dominated population and a very weak economy, while Britain's National Health Service faces widespread public pressures to sharply expand its resources and services.

Health-care costs present special problems for the United States, too, despite its more favorable demographics, because Americans already spend a much greater share of their national income on health care than others. Americans also are strongly committed to using the latest technologies, many of them developed there. On top of that, the next U.S. president and Congress will very likely move to provide insurance coverage for the fast-growing numbers of people who lost theirs to the sky-high prices. Most important, the United States virtually alone has few price controls or global budgets to keep costs in check. All of these factors together will drive those overall costs to $10,000 per person within a few years, and the ensuing political clashes could come over the higher taxes needed to cover the exploding medical bills for the elderly, or over how much medical costs are holding down working people's wages.

Every advanced country has dealt with financing problems in their health-care systems before, but nothing like what they will face over the next decade. In Japan, America, and across Europe, people's taxes, premiums, and co-payments will rise, and the quality of their medical care will slowly degrade as access to some of the latest treatments becomes increasingly limited.

China faces a health-care crisis of a different kind. There, the central issue will not be the high costs of the latest generation of drugs, devices, and procedures, but the more expensive challenge of restoring basic care for some one billion Chinese. In a society growing more prosperous every year, but only for about one-third of the people, with a government undertaking scores of expensive public projects for modernization, the millions of Chinese who now protest over wages and corruption could well demand the restoration of universal medical care.

Energy costs could present nearly as great a burden for many societies over the next decade as health-care costs. The dimensions and urgency of this problem come from an intersection or collision of a number of economic and political developments, as well as inexorable forces of nature. The first two factors here are the standard economic ones—fast-rising worldwide demand for energy, driven by the rapid modernization in China and other developing countries committed to globalization, and the increasing costs of securing the energy supplies to meet the new demand. The same economic processes now enabling China, India, and others to develop and grow faster than any society before them have already pushed oil prices from $20 to $30 a barrel in the first years of this decade, to $50 to $80 a barrel and more for the last several years, taking slices of varying sizes out of growth everywhere. The next ten to fifteen years may well include periods when energy prices fall sharply; but it's even more likely that the price will reach $90, $100, and more per barrel.

On top of supply and demand, the terrain of "petropolitics" has become more treacherous. Oil supplies are increasingly concentrated in politically fragile Middle Eastern and African countries, along with Russia—and the stability of the Middle East in particular may be more problematic than at any time in decades, as the United States extricates itself from Iraq and the rising regional power of Iran puts additional pressures on its Saudi neighbor. Energy will complicate the geopolitics of the next decade in other ways. China's growing network of investments, joint ventures, and long-term contracts with oil states in the Middle East, Africa, Latin America, and Central Asia could increasingly place it at odds with Russia, Japan, and the United States. And Russia will continue to try to wring geopolitical advantages out of Europe's growing dependence on Russian oil and gas. The American superpower will try to oversee all of this, maneuvering to contain China's ambitions, deflect those of the Russians, and maintain the Middle East as a stable energy supplier for the world.

The new factor is the growing worldwide concern that the environmental effects of all this energy use on the global climate may soon reach a serious tipping point. These concerns will send worldwide energy prices even higher, first by reducing investment in carbon-intensive ways of generating power. Already, for example, the prospect of future regulation is slowing new investments in coal-fueled power plants; and five years from now that slowdown will push up European electricity prices. The positive side is that these concerns are also fueling greater interest in renewable energy sources such as wind, solar power, and biomass, which by one estimate comprised 30 percent of new power-generation investments worldwide in 2005.[1]

Much larger costs will come as the advanced countries actually regulate the greenhouse gases that come from fossil fuels, an increasingly likely prospect over ten years. A solid consensus has formed in Europe, Japan, and America that the world's current trajectory for producing CO_2 and other greenhouse gases will eventually do serious damage to climate conditions everywhere. Still, the proposed responses will be divisive, from top-down regulation to carbon taxes and complex cap-and-trade systems. However, there's no sign that China, India, and other developing countries with very fast-growing emissions share the consensus. Any prospect of their going along will require the kind of effective international leadership from the United States that has not been seen since the first Gulf War and the WTO's successful launch.

HOW MUCH COUNTRIES WILL HAVE TO SPEND ON HEALTH CARE

Different societies spend greatly varying amounts on health care. China un-surprisingly spends much less than any advanced country, with its govern-ment dedicating less than 2 percent of GDP to health care and its citizens contributing another 2.5 percent on their own. By any measure, America spends the most—about 16 percent of its GDP at last count in 2006 or about $6,700 per person, compared to between 10 and 11 percent of GDP in France and Germany, and about 8 percent in Japan and Britain.[2] That makes Japan and Britain's financial commitment to health care considerably closer to China than to the United States.

No one knows how much difference these disparities make in the actual quality of the care that people receive, especially across advanced countries. In 2004, the Commonwealth Fund issued a long-term study of medical care in Britain, America, Canada, New Zealand, and Australia.[3] While spending twice as much per person, the United States did outpace the U.K. in survival rates for breast cancer and cervical cancer—but British survival rates were higher than America's for a number of other cancers, including colorectal, childhood leukemia, and non-Hodgkin's lymphoma, as well as kidney and liver transplants. But the study couldn't say, for example, whether more Britons survive transplants because their care is better or because Americans perform transplants on sicker patients.

One clear difference is that Britons experience many fewer financial barri-ers to care. That's because the U.K. National Health Service picks up more than 85 percent of the nation's total tab for medical care; and Germany, France, and Japan aren't far behind, with government financing 75 to 81 percent of their people's total health-care spending.[4] What that ultimately means is that people pay for each other's care through their taxes, which is a clue to why these countries spend so much less than America: The hard po-litical fact is that taxpayers will tolerate only modest and occasional tax hikes to keep the system going, and so Europe and Japan apply price controls and some form of global budgets to their health-care sectors.

America obviously has very different arrangements, with the government picking up 45 percent of the national tab to cover low-income and retired peo-ple (and its own employees), while most companies provide private insurance for their employees. Americans blame some of their high costs on this system's inefficiencies, and it does involve high administrative costs. But Britain or Germany's huge public health-care bureaucracies also are notoriously inefficient—and in any case, the greatest "inefficiencies" come from the uncertainties inherent in most medicine. The most important reason why America spends so much more on health care is simply political: While Europeans and Japanese can force their politicians to find ways to deliver care without raising their taxes too much, Americans unhappy with their rising costs or shrinking coverage cannot do much about it except try to find another job.

Americans spend so much on health care that the per capita taxes that support the elderly and poor are greater than the per capita taxes the Germans, Japanese, and Britons pay to cover everybody.[5] On top of those taxes, Americans have to pay for their private insurance plus out-of-pocket costs, which together are greater per person than all per-person medical spending in Britain and Japan.[6] Americans have put up with it because their incomes are higher than Japanese or most Europeans, and there's been little they could do about it.

Still, medical spending has been rising rapidly in Europe and Japan, too, because price controls cannot stop people from aging or arrest the technological advances being developed to treat the illnesses of old age. From 2000 to 2004, per capital health-care spending rose by about 4 percent a year in Japan, 5 percent a year in Germany, almost 7 percent a year in France, and more than 9 percent a year in Britain.[7] Britain's increases even outpaced the United States, where the costs grew by a little over 8 percent a year.

This is just the beginning. As the boomers start to retire from 2010 to 2020, the share of the population age sixty or older will rise in France from 23 to 27 percent, in Germany from 27 to 29 percent, and in Japan from 30 to 34 percent.[8] The United Kingdom is in a little better shape, with its share of people sixty and older going from 23 to 25 percent, while America's older cohort will rise from 18 to 22 percent.

This rapid aging is a problem, because the older we are, the higher our medical costs, with most people's final illnesses the most expensive of all. In the United States, the average cost of a person's medical services jumps 70 percent from ages 65–69 to ages 75–79, and another 35 percent over age 80.[9] These increases are less dramatic when the government controls health-care prices, but still substantial. The average Japanese person's medical costs rise 62 percent from ages 65–69 to ages 75–79; in Germany, those costs rise 39 percent; and in Britain, 37 percent.[10] And from 2010 to 2020, the share of the population age 75 and over will reach 15 percent in Japan, 12 percent in Germany, and 10 percent in France.[11] Here, too, the demographic outlook is better for Britain and America, with 75-year-olds and older accounting for just 8 percent of all Britons in 2020 and 6.4 percent of Americans.

The costs rise so much because older people are so much more likely to have heart disease, cancers, and other conditions that have become very expensive to treat. More than 80 percent of all Americans who die from coronary disease are age 65 or older, for example. Cancer deaths also are highly concentrated in old people in every advanced country, with more than 70 percent occurring in people age 65 or older and the most common age group for cancer deaths being 75 to 84 years old.[12]

While the aging of the boomers and their parents will play a large part in future health-care costs, the facts are that those costs rose twenty-sixfold over the last thirty-five years, going from $75 billion in 1970 to $2.1 trillion in 2006, increasing 4 to 5 percent a year after inflation even when the boomers entered their generally healthy twenties and thirties.[13] Looking both back and ahead, the most powerful cost driver in health care is innovation, the same force that helps make an economy strong. Expensive new drugs receive most of the political attention, but the largest cost-driver in health care is the proliferation of new procedures such as angioplasties, joint replacements, transplants, arthroscopic surgeries, new medical devices from MRI and CT scanners to implanted defibrillators, and the doctors' time and hospital care required to carry out the procedures and use the equipment. In 2006, while Americans spent $214 billion on prescription drugs, they shelled out $421 billion for physicians and $652 billion for hospital care.[14]

Most of the new procedures and technologies are worth the cost. Heart disease and heart attacks have long been the leading causes of death in advanced countries. But since 1980, their U.S. mortality rate has fallen by nearly half, and medical innovations are the main reason.[15] In the 1970s, when heart attacks killed about 3.5 people per thousand Americans every year, people surviving an initial attack went to cardiac care units where they usually received lidocaine for irregular heartbeats, a short regimen of beta-blockers to lower blood pressure, a "clot buster" drug, and in some cases, artery bypass surgery. By the mid-1980s, angioplasties replaced many of the bypasses, and beta-blockers became a long-term maintenance therapy for high blood pressure. In the 1990s, new drugs came on the market that prevent clots from forming, along with stent and revascularization procedures to keep blood flowing, and implanted defibrillators. The number of heart attack patients who received surgery soared, from about 10 percent in 1984 to more than 50 percent in 1998,[16] and total spending on heart attacks rose 50 percent—but the number of Americans dying from heart attacks fell to 1.9 per thousand.

These improvements and comparable ones for conditions that only recently were considered untreatable—AIDS, renal disease, many cancers, and more—raise costs not only because the new drugs and equipment are expensive and the doctors and nurses who use them have to be trained and paid accordingly; in addition, more people survive their crises and then go on to expensive long-term treatments, more people have to be treated for secondary conditions, such as anemia in dialysis patients, and more people survive to eventually develop other conditions that also require expensive, high-tech treatment.[17]

According to nearly everyone who looks closely at health-care costs, these factors account for most of the real growth in medical spending everywhere.[18] And these costly technological advances and innovations are actually accelerating. In 2005, U.S. drug, biotech, and medical equipment companies spent $65 billion on research, the U.S. government spent another $40 billion, and universities and foundations another $10 billion—all to develop the new treatments, procedures, and equipment of the next decade.

A case study of how new technologies push up medical costs, involving the development of a new procedure for over two million Americans with a condition called atrial fibrillation, or rapid and irregular heartbeats, was recounted in *The New York Times*.[19] A decade ago, the condition usually went untreated; but after scientists a few years ago determined that it can lead to strokes or serious heart deterioration, industry and academic researchers developed a new treatment called catheter-based ablation. Surgeons thread long catheters through a patient's veins and into his or her heart, with tiny devices attached to the end that permanently neutralize the part of the heart muscle that produces the abnormal electrical impulses. The procedure costs $25,000 to $50,000 because it requires very specialized equipment that costs $5,000 and can be used only once, four hours of a cardiac surgeon's time along with his or her team, and the hospital's facilities. The FDA hadn't even approved the procedure as of summer 2007, but the American Heart Association and American College of Cardiology support it, and the surgeons and hospitals that have already done it thousands of times simply bill insurers and Medicare using the codes for other cardiac procedures.

Once the treatment is formally approved, as it almost certainly will be, its use will grow. The hospitals and cardiology clinics that will offer it will buy a lot of high-priced equipment beyond the catheters and their attachments, including high-tech diagnostic, computerized mapping, and robotic guidance systems. Industry experts figure that the equipment alone will generate sales of $1.5 to $5 billion over the next decade; and the surgical personnel, hospital, and associated costs will come to several times as much. If the patients using the procedure avoid strokes and heart deterioration, it will save money over a long term—and so long as it's only used for cases in which an alternative $1,000 drug therapy doesn't work. Yet doctors and hospitals generating large revenues from catheter-based ablations have little incentive to make that distinction, and most patients will follow their advice.

American health care has no central authority that can limit the use of the treatment or set its price, but it does have some indirect, built-in constraints. Lack of insurance effectively denies many treatments to about 15 percent of people. Private insurers and the federal government in its Medicare and Medicaid programs also apply a loose form of price controls by usually paying hospitals, doctors, and pharmacies less than their retail rates. The private insurers that cover 60 percent of Americans also withhold coverage for procedures or drugs they consider "experimental" or outside the norm; and in about 10 percent of cases, they deny coverage to some patients, often for "preexisting conditions." These practices highlight the inequities that emerge inevitably when the pursuit of profits in private health care collides with fast-rising medical prices and strong resistance to higher premiums. But compared to the protocols and prices set by the British, Japanese, German, and French health systems, these practices have had modest effects on prices and access to high-tech medicine for most Americans.

Another reason why U.S. health-care costs are so much higher is that the systems in other countries tacitly shift some of their costs to the United States. While America sticks with largely uncontrolled prices for new medical equipment, drugs, and other treatments, most of them are also used in Europe and Japan, where price controls force the U.S. developers to accept less for each sale. So, for example, the same prescriptions drugs cost between 43 and 50 percent less in Germany, Britain, and France than in the United States.[20] Since the companies producing these products price them to generate an overall return set by financial markets, European and Japanese price controls drive up prices inside the American market.

The other big problem is that some of the most cost-effective technologies are underused and others that are much less cost-effective are overused.[21] Coronary bypass surgery works well for people with heart disease of the left main coronary artery, but not for those with single-vessel disease—although many of them also have the surgery. Drugs that lower cholesterol work well for those with heart disease or multiple risk factors for it, but not for millions of others who also take them. And Dr. Alan Lotvin, a cardiologist who now heads the largest medical continuing education company in the United States, estimates that half of current angioplasties don't have to be done. Further, some new treatments provide only modest benefits at large cost. Five years ago, the two principal drugs for colon cancer cost about $500 per regimen and produced an average survival rate of eight months; current chemotherapies for colon cancer cost $300,000 to $500,000 with a survival rate of thirteen to twenty months. And how doctors and hospitals are paid can affect their use of expensive procedures: Patients with fee-for-service coverage are 50 percent more likely to have electrocardiograms and 40 percent more likely to have chest radiographs, for example, than others like them in all other respects.

The Kaiser Foundation estimates that on the current path, U.S. health-care expenditures will top $4 trillion by 2015, which will work out to $12,200 per American,[22] and $5.7 trillion by 2020. America could follow the rest of the world and cut into these costs with price controls, either in a system that keeps private insurers in business or through a new national health-care system. The unavoidable result will be a little more of what lies ahead for everyone: People's medical care will gradually become less high-tech, at least for those who can't afford to pay a large premium over everyone else. People's initial treatments will become the least-expensive options, insurers or the government will approve fewer expensive new technologies, drugs, and procedures for general use; and their development eventually will slow down.

THE COMING CRUNCH IN FRANCE, GERMANY, AND THE UNITED STATES

Prices are higher in America, but every advanced country will face much the same health-care crisis over the next decade. This may be especially so for France, Germany, and the United States, because their health-care systems are close cousins. To begin with, all three systems give people broad autonomy to manage their own health care. Everyone in France and Germany can choose his or her primary doctor, as can most Americans, with the notable exception of uninsured people who have to accept whomever they get in hospital emergency rooms or charity clinics. Seeing a doctor is also fairly easy, with the French and Germans visiting physicians an average of eight to nine times a year, and Americans an average of five times a year.[23] All three countries also provide most care on a fee-for-service basis, with Americans and French paying and then being reimbursed while German providers are paid directly by the government.

All three systems also combine public and private elements, although in different ways. In France, the government picks up 70 to 80 percent of everyone's costs for most services, drugs, and devices, regardless of age or income, and nearly all French people also carry private insurance for the other 20 to 30 percent.[24] In Germany, the public-private split is based on income: The government covers nearly all health-care costs for everyone earning up to $60,000 a year (including "nature healing," acupuncture, homeopathy, glasses, teeth replacements, health spa visits, and some cosmetic procedures). The top-earning 15 percent of Germans also can take the public coverage, but about two-thirds opt out and go private.[25] Finally, in the United

States, the government finances care for elderly and low-income people, or about one-quarter of the country, while 80 percent of the rest hold private insurance through their jobs. Those left, about 15 percent of Americans, fall somewhere in between—too young and not poor enough to qualify for the government programs, and working in jobs that don't provide private coverage and pay too little to buy it. The essential difference affecting costs, however, is that Germany and France have systems in which the government determines what medical services, products, and procedures are reimbursed, and sets their prices. These regulations amount to not only price controls on most drugs and procedures, but also wage controls on doctors: French physicians earn roughly twice the average income in France, and German doctors even less, while American doctors take home five times the U.S. average income.[26]

All three systems are straining to contain their costs. Angioplasties, transplants, third and fourth rounds of chemotherapy, and thousands of other expensive procedures are a little less widespread in Europe, and cost less in Paris and Berlin than in New York or Miami. But all three systems embrace modern medicine and leave most decisions to doctors and patients, with little to constrain their demand. The result is that the United States devotes the world's biggest share of its economy to medical care, and France and Germany are numbers three and five. Moreover, while American health-care costs have been rising by 8 percent a year since 2000, those costs have been rising 5 percent a year in France and Germany. Take account of the rapid aging facing France and Germany, and, on their present paths, all three systems could see health-care costs double or more between now and 2020.

All three countries almost certainly will respond in much the same way as they have in the last decade, but with greater cumulative impact on the quality of people's care. The first line of defense by the French and German governments, U.S. insurance companies, and America's Medicare and Medicaid programs is to cut reimbursements to hospitals, doctors, and drug companies. That probably works least well in France and in caring for the elderly in America, since in both cases private insurers offer supplementary coverage to make up most of the difference. Other limits to this approach also have become apparent recently. For example, French and German physicians are unionized, and both conducted successful strikes in 2005 and 2006 to raise their fees or salaries. Over time, this strategy also will inevitably degrade the quality of care, or at least slow its progress, because it targets the most expensive services, drugs, and procedures, which are usually the most recently developed and advanced. And if it's meant to produce major savings, it has to be targeted to those with the most serious problems: In France, 10 percent of the people account for three-fourths of health-care spending,[27] and the proportions are comparable everywhere there are advanced medical systems.

The second line of attack on costs is to make it more expensive or more difficult to get treatment. In 2003, Germany ended its long tradition of free care and introduced modest patient co-payments of five to ten euros for each doctor's visit, drug prescription, and day spent in a hospital or rehab facility. The payments are still considerably less than the co-pays of 20 to 30 percent in France and Japan. In France, the higher co-payments are passed on to people's supplementary coverage, where premiums naturally have been rising. Americans pay the highest and fastest-rising out-of-pocket costs: In 2004, they averaged $255 per person for those with private coverage, with the top 10 percent spending more than $1,500, and almost $600 per person including those without coverage.[28] The big limitation of this approach is its effect on people's health, because when co-payments and out-of-pocket costs rise sharply, people get less care. Even so, it doesn't come close to the root causes of the fast-rising costs, since most of those costs come from advanced treatments for people with serious illnesses, and they're not deterred by co-payments.

The third response to these rising costs is to pay them by raising taxes or increasing insurance premiums. In Germany, Prime Minister Angela Merkel announced plans in 2007 to raise "contributions" to the government system by half a percentage point, to 7.1 percent of a worker's salary and 8 percent of a company's payroll costs, or a total of some 15 percent of most working families' incomes. There are also plans in Germany to raise taxes on wages and pension income to limit looming deficits for long-term care. In France, the government replaced the 6.8 percent payroll tax for employees with a 5.25 percent tax on all income—the Social Democrats have called for the same shift for Germany—plus the 11 percent employers' tax on payrolls.[29] In the United States, the big increases have come in the private insurance premiums paid through work, which rose 87 percent from 2000 to 2006, compared to overall inflation of 18 percent.[30] In 2006, those premiums averaged $11,500 to cover a family of four, including $3,000 paid directly by workers.[31]

This approach, like the others, ultimately works only on the margins. As every major country watches its population age rapidly over the next ten years, and the costs of the treatments they will need increase even faster, they won't be able to raise taxes or premiums enough to cover them. This tact will probably reach its limit even sooner than cuts in reimbursement and increases in co-payments, because taxes and premiums already are so high. With overall tax burdens of 45 to 50 percent across most of Europe, how much higher taxes will European workers be willing to pay to finance primarily the rising medical bills of retired people? And with insurance premiums paid through the workplace averaging $11,500 for an American family of four, how much higher can they go?

There are still a few comparatively easy ways left to cut costs in every system. The Germans almost certainly will require that people who want to see a specialist get a referral from their primary doctor, as everyone else already does. The French, who use about twice as many pharmaceuticals as Americans or most other Europeans, will take steps to use more generics. And the U.S. government could force doctors, hospitals, and insurers to cut their administrative costs by adopting standardized billing and reimbursement procedures, perhaps when it finally takes steps to cover uninsured people.[32] Even so, these changes and everything else will still leave growing chasms of costs. Regardless of what else happens, by 2020 most people's access to the most expensive and technologically advanced treatments and procedures of the time will be reduced substantially. And over time, those reductions will slow the pace of additional medical innovation.

THE HEALTH-CARE CRISIS IN JAPAN AND THE UNITED KINGDOM

Japan and Britain already have arrived where Germany, France, and the United States are headed. The ability of Japan and Britain to keep their national health-care spending to about 8 percent of (2005) GDP—about half the share that America spends, and three-quarters the share in France and Germany[33]—seems remarkable. What sets them apart and has made such a large difference, simply stated, are much stricter government controls on the prices of everything in health care.

In Japan, doctors' fees, the salaries of nurses and other health-care personnel, hospital charges, prices for drugs, medical equipment and medical supplies, are all determined, after some negotiation, by the Central Social Medical Care Council of the Ministry of Health, Labor, and Welfare. This council has set these prices at such low levels that, from 1980 to 2000, Japanese health-care costs actually grew more slowly than overall consumer prices.[34] One reason is that when demand for a health-care service goes up in Japan, the government keeps the lid on spending by cutting the reimbursement. When the use of MRI scans jumped in 2000 and 2001, for example, the Japanese government cut the fees it paid to doctors and hospitals from $150 to a little over $100—the same scan costs from $650 to $1,200 in America, and not much less in Germany and France—and Japanese doctors and hospitals started using them less.[35]

Japan's strict price and wage controls in health care inevitably cut into the supplies of doctors and the most expensive services that require the most expensive equipment. There are substantially fewer doctors per thousand people in Japan and Britain than in America, Germany, or France, because they earn so little. Since anyone can see any doctor or specialist in Japan without a referral or paying anything beyond a standard co-payment, Japanese also see their doctors very frequently—although it's often said that visits to Japanese doctors consist of "three-hour waits and three-minute contacts."[36] Japan also uses very costly high-tech medicine less broadly, with less than half as many ICU and coronary care beds as the United States per thousand people, and roughly one-third the surgeries.[37] Instead, drug therapies are more common—drugs account for 20 percent of all health-care spending or nearly twice as much as almost anywhere else except France—because they're cheaper than surgery and Japanese doctors can sell drugs at a markup. So the catheter-based ablations described earlier will never be used as widely in Japan as in the United States.

While Japanese controls effectively ration high-tech medical care where they can, Britain's price controls produce shortages, especially of doctors and their time, and bottlenecks for hospital procedures. Overall, Britain has one of the world's simplest government-run medical systems. The British National Health Service (NHS) owns the hospitals, employs the doctors, sets the fees for all inpatient and outpatient services, and then pays the costs without co-payments, except for modest charges on about 15 percent of prescriptions and larger fees for dentistry and optical services.[38] Britain's low fees, combined with free access to personnel and facilities, have led to the system's now-famously long waits and quality problems. The shortages in British health care begin with the lowest number of doctors per thousand people of any developed country.[39] The NHS is such an unattractive prospect for budding doctors that it has to fly in MDs from other European countries to be on call on weekends,[40] and the British Medical Association reports that two-thirds of the doctors who joined the health service in 2004 were trained outside Britain, mainly in Africa. The longest waits, averaging two months or more, involve getting to see specialists and, if necessary, receiving hospital treatment.[41] After a major push and large funding increases by then–Prime Minister Tony Blair, the numbers of people waiting more than six months for hospital procedures fell—by how much is a matter of dispute—but the push had little effect on the large numbers who at any time wait for up to five months. Unsurprisingly, those who can afford to opt out of this system generally do so, with 12 percent of Britons now purchasing private insurance and access to doctors practicing outside the NHS and to hospitals not owned by the NHS.[42]

One reason they're opting out is public reports and private studies showing how the low levels of support have affected quality. The National Audit Office reported that Britain has the worst record in Europe for hospital-acquired infections, with 9 percent infected at any time and causing five thousand unnecessary deaths a year. Similarly, the OECD found that fatalities in the first seven days from stroke are twice the average in Britain as other OECD countries. And much like Japan, British doctors and hospitals perform heart procedures at one-third to one-half the rates of France and Germany, despite having the world's second-highest incidence of heart disease.[43]

So many factors affect the health of a nation's population that it's possible to characterize a health-care system in almost any way. Against those data, for example, are the reports that Britain's neonatal mortality rate is lower than Germany's, and its maternal mortality rate is lower than France's—but its death rate from communicable diseases is higher than other advanced countries.[44] Similarly, a recent report from a conservative think tank in London announced that almost 31,000 Britons die every year because the NHS doesn't match the average standard of care in EU countries for heart disease, respiratory disease, and cancer.[45] But the report's data show that British death rates from two of those causes, heart disease and cancer, are well below the EU average.

Whatever the verdict on British public medicine, widespread public dissatisfaction has forced London to double government spending on health care from 2000 to 2008. As Britain's boomers begin to retire, the U.K. will experience the same crunch facing Germany and France: In 2003, the average cost of medical care for a Briton under age sixty-five was 440 pounds a year; over sixty-five, it jumped to nearly 1,000 pounds a year, and much higher than that after age seventy-five or eighty-five.[46] Britain's baby-boom bulge is smaller than many other advanced countries, but the number of elderly Britons will still increase nearly 23 percent from 2005 to 2020.[47]

The British system is financed by substantial payroll taxes, plus additional funds from the income tax and the 17.5 percent VAT. By the British government's estimates, keeping pace will require increasing funding between now and 2020 by about two-thirds.[48] That's a daunting prospect politically. Raising Britain's VAT by ten percentage points, for example, would take care of just one-quarter of the projected shortfall. But when the president of the Royal College of Surgeons, Bernie Ribeiro, recently urged the introduction of the substantial co-payments seen everywhere else, the head of the largest health workers union, Unison, denounced it. Nevertheless, over the next ten to fifteen years, it's virtually inevitable that British health care will not only introduce co-payments, but also higher taxes and less use of expensive new medical technologies, whatever the consequences for quality.

Japan's looming problem is that its low-tech cost containment strategy is colliding with its demographics. Japan has been aging faster than anywhere else for over a generation; and by 2020, more than one-third of Japanese will be age sixty or older.[49] The aging of the boomers is an issue everywhere, but nowhere else will it affect the health-care system as much. Japan spends five times as much per person on care for elderly people as for younger Japanese,[50] because its low-tech medicine has fewer applications for the conditions that affect older people, including the heart disease and cancers that cause two-thirds of all deaths in Japan.[51]

Over the next ten to fifteen years this problem will become even more serious. A recent analysis by Harvard health economist David Cutler found that despite Japan's tight hold on prices, the rising costs of the overall system can be traced equally to its aging boomers and technological changes.[52] Moreover, much of the current development of expensive new equipment, treatments, and drugs is focused on heart disease and cancer. By 2020, according to the best estimate, technology and aging will increase health-care spending in Japan by some 75 percent or more—despite the strict controls—and expand the share of Japan's GDP devoted to health care by four percentage points or 50 percent.[53]

Japan finances its health care the same way almost everyone else does—by taxes and co-payments. In the last fifteen years, the government has already doubled the wage-based taxes paid by workers, which account for 70 percent of health-care revenues. It also raised co-payments for the nonelderly from 10 to 30 percent in two steps, with no apparent effect on demand for medical services.[54] Then in 2000, Tokyo compounded its long-term problem by creating universal coverage for long-term care, in a benighted attempt to save money in the short term by shifting people from hospitals to nursing homes. (It didn't work.)[55] So in 2006, the government also raised co-payments by the elderly to as high as 30 percent, based on age and income, introduced a new 20 percent co-pay for the medical costs of young children, and cut fees again to hospitals, clinics, doctors, and pharmacies.

Japan will have very few options left as its health-care costs rise sharply over the next decade. Some Japanese experts estimate that the costs will require another doubling of the taxes paid by working people.[56] The actual crunch could be even worse if Japan fails to open up its economy in ways that ultimately can boost its growth and tax base for its health care. All of these forces make a major political battle in Japan over health care almost certain. It's safe to assume, even for Japan, that no elected government can follow up the steep tax increases of recent years with another 50 percent hike in the payroll tax, topped off by substantial increases in the income, VAT, and municipal taxes that also help fund health care. Nor can co-payments go much higher without cutting into universal coverage. And the draconian price controls that would be necessary to close the gap will cut off Japan from most medical advances.

In Japan, as almost everywhere else, the most likely developments over the next ten to fifteen years will be new medical protocols dictating lower-tech medicine, even when a person's life lies in the balance, as well as higher taxes, higher co-payments, and lower fees for higher-tech treatments. Wealthy people in Japan and everywhere else will be able to buy personal access to the latest technologies and biologic drugs. For most other people, the long trend of improvements in health care will likely slow and could even stop.

CHINA'S VERSION OF THE HEALTH-CARE CRISIS

China will face its own, unique health-care crisis in the next decade. As the only society in history to unravel a system of basic, universal coverage, China now faces the prospect of having to build a modern medical system for more than 1 billion people in a short time—and divert enormous resources from its core modernization program to do it.

In the 1970s, virtually all rural Chinese had access to very basic medical care through a vast network of clinics operated by the agricultural communes, while almost everyone else worked for the government or state-owned enterprises that provided coverage at city and town hospitals. With the end of the commune system in the 1980s, most of the rural clinics closed down and free access ended; and as most state-owned enterprises have phased out in the last fifteen years, coverage for their employees has disappeared, too, including for millions of retirees.[57] What's left is a system that provides coverage for about 20 percent of Chinese—the state's soldiers and civil servants, and those working for the remaining state-owned enterprises or foreign-owned companies.

With almost no primary care available through independent doctors, hospitals and government clinics serve all of China's medical needs,[58] at least for the fraction that are insured or can pay on their own and who also live in a large city with modern hospitals. Outside the cities, according to Dr. Michael Moreton, an American obstetrician who practiced at China's only privately owned western-style hospital, Beijing United Family Hospital, "hospitals have little in the way of modern equipment, or even modern plumbing."[59] And in China's villages, the World Bank reports, 70 percent of "doctors" have no more than a high school degree plus an average of twenty months medical training.[60]

For the nearly 80 percent of Chinese without insurance or the private means to pay, doctors won't see them and hospitals won't admit them, regardless of how sick or injured they are. A young American of my acquaintance, teaching with her husband in Tsingtao, had a baby in 2007, prepaid, and when she asked for an epidermal injection, her husband had to run to their apartment to fetch 500 remembi before her obstetrician would provide it. *The New York Times* recounted a more serious but also typical instance of pay-before-care in early 2006. Jin Guilian, thirty-six, had worked for ten years as an orderly at a hospital that refused to treat him when he developed a heart condition and arm infection, because he couldn't pay.[61] His family traveled five hundred miles to take him home to Fuyang, where they could afford to cover four days in the local hospital at $15 per day. From there, Jin went to an unheated clinic "where stray dogs wandered the grimy, unlighted halls," and then home to die of conditions that could have been treated easily and successfully where he had worked.

The policy of "pay up front or take your chances" extends even to vaccinations for children and communicable diseases. Measles and tuberculosis, for example, are still epidemic across China. The World Health Organization estimates that 30 million Chinese will develop TB in this decade,[62] and that every year, 250,000 Chinese will die from it, along with comparable fatalities from hepatitis.[63] The government's abdication of health care is also evident in World Bank data showing that in China, people's out-of-pocket costs now account for nearly two-thirds of all health-care spending.[64] Even among the 20 percent with insurance, one-third of hospital costs are picked up by patients and their families.[65] And the government's own estimates are that half of all Chinese who are ill or injured go untreated because they cannot afford it.

These prospects also are a principal reason why Chinese save some 30 percent of their incomes—although most can never save enough to cover care for a serious condition. In 2003, the average hospital stay cost the equivalent of 43 percent of an average family's yearly income.[66] This problem is especially acute among elderly Chinese, 90 percent of whom are uninsured and, according to a recent study from the Elderly Health and Family Research Center at Beijing University, generally live in poverty.

The social and political problem for China's leaders is how long they can justify what the British medical journal *The Lancet* calls a "pay or die" policy, when hundreds of millions of Chinese can remember free access to care and know that their country has grown much more prosperous. In 2002, the Politburo promised to rebuild universal access by 2010 and began a series of pilot programs in fifty cities and two thousand rural counties. So far, however, all have met with little success. The urban programs are built around private, tax-subsidized individual health accounts, but few people without coverage can afford them. The program for rural areas suffers from the same problem. It's a voluntary system to cover serious illnesses, and charges individuals ten remembi a month to join, matched by ten each from the local and central governments. Since it's voluntary, the program cannot avoid adverse selection, where those most likely to need serious treatment are most likely to join. That's a serious issue, because the three payments combined raise only about $4 per person per month.[67] Relatively few Chinese have joined, because such meager funding still requires large out-of-pocket expenses for serious illnesses and no coverage at all for lesser problems.

Providing meaningful coverage to more than 1 billion people over the next decade would cut deeply into the Chinese government's modernization programs, because it will involve more than the costs of treatment. To provide that treatment, China also will have to build, equip, and staff thousands of new clinics and hospitals across the country—and at the same time face the same forces pushing up health-care costs everywhere else. From 2005 to 2020, the number of Chinese age sixty or over will go from 144 million to 244 million, as their share of the population rises from 11 to 17 percent.[68] The additional burden of caring for people of the age when the most serious and expensive problems usually strike—heart disease and cancer account for more than half of deaths in China[69]—will involve not just the 100 million new elderly, but 230 million or so, since only 10 percent of the current elderly have coverage. The costs of treatment for younger Chinese who also have no coverage will be less per person, but providing even basic care for another billion people will strain the government's finances and depress domestic investment in other areas.

Chinese health care is low-tech next to what's available in America and Europe; but it is not impervious to pressures to expand the most costly services. Chinese hospitals and rural health centers operate under what's called a "Management Responsibility System" that provides fixed payments to hospitals and clinics plus the right to charge patients, with or without insurance, additional fees on a fee-for-service basis at rates set by a central price commission. As everywhere else, most of the services recommended are the higher-tech ones with the highest fees, especially since China's poorly paid doctors earn bonuses based on the revenues they generate. The same rules apply to prescription drugs—the government sets the prices that hospitals and doctors pay, but they can charge patients more for them. The ironic result is that even with 80 percent of China's people having almost no access to serious medical care, much of the care and medications that the rest receive are unnecessary.[70]

China's leaders are certain to miss their 2010 deadline and unlikely to achieve broad coverage by 2020. But the current arrangements are also untenable. In late 2006, two thousand people in southwestern China ransacked a hospital over high medical fees and the shoddy care provided a three-year-old boy who died from accidental poisoning,[71] and *The Lancet* reports that numerous protests of significant size have occurred over refusals of care.[72] A serious effort to extend coverage and care also would pay large long-term economic dividends. But unless social unrest forces the leadership's hand, real progress in this area will continue to play second fiddle to the investments for industrialization.

THE WORLD'S ENERGY PROBLEMS START BUT DON'T END WITH SUPPLY AND DEMAND

Much like health care, how much energy prices rise or fall over the next decade, and the resulting effects on national economies, could depend as much on politics as on simple supply and demand. Globalization will continue to drive the demand for energy at faster rates than the supplies available at current costs, as it has for the last five years. And while the world has ample sources of energy for any conceivable level of demand over the next decade, tapping many of those potential sources—tar sands and oil shale, solar power, biomass, deep-sea natural gas—will be much more expensive than most people are used to. How these energy price increases will affect countries, companies, and households, however, will depend on whether they happen gradually, as they usually do when markets determine prices, or abruptly, as often happens when political developments drive up the price.

When prices rise gradually, people can find ways to use a little less energy, and companies can absorb or pass along their cost increases in increments that most shareholders and consumers can tolerate. So, as oil prices gradually doubled over the last three years, most households and companies successfully adjusted their other spending and investments, little by little. At a minimum, the next ten years will see a good deal more of this belt tightening by American, European, and Japanese businesses and households. Not only will globalization force everyone to tap into increasingly expensive sources of energy, but responses to climate change will almost certainly further push up prices. But it may not only happen gradually. The Middle East continues to be a war zone, and the governments of pivotal producers such as Saudi Arabia, Iran, Russia, Nigeria, and Venezuela all remain politically fragile. Price shocks from sudden disruptions in oil flows, therefore, are also significant possibilities for the next decade. The good news is that periods of energy price declines also are likely, as higher prices bring online new energy sources and more energy-efficient technologies.

Worldwide energy demand will increase by more than one-third between now and 2020, according to the most recent U.S. government analysis. Most of that increase will come from developing countries, with an 80 percent jump predicted for China, India, and the rest of developing Asia.[73] While energy demand in Europe, Japan, and the United States is expected to rise just 14 percent, everyone will pay the higher prices driven by Asian fast-rising energy use. And by 2020, all developing countries will consume half of all worldwide oil production, 55 percent of all natural gas, and nearly 70 percent of coal.[74]

The huge increases in energy demand coming from developing countries reflect not only their rapid economic growth, but also how energy-intensive their economies are. China, India, and almost all other developing economies focus almost entirely on manufacturing and agriculture rather than services, and their manufacturing and farming operations are usually both relatively energy-inefficient and concentrated in sectors that require particularly large amounts of energy, such as smelting, cement, iron and steel, meat and dairy products, and brewing.[75] There are big differences in the "energy intensity" of the economies of the advanced countries, too—for example, Ireland uses one-fourth the energy to produce one euro of GDP as France or Germany, because so much of Irish GDP comes from IT and financial services[76]—but nothing approaching the energy intensity and inefficiency of China, India, or even Argentina.

No one knows what all this new energy demand will mean precisely for oil prices one year from now or in ten years, and one of the reasons is that most oil-producing nations don't know how large their own reserves are (or if they do, they won't say). Kuwait, for example, claims oil reserves of 99 billion barrels while *Petroleum Intelligence Weekly* estimates them at half that level. Angola has leaned in the other direction, claiming 5.4 billion barrels of reserves both before and after four large, new oil fields were discovered; and without explanation, Mexico first reduced its claimed reserves from 27 billion to 13 billion barrels in 2002 and then upped the number to 16 billion barrels the next year. Even the data on Saudi Arabian reserves are unreliable, as the government there has hinted that the country's reserves might be twice the 260 billion barrels it has long claimed. On balance, however, the world's confirmed oil reserves jumped 60 percent over the last 20 years and known natural gas reserves doubled.

Whatever worldwide reserves actually are, over the next ten years, the world will meet its new energy demand at prices likely to reach $100 per barrel or more for oil. One key reason is that investments to develop new fields and new areas of existing fields slowed sharply over the last ten years, in important part because low oil prices from the mid-1980s to the late 1990s convinced OPEC that oversupplies are the norm. Investment is also down in Russia, which accounted for two-fifths of all oil production increases from 2000 to 2005, following President Putin's decisions to seize the Yukos energy conglomerate and further restrict western investment in Russian oil operations. The other reason for low investment in the face of rising demand is that the world's ten largest oil companies are all monopolies owned by the governments of, in order, Saudi Arabia, Iran, Russia, Iraq, Qatar, Kuwait, Venezuela, the United Arab Emirates, Nigeria, and Algeria.[77] (The biggest private oil company, ExxonMobil, ranks number fourteen.) The energy group at McKinsey and Company estimates that in order to meet the rising demand of the next two decades from conventional sources, capital investments in Middle Eastern and African oil production would have to jump from their current levels of $8 billion a year to some $45 billion per year.[78] But the state-owned oil companies that control 90 percent of worldwide reserves and 80 percent of global production continue to resist either making those investments themselves or allowing foreign private companies to fill the gap.

As energy prices march up, the effects will be felt most in countries where the economies are most energy-intensive and the demand for energy is growing fastest—which means China and India, Brazil and Argentina, and parts of Central and Eastern Europe. Scandinavia, Germany, France, and the rest of Western Europe will be affected least, because their high energy taxes and dense populations have made them very energy-efficient places. The effects on the United States and Britain will fall somewhere in between, because they're less energy-efficient than most of Europe, and in the American case, tied closely to the major developing economies.

The last decade of energy underinvestment also has produced a public debate about "peak oil," or the point in time when worldwide oil production will top out and begin to decline, putting even more upward pressures on prices. Energy scholar Daniel Yergin sensibly calls this discussion one between "geological pessimists" and "technological optimists."[79] The pessimists focus on oil that can be tapped in conventional, inexpensive ways, while the optimists point to the huge amounts of "unconventional" oil locked in tar sands, shale, and below deep ocean shelves. The Athabasca region in Alberta, Canada, for example, has 54,000 square miles of oil sand deposits rich in bitumen, a tarlike substance that can be mined and processed into oil.[80] It costs about $40 to mine and upgrade enough bitumen to produce a barrel of oil, compared to a worldwide average for conventional crude of $12 to $15 per barrel. Based on extracting one barrel of oil from each two tons of sand, these deposits could give Canada the world's second-largest reserves. At a minimum, the Canadian tar sands could produce 4 million barrels a day by 2020, or the equivalent of one-third of current Saudi production—at three to four times the cost of conventional oil. Many other countries hold large quantities of oil shale, which are rocks that when heated release forms of refinable petroleum liquid called kerogens. Geologists estimate that the oil shale deposits in Wyoming, Colorado, and Utah hold the equivalent of 800 billion barrels, or three times Saudi Arabia's proven conventional reserves; and companies in Estonia, Brazil, and China already produce oil from shale on a modest scale, for a little less than oil from tar sands. As global oil demand continues to rise faster than production of conventional crude oil over the next decade, the world will extract more oil, at higher costs, from both tar sands and shale.

The world also has huge natural gas reserves, both conventional and unconventional, to help meet energy demand. Most of the cost issue is tied to the expense of moving the gas from the countries that produce it to those that consume it; and over the next decade, governments and companies will spend tens of billions of dollars building new natural gas pipelines and special tankers for liquefied natural gas. As a matter of economics, such large investments will still be less expensive than making the use of more coal environmentally acceptable in Europe or America, by developing and applying new chemical processes to capture and sequester the carbon, or building new nuclear power plants and disposing of their wastes safely. At the price points where those options become viable economically, renewable energy from the sun, wind, biomass plants such as switchgrass and corn, and geothermal sources, all could be developed on significant scales.

The next decade will prove both sides of the "peak oil" debate right. Demand driven by globalization will drive oil prices past $100 per barrel; and at those prices, tar sand and shale oil will become competitive, as will new tanker fleets for liquefied natural gas. According to U.S. government estimates, at those prices, western businesses and people also will shift more of their energy use to renewables, as well as natural gas.[81] The higher prices also will bring into play more energy-efficient technologies and energy-saving lifestyle changes that can also ease the crunch. When oil prices soared in 1977 and stayed high until the mid-1980s, American companies, households, and drivers altered their patterns and habits enough to cut the U.S. economy's overall oil intensity by more than 5 percent a year. If renewables come online on a major scale this time, those new uses also will likely hold. Amory Lovins, a leading exponent of non-oil-based energy, notes that it took less than a generation for automobiles to displace carriages and for diesel to displace steam in locomotives—and nobody ever went back.[82] Even the U.S. government forecast sees renewables accounting for almost 10 percent of American energy use by 2020.

These various shifts also will produce a bit of a roller coaster in energy prices, going way up as demand meets the limits of conventional production, and then down as new sources come online, high prices cut into demand, and the costs of new energy technologies decline.[83]

OIL SHOCKS AND ENERGY SECURITY

The energy price increases that worry most people, or ought to, are those that happen suddenly, because they force everyone to revise their plans all at once and scale back sharply. Sudden price hikes that last for at least six to twelve months also push up other prices, reducing a country's investments and over-all growth. With a growing share of the world's energy coming from coun-tries with fragile or unstable governments, and some evidence that storm systems are worsening, a disruptive price spike triggered by political or natu-ral events is likely in the next decade. Even if the regimes in the major oil ex-porting countries all manage to hold on, most OPEC members are located in the midst of or nearby the world's main war zones and terrorism locales. Large production and shipping operations are especially vulnerable to attacks, and particularly in Saudi Arabia, with the world's largest oil fields at Ghawar, largest processing complex at Ras Tanura, and largest tanker facilities at Sea Island. Nearly half of all Saudi oil goes through Ras Tanura and the two huge terminal complexes at Sea Island; and by one calculation, an attack on those facilities could choke global oil supplies for two years.[84]

More generally, the International Energy Agency calculates that 80 per-cent of Middle East oil exports are shipped through three channels, "all of which are predisposed to accidents, piracy, terrorist attacks or war"—the Straits of Hormuz between Iran and Oman, the Bab el-Mandab connecting the Gulf of Aden and the Red Sea, and the Suez Canal and Sumed Pipeline connecting the Red Sea and the Mediterranean.[85] Should violence disrupt the production or shipments of a major OPEC producer for a sustained period in the next decade, prices would shoot up fast to economically damaging levels in almost every country in the world.

The most serious scenarios involve the world's two dominant oil exporters, Saudi Arabia at about 10 million barrels per day and Russia at 6 to 7 million barrels. Everyone else is far behind—Norway and Iran export a bit under 3 million barrels a day, and Nigeria, the United Arab Emirates, Kuwait, and Venezuela each ship a little over 2 million barrels a day. A catastrophic event that sharply cut back exports from the Saudis or the Russians, or a political decision by either with the same effect, would send worldwide oil prices soaring. Of the two, Saudi Arabia is more likely to suffer such an event, while Russia is more likely to take action itself.

For at least two decades, Saudi Arabia has used its oil power mainly to manage OPEC and police its members. It has shaped and led OPEC's policy of allowing world oil prices to gradually double since 2004, by reducing OPEC production quotas whenever oil inventories rise in the major consuming nations. And the Saudis don't tolerate other members deviating from their quotas. In 1998, when Venezuela pushed its exports well past its quota, Riyadh ramped up its own production and let the price fall until Caracas finally gave in at $9 per barrel. (The episode discredited then-President Rafael Caldera Rodríguez and his prowestern party, and helped elect the anti-American firebrand, Hugo Chavez.) For their own security, the Saudis are unlikely to engineer a sudden upward price shock in the next decade, especially one that would injure the United States. Al Qaeda and its associates and imitators are firmly entrenched on the Saudis' northern border in Iraq, and a devoutly Shia and militant Iran is less than one hundred miles away across the Persian Gulf. For the foreseeable future, the Saudi regime, along with the governments of the United Arab Emirates, Kuwait, and Qatar, will continue to rely on U.S. military protection.

Long after the United States leaves Iraq to its own fate, America will be actively engaged in doing what it can to ensure stable oil flows—which will include protecting Saudi Arabia and the regime of its ruling family. America's commitment to the Middle East is not about its own access to oil. The U.S. imports almost 60 percent of its oil, but 80 percent of that now comes from outside the Persian Gulf and only 13 percent comes from Saudi Arabia.[86] Japan has much more at stake directly in the Persian Gulf, with almost 90 percent of its oil imports coming from the Middle East. So does China. It currently imports much less oil overall than America, but almost 60 percent of it comes from the Persian Gulf, and by 2015 the Gulf region will account for 70 to 80 percent of Chinese oil imports.

If the United States didn't import one more barrel of oil from the Gulf, its economic and geopolitical interests there would still be compelling. A supply shock originating in the Middle East will directly affect the price of U.S. oil imports from Mexico, Venezuela, Canada, and elsewhere. Moreover, globalization vastly increases America's economic stake in the Gulf's stability, since the United States is the largest participant and owner in the global economy, with annual trade flows of nearly $4 trillion a year and foreign investments of some $13 trillion.[87] No U.S. administration will ignore the prospect of an oil shock rocking the Chinese and Japanese economies—with incalculable political effects in China—and beaching the global economy. As the only country with the capacity to provide a measure of global security, the United States also will act—hopefully more wisely and effectively than in Iraq—if political upheavals in the Gulf or African oil states undermine the stability of those governments. For the next decade and beyond, the American military also will continue to provide worldwide maritime security for the 40 million barrels of oil that cross the world's oceans on tankers every day. By 2020, daily ocean oil shipments will reach 67 million barrels, plus millions of tons of liquefied natural gas, and the U.S. Navy will have the only force capable of maintaining the security of the global sea lanes.

Europe, Japan, and the United States are much better prepared for an oil shock sometime in the next decade than they were in the early 1970s. Following the 1973 Arab oil embargo, together they created the International Energy Agency, headquartered in Paris, to monitor worldwide energy markets. They also established national stockpiles of oil that were used just before the first Gulf War and following Hurricane Katrina, pushed national conservation efforts, and prepared plans to share their supplies during future disruptions. They agreed to take steps to diversify their own energy imports; and to deal with potential domestic disruptions, they expanded refining and storage capacities, extended their distribution systems, and stockpiled equipment. One reason that Katrina didn't tank the U.S. economy is that America now has more than 100 refineries, 2,000 offshore platforms, 160,000 miles of oil pipelines, facilities to handle 15 million barrels per day of imports and exports, 410 underground gas storage fields, and 1.4 million miles of natural gas pipelines.[88]

For all these steps and preparation, a serious disruption in the next decade will send oil prices past $100 to perhaps $150 per barrel, and then panic will likely tank the world's stock and bond markets. If the disruption persists for even a few months, it will drive the world into a serious recession, and in a period when many countries will be especially vulnerable. In Europe's major economies, a serious recession will depress revenues but not health-care and welfare costs, putting significant pressures on the unity of the European Union. And if the recession's severity varies significantly from EU member to member—and it most likely will—it could strain the allegiance of Southern Europe to the euro, including that of Italy and Spain. A serious recession in Japan, after a decade of generally very slow growth, could finally end the LDP's enfeebled political hold on power and break open Japanese politics. The United States will be more exposed to an initial oil shock, since it is less energy efficient than most of Europe or Japan; but its more market-based economy will adjust more quickly and efficiently to the new conditions, reducing the downturn's impact.

The big question is the effect of an oil shock and global recession on China, and how it will respond. More than any other major country, China's growth depends profoundly on its exports and investment, and a global recession will hit both hard. A sharp slowdown in growth also will force Beijing either to suspend its program to unwind state-owned enterprises, at least for a while, or allow unemployment to rise sharply. The most serious issue for China's leaders—and for the rest of the world—is whether a serious downturn in China will undermine the leadership's strategy of maintaining its monopoly on political power by providing continuous economic progress. If large-scale unrest were to break out, Beijing could be forced to crack down again, as it did in 1989, or finally introduce the first elements of modern political liberty.

THE PETROPOLITICS OF CHINA AND RUSSIA

It is sometimes claimed that the changes in energy markets are creating a new international "petropolitics" that will have far-reaching effects.[89] While the implications are often exaggerated, energy issues will pose serious challenges for American policy in the next decade and could enhance the geopolitical clout of China and Russia.

Beijing's approach to the Middle East and international energy politics has been avowedly nonideological, even opportunistic, since its first moves towards economic reform in the late 1970s.[90] In the 1980s and 1990s, China sold arms to both sides of the Iran-Iraq War, collaborated with Israel on the F-10 fighter while selling intermediate-range missiles to the Saudis, and considered missile sales to Syria and Libya before backing off under U.S. pressure. In recent years, China blocked UN action against Iran, its energy partner, over its nuclear program in 2006, and then agreed to milder sanctions in 2007. The constant element has been China's determination to secure reliable access to the enormous quantities of oil and natural gas it requires for its modernization program. From 1990 to 2002, Chinese demand for oil soared 90 percent while domestic production grew 15 percent, and that demand has accelerated even more in the last five years.[91] To meet these energy needs, China has spent fifteen years courting OPEC members, starting with Oman and Yemen in the early 1990s and then moving on next to Kuwait, the United Arab Emirates, Algeria, Egypt, Libya, and Sudan.

By the turn of the century, Beijing focused its energy attention on Saudi Arabia, Iraq, and Iran, and became a world player in energy. In 1999, China formed a "strategic oil partnership" with Riyadh, with the Saudis selling crude and natural gas and China opening its refining sector to Aramco, the Saudi state oil giant. Today, Saudi Arabia is China's leading oil supplier and the leading investor in Chinese refineries. Furthermore, in 2002, China adopted a new "going out" policy of encouraging its three, huge state-owned energy companies—China National Petroleum Corp., China National Petrochemicals Corp., and the China National Offshore Oil Corp.—to purchase ownership shares in oil projects around the world. It also directed other state-owned enterprises to expand their exports to and investments in countries with state-owned oil companies. (India recently adopted the same approach.) At the same time, Beijing signed a twenty-two-year production-sharing arrangement with Saddam Hussein to develop the country's second-largest oil field, once UN sanctions were lifted. Its current status is still uncertain.

China also has invested tens of billions of dollars in Iranian oil operations over the last decade, much of it in "buy-back" deals in which China pays to develop the field and then claims a share of its production. In the next decade, these ties are likely to expand and deepen. Iran will need even more foreign investment, technology, and expertise, because its current production relies heavily on a small number of large, technologically outdated oil fields. And if Iran's nuclear program continues to isolate it from western companies, it almost certainly will turn to China's state oil companies, which themselves have gained access to the latest technologies and methods through their joint ventures with western companies. Beyond Saudi Arabia and Iran, China also has energy projects in virtually every oil-producing country in the world— Algeria, Angola, Azerbaijan, Canada, Chad, Ecuador, Egypt, Kazakhstan, Indonesia, Mongolia, Myanmar, Niger, Nigeria, New Guinea, Peru, Russia, Sudan, Syria, Thailand, Venezuela, and Vietnam, and in some cases in joint ventures with Italian, Indian, Australian, and Saudi companies. And the direct economic import of all this Chinese energy activity has been positive for the United States, Europe, and everybody else, since it helps expand global oil supplies.

All of these projects and approaches are part of a political agenda formed at the highest levels of the Chinese government. The three major state-owned energy companies that comprise nearly the entire industry in China are run by executives named by the Central Committee of the Communist Party, their strategies are set by the National Development and Reform Commission, and the Standing Committee of the Politburo has the final say on all major decisions.[92] At least thus far, those strategies and decisions are driven primarily by the energy imperatives of rapid industrialization, not geopolitical ambitions. Since 2003, China has been the source of 40 percent of all the growth in worldwide oil use; and its energy demands will increase even faster over the next decade as the program to modernize agriculture takes hold. China also is in the early stages of exploding automobile ownership, which will further increase its energy needs. In 2005, there were about 23 million cars on China's roadways, roughly as many as Great Britain, owned by 5 percent of China's population. By 2020, the number of cars in China is expected to reach 90 million, and then 130 million by 2030, or roughly the number on American roads in the mid-1970s.[93] These needs, on top of industrialization, are expected to expand Chinese oil imports to 7 million barrels a day by 2020 and 10 million barrels a day by 2030, or as much as the United States imported in 2006.

While geopolitics has not set China's course in world energy markets, the ties it is forming will have geopolitical effects. By 2020, the Saudis will sell more oil to China than to the United States; and while the Saudi government will almost certainly still be dependent on the United States, it could call on Beijing's support when it finds itself disagreeing with Washington. China's economic ties to Iran also will have political implications. Beyond oil, Chinese state companies have set up automobile and television plants in Iran, they're building an extension of the Tehran subway system and a national broadband network, and in a number of areas Chinese companies stepped in when European business pulled back under threats of U.S. sanctions in 2006 and 2007. At a minimum, Iran will turn to China for military purchases, a potentially serious matter as America recasts its position in Iraq.

China's deteriorating relations with Japan also are bound up in energy. The most serious conflict so far unfolded when Japan granted a Japanese company, Teikoku Oil, rights to explore a vast undersea natural gas field lying in the East China Sea between China's central coast and Japan's Ryukyu island chain. The sea between the two countries is 360 miles wide, and Japan offered to set the boundary midway—which would give Teikoku Oil access to the field. China insists that its territory extends to the edge of its continental shelf, which takes in the entire field and would actually bring Chinese territorial waters near the shores of Japan's islands. When the Teikoku vessels arrived in the disputed area, China dispatched missile-equipped destroyers and frigates to its own nearby drilling platform—and then, as Tokyo tells it, a Chinese ship turned its gun turrets towards a Japanese plane monitoring the activity. The incident sparked state-orchestrated anti-Japanese demonstrations in Beijing, and then counterdemonstrations in Tokyo after China began drilling in a field near the disputed borderline. Two years later, an uneasy standoff still persists. Eventually, it could corrode Sino-Japanese relations, with both sides looking for America's support; or it might end in two joint development projects, one along the median and another near Japan's islands.

China's energy activism in the Middle East, Central Asia, and Africa eventually could make it a competitor of the United States for influence in those regions—but not for some time. For at least the next decade, China will have compelling reasons to keep any such competition in check. The United States is not only the indispensable economy for China's modernization; its blue-water navy is also the only guarantor of the sea lanes through which millions of barrels of foreign oil will flow to China every day for the foreseeable future. Moreover, China's energy strategy suggests that its next step will be joint ventures with U.S. oil companies in Africa and Latin America, and perhaps in Central Asia and the Middle East as well—much as it has with the Italian oil company Agip and with U.S. and other western oil companies inside China. This all will likely change when China becomes a major military and naval power, but that won't happen by 2020 and for some years beyond.

While China moves aggressively to nail down the energy resources it will need to continue modernizing itself, Russia seems set on using its energy resources to try to regain a measure of its former geopolitical power. Through the 1990s and into the early years of this century, Russia slid from world power to a midsize country with an ailing and shrinking economy, fighting a brutal internal war in Chechnya, and struggling to hold itself together. It could not stop America from corralling its former client states into NATO, abrogating the ABM treaty, waging two wars in the Middle East and two military interventions in Eastern Europe, setting up new military bases across nearby Central Asia, and even supporting anti-Russian "color revolution" leaders in countries that had been part of the old USSR.

The doubling of oil and gas prices since 2004 has given Russia its first geopolitical leverage since the Soviet Union collapsed. As the world's second-largest oil producer and exporter after the Saudis, and the world's largest producer and exporter of natural gas, Russian fields produce more than 8 million barrels of oil a day (more than twice Iran's output) and nearly 2 billion cubic meters of natural gas a day (20 percent more than the United States). More important, the country exports 6 million barrels of oil a day, mainly to Europe—which buys 4 million barrels of Russian crude and 2 million in refined products every day. Russia also exports 600 million cubic meters of natural gas every day, meeting 43 percent of Germany's needs, 26 to 30 percent of Italy and France's needs, 60 to 70 percent of the natural gas consumed in Austria, Hungary, and Turkey, 80 percent of the gas used in the Czech

Republic and Ukraine, and virtually all the natural gas used in Greece, Finland, Bulgaria, Georgia, Slovakia, Belarus, and the Baltic states.[94]

Communism may be dead, but the Russian government has consolidated effective control over almost all oil and gas production in the country, so their exports and domestic use can be instruments of state policy. The Putin administration not only seized and broke up Yukos, the giant domestic oil company, it also strengthened the effective monopoly of Gazprom, the state-owned gas conglomerate that produces 93 percent of the country's natural gas, by putting Sibneft, Russia's largest oil company, with reserves of 119 billion barrels, or more than any nation except Saudi Arabia and Iran, under Gazprom. Gazprom is also developing the vast Shtokman gas field in the Barents Sea to produce liquefied natural gas to ship to the American market by 2012. The Kremlin also is extending the reach of Transneft, the state-owned company that operates the world's largest oil and gas pipeline networks, by building a new pipeline under the Baltic Sea to Germany and eventually Britain.

The Russian government uses its vast energy assets, first, to take some of the domestic edge off the country's often disappointing economic performance, by selling Gazprom oil and gas to Russian businesses and households at roughly one-fifth world prices. The upshot is that even with record energy exports at record energy prices, the Russian oil and gas industry barely breaks even and has very few resources to upgrade its equipment and operations. The Russian government also sets politically determined prices for its energy exports. When Belarus was a Kremlin favorite, it paid Gazprom about $50 for the same gas that cost the new EU members Latvia and Estonia $110 and cost Poland, a frequent critic, as much as $200. Russia also hasn't hesitated to enforce its political prices: In January 2006, when Ukraine balked at a fivefold increase in its Gazprom prices following its "Orange Revolution," Putin cut off natural gas flows to the country—and when the European Union objected, the Kremlin suggested it might reroute the natural gas headed for Europe to China.[95] Selling natural gas from fields in western Siberia to China instead of Europe could cost Russia $50 to $70 billion—even more than it would have cost Europe to replace it—but for the Kremlin, it's all about flexing geopolitical power. Similarly, Moscow's decision to extend the Siberian pipeline to Nakhodka on the Sea of Japan instead of Daqing in China, while Japan and China were feuding over oil in the South China Sea, was all about Asian politics.

While the Kremlin will almost certainly continue to use access to its oil and gas and the prices it charges to secure political concessions, its leverage will still be fairly limited. Gazprom jacking up its prices didn't roll back the political changes in the Ukraine; and its western European customers have other alternatives. Russia may be able to intimidate small countries deeply dependent on its oil and gas, but even many of them have other potential sources, including Iran—and in a few years, perhaps Iraq as well. Moreover, Russia's drive to consolidate state control over the country's energy production and exports has sharply reduced its own access to the foreign investments, technologies, and expertise needed to expand those resources, making Russia one of the world's least-efficient energy producers. Western analysts expect Russian oil and gas production to truly peak by 2010 and then decline—even official Russian forecasts see production leveling off in the next decade—and production at the three giant natural gas fields in western Siberia that account for 70 percent of Gazprom's output already is declining. And any period of global oversupply—or a global recession that cuts demand—will cripple the Kremlin's strategy and send the Russian economy reeling.

HOW THE WORLD WILL ADDRESS CLIMATE CHANGE

The dark side of the world's fast-rising energy use—even darker than Kremlin plotting—is the pollution it's producing and its impact on the climate. These concerns first went global in the 1990s, when the United Nations first tried to address growing unease about the greenhouse gases released by fossil fuels. The resulting Kyoto protocols have failed to make a serious dent in any country's emissions. But the next ten years will likely see another concerted effort to build new international arrangements to control emissions of carbon dioxide (CO_2), the principal greenhouse gas. The next effort will probably use an approach different from Kyoto, since in order to succeed, it will probably have to be led by the chief antagonist to the Kyoto protocols, the United States government, and include China and other fast-developing countries that have vowed to never join a Kyoto-type program.

The basic science of climate change is now internationally undisputed. The greenhouse gases produced mainly by burning fossil fuel accumulate in the earth's atmosphere—they don't break down for about a century—absorb infrared radiation that otherwise would travel back into space, and radiate some of it back to the earth. So, as CO_2 levels have increased in the atmosphere, so have the temperatures of the earth's oceans and air. Over the last 150 years, atmospheric CO_2 levels rose from 271 parts per million to 370, or by 37 percent. Among the more visible effects observed in the last decade and popularly attributed to the consequent temperature increases have been melting glaciers and snow caps, disappearing lakes, successive years of record high temperatures on most continents, and stronger and more frequent hurricanes and tornadoes in the United States, monsoons and typhoons in Asia, and floods in Europe, Africa, and Asia.[96] Scientists don't agree on how quickly or how much these effects will worsen, or at what concentrations the changes will be irreversible. A rough consensus has emerged, however, that 450 to 500 parts per million may be the tipping point for serious, permanent climatic effects, and that without serious steps CO_2 levels will reach 600 parts per million by 2050.[97]

If the world's major governments intend to avoid that nasty scenario, they will have to figure out in the next few years how to keep emissions below their current levels even as worldwide energy use doubles over the next twenty years.[98] The challenge is partly technological, since achieving the goal will require the development and spread of much more energy-efficient equipment and materials, and probably new alternative fuels. The challenge is also an economic and political one, since every plausible response involves higher energy prices that will take a piece out of growth—and most politicians are reluctant to do that in order to head off even large costs at some future time. So solving climate change will be one of the daunting political problems of the next decade, with many governments trying to convince their electorates to accept even higher energy prices along with other changes in their lives.

The greatest challenge in climate change could well be geopolitical. The European Union and Japan will ramp up pressure on Washington to take the lead, because the relatively high energy-intensity of the American economy and the automobile-centered lifestyles of most Americans dictate that if worldwide emissions are to come down, the United States will have to absorb larger costs than anyone else. Moreover, an effective effort also will have to include China, India, Brazil, and other large, fast-developing countries that see climate change as a problem that should be solved by the advanced countries that produced most of the greenhouse gases. That may be true, but it also will soon be beside the point, since developing nations will soon produce as much greenhouse gases as the advanced economies.

Given these hurdles, some observers see little prospect for a meaningful global compromise on climate change. The prospects for the World Trade Organization looked bleak for a time, too; but it gained momentum with effective U.S. and European leadership and a growing recognition that its costs would be less than feared. American leadership isn't there yet for climate change, but will likely emerge in the next few years. The public education and media blitz led by former Vice President Albert Gore has shifted American public opinion in support of action; the Democratic majority in Congress also has called for action, and almost every serious candidate for president in 2007 did so as well. Moreover, in 2007, ExxonMobil, Chevron, and other large U.S. oil companies for the first time called for a public debate not over whether climate change is a serious problem but on the best ways to address it.

There's also growing evidence that the adjustments of a serious program may cost less than many fear. A 2007 study by the McKinsey consultancy, for example, analyzed the costs of the available options if the major countries agreed to caps on CO_2 emissions that would limit atmospheric concentrations to 450 parts per million in 2030.[99] The study found that achieving that goal would entail changes in most aspects of modern energy use, but not very drastic ones—and that's without startling technological breakthroughs to ease the effects down the road. A good part of the effort would involve steps to improve energy efficiency that over time wouldn't cost people anything. Shifting to hybrid automobiles, improving the insulation in buildings, plants, and homes, and using more energy-efficient heating, cooling, and lighting systems would generate savings through reduced energy use that would offset the initial higher costs. By one estimate, better insulation and more efficient heating, cooling, and lighting alone could reduce the growth in worldwide electricity demand by half over twenty years. The next, least expensive steps involve planting more forests and slowing deforestation—vegetation absorbs CO_2 and produces oxygen, so the more, the better—and changes in livestock practices. After that come the more costly steps that involve the measures that many people think of first—shifting to alternative energy sources that produce less greenhouse gases, such as wind, solar, and carbon-capture technologies—followed by even more expensive measures, such as using biodiesel fuels and clean coal in manufacturing.

Progress in arresting climate change over the next 10 to 15 years could be less painful than many people assume, but only with a genuine worldwide effort. The geopolitical challenge is that the advanced countries most likely to take serious steps cannot achieve the global goal on their own, because the majority of the less-expensive options for containing greenhouse gas emissions involve action by developing countries. It's cheaper, for example, to outfit new homes, buildings, plants, and automobiles with more efficient and clean technologies than to retrofit existing structures and cars—and over the next ten to fifteen years, most of the world's new homes, buildings, and plants will be built in fast-growing developing countries, principally in Asia. But this will be hard to sell to developing countries, since it's also more expensive initially to use more energy-efficient and clean technologies, unless the advanced countries create a global fund to help offset some of those initial costs.

In addition, most of the potential climate-change gains from forestation and livestock practices are located in Latin America and Africa, and these steps will involve initial costs to correct.

The issue of the CO_2 emissions from automobiles is different. Americans, Europeans, and Japanese will still buy the majority of new cars in 2020, and probably 2030, too, because the national fleets in advanced countries turn over every six to eight years. But China, India, Indonesia, Bangladesh, Brazil, and other large developing countries also will vastly increase their auto purchases over the next ten to fifteen years, and hopefully they, too, can be moved to hybrids or "plug-in," battery-powered vehicles.

A dilemma for now is that China, India, and the other fast-growing developing countries say they will never join a system that caps their emissions. The onus of changing their minds will fall to the United States, because it is virtually the one country with the leverage to induce them to participate. (Saudi Arabia probably has the requisite leverage, too, but not the incentive.) Two years ago, U.S. leadership on climate change seemed fanciful. In 2007, President Bush for the first time agreed that international action on global warming is required; and in 2009, serious climate change talks could be part of the new president's efforts to rebuild international support for American leadership.

The real question for China, India, South Korea—and, for that matter, the United States, Germany, and Britain—is how much it will really cost to do what's required to reduce the risk of global warming. Economists who ask that question generally conclude that the average cost of all the measures needed to keep those future greenhouse gas concentrations within what's thought to be a safe range, twenty and thirty years from now, would be about 40 euros or $50 per ton of CO_2 reduced.[100] That's not trivial, but it's also not backbreaking: At current rates of energy use, it would cost about $300 a year per American (today's dollars), about $100 per year for Europeans (they consume less energy because their energy taxes are four times higher than America's), and about $25 a year per person in China. Those per-person costs would be higher in 2020, especially for China, because people will use more energy—but it also should be more affordable for most people, since average incomes will also be substantially larger by then in China, America, and most other places.

The catch is that most people will resist changes that entail spending more today, say for a hybrid car or a highly energy-efficient washer, to head off the larger costs over time of consuming more high-priced energy. One way to overcome this resistance in some areas will be government mandates and regulations requiring higher fuel efficiency standards for automobiles and higher energy efficiency standards for appliances, lighting, and insulation. These regulations will cost consumers and builders more at first, and save consumers and homeowners even more later on. But those measures are only enough to get half of the job done, and only if everyone in the world was part of it. If the United States, China, and the rest of the world can agree to seriously address this problem in the next decade, people and businesses will find themselves dealing with either a "cap-and-trade" regime such as Kyoto, or with new carbon taxes—and both approaches will lift energy prices even higher than global supply and demand have in store.

Under cap-and-trade, each country would agree to cap its annual CO_2 emissions at an agreed-upon level and then allocate the total in permits distributed to its major greenhouse gas producers—utilities, manufacturers, agribusinesses, and oil and gasoline producers or distributors—based on some fraction of their emissions in some previous year. Those that can come under their caps at relatively low cost will do so and sell their excess permits to others who can't meet their own caps, so, in theory, the overall emissions cap can be achieved each year at the lowest overall cost. Consumers and businesses ultimately will shoulder those costs. Utilities or manufacturers who make changes to come under their caps will pass on the costs of their changes, as will those who have to buy more permits. If everything is passed along, it should discourage people from using energy produced by those least effective at reducing the carbon content. Most American and European politicians who want to address greenhouse gas emissions currently favor cap-and-trade. It's not hard to discern its political appeal, since the inevitable energy price increases will appear to come from the market, not the government.

However, most economists who favor serious action don't favor cap-and-trade—and for good reasons. They see that any system of preset caps will make energy prices even more volatile than they are already, and that's bad for investment and the overall economy. They remember the late 1970s, when the U.S. Federal Reserve set a cap on the money supply, and the price of money—interest rates—shot up. In this case, the prices for permits and the underlying energy could soar whenever energy demand is greater than was expected when the caps were set—for example, because the summer is unusually hot, the winter unexpectedly cold, or the economy grows faster than predicted. Cap-and-trade also seems vulnerable to corruption and manipulation. The permits will be worth tens of billions of dollars, which governments both honest and corrupt will have to distribute and then monitor after every trade.

The system will be a boon to the financial firms that will trade and speculate in the permits—financial markets for these permits have been created already in Chicago and London—and their commissions and trading profits ultimately will raise permit prices as well.

Instead, many environmental economists favor carbon fees or taxes over cap-and-trade, because they like the clear and direct price signal that a carbon fee provides to favor less carbon-intensive fuels. Moreover, carbon fees don't make energy prices more volatile when demand rises unexpectedly, as cap-and-trade does. Such taxes or fees also certainly should reduce the use of the most carbon-intensive forms of energy, as high utility and gasoline taxes in Europe have moved businesses and households there to be more energy-efficient than in America. And everyone agrees that carbon taxes would be much simpler to administer than the permit system under cap-and-trade, and less open to corruption. A carbon tax system also could be designed to encourage people to favor less carbon-intensive fuels and products while protecting them from high short-term costs, for example, by using the revenues from a steep carbon tax to reduce their payroll taxes. The drawback is that carbon taxes will allow emissions to increase with unexpected demand, as happens today. Those increases could be offset by increasing the carbon tax rate or fee the following year, if politicians are willing to raise taxes for the planet's long-term health.

Whether or not the world comes to grips with climate change over the next decade will ultimately depend on the United States and China. America is the largest producer of greenhouse gases, by far; and by 2020, China will be closing in fast. The United States withdrew from the Kyoto cap-and-trade arrangement when it became clear that German, British, and Russian gamesmanship with their caps would relieve them of any significant burden, leaving nearly all of the worldwide adjustments to the United States, Australia (it also withdrew), Japan, and Canada (they stayed but reinterpreted Kyoto's terms to avoid most costs). If the United States leads a second global effort, it will try to use its superpower leverage and resources to persuade China to participate—perhaps with pledges of joint ventures to develop and produce alternative fuels and new energy-efficient technologies. And if China reverses its position and joins the United States, Europe, and Japan in a global approach to greenhouse gas emissions, it will be difficult for other large developing nations to hold back.

Apart from what the United States might want, China's incentives to address climate change will increase over the next decade. The World Health Organization reports that seven of the world's ten most polluted cities are in China; and the World Bank estimates that pollution currently costs China the astounding equivalent of 8 percent of its GDP each year, mainly in health problems and damage to crops and structures. Without serious policy changes, China's pollution is projected to increase another fourfold by 2020, creating burdens that "China isn't able to withstand," as the head of China's State Environmental Protection Administration, Zhang Lijun, has put it.[101] Perhaps most appealing of all, Beijing may come to see that support for climate change reforms will help slow the growth of China's energy needs and oil dependence.

No one can know what a new global approach to climate change will look like. If the commitment is serious—which still is far from certain—carbon taxes will meet the need better than a Kyoto-type cap-and-trade. Carbon taxes also have one very valuable political advantage: They generate large amounts of revenue that governments can use as they choose. China's leaders may come to see carbon taxes as a growing funding source for health care, while a future U.S. president is more likely to use the revenues to cut payroll or other taxes. Cap-and-trade will generates tens of billions of dollars, too, but the funds go to energy producers and other businesses with excess permits to sell, to the financial institutions that will manage the permit trading, and to speculators and investors who correctly guess the short-term direction of permit prices.

Whether the world's major countries agree in the next ten years on harmonized carbon taxes, some other way of reducing greenhouse gas emissions, or nothing at all, energy prices will keep rising. That, in itself, will push governments and businesses to take helpful steps. By legislation, regulation, or simply because gasoline prices reach $5–$6 per gallon in America and, much higher than that in Europe and Japan, a growing share of the world's automobiles and trucks will run on electric fuel cells, cellulosic ethanol, and other biofuels. Brazil already produces hydrogen ethanol for less, per gallon, than gasoline, and Volkswagen and General Motors have introduced "flex cars" that can run on pure gasoline, pure ethanol, or any combination of the two. While higher prices for electricity and heating will not drive many people to retrofit their homes or replace their appliances, higher energy prices will eat into most people's energy consumption, which in turn will reduce emissions on the margin.

Ultimately, neither the world's energy problems nor the prospective changes to the climate can be settled on those margins. Much like the financing of health care as millions of boomers reach the age when they will need expensive medical treatments, the issues of energy and climate change will become more heated and serious over the next decade and require large and serious reforms.

CHAPTER EIGHT

HISTORY'S WILD CARDS: CATASTROPHIC TERRORISM AND TECHNOLOGICAL BREAKTHROUGHS

NATIONS CANNOT CHANGE THE SINGULAR GLOBAL FORCES now shaping their paths. Yet no outcomes are ordained, and societies always have leeway to affect the consequences of large forces that they cannot control directly. America's comparatively open approach to immigration will lessen the economic costs of its baby boomer baby-bust cycle, and other societies could adopt a similar attitude. China's profound economic reforms position it to develop and prosper through globalization in ways currently beyond India's reach, but with aggressive leadership, India could conceivably undertake comparable reforms. Nor is it beyond imagining that Germany and France—especially France of late—could reverse their slow economic declines by recasting their welfare states more along the American or Irish models. There is no scenario that will alter America's position as sole superpower for a long time, but China and Russia could find their way to an alliance that might bring back elements of balance-of-power geopolitics in Central and Southern Asia.

The history of just the past generation also is replete with utterly unexpected events, especially the Soviet Union's implosion and China's embrace of capitalism, and the next ten to twenty years will include many unforeseen developments. Even so, there are very few that could materially affect the three forces examined here. But there are at least two wild cards—one dark and one much more promising—that just might. The powerful, dark possibility is catastrophic terrorism. America's historically disproportionate military superiority has led its opponents to try unconventional tactics such as guerilla insurgencies and unconventional weapons from improvised explosive devices (IEDs) to biological weapons. America's sluggish response to these well-understood developments already may have cost the United States its aura of invincibility. While it is still very unlikely, terrorists could successfully escalate their unconventional tactics and weapons to such extremes in the next decade, that the effects could alter geopolitics and globalization.

A terrorist attack that devastates Saudi oil facilities, for example, would sink equity and bonds markets around the world, set off a deep global recession, and slow or possibly stall out China's modernization program, with large and unpredictable political effects. Terrorists also might manage to play a number of nuclear cards. The easiest to pull off would be an attack on a nuclear power plant or setting off a radiological weapon, although the concrete economic and political effects would probably not be nation-changing. A more dreadful possibility—although also a much slimmer one—is the detonation of a stolen "loose nuke" or improvised nuclear device in a large American or European city. This scenario is usually contemplated safely in thriller movies; but if it ever happens, its effects could change much that we now take for granted.

The more upbeat, historical wild card is unexpected technological advances that could relieve or even solve intractable economic and social problems. This happened recently on a historic scale when a generation of advances in computing power and capacity helped drive globalization and, with it, the economic ascendance of China and the strongest sustained period of worldwide growth on record. It's not likely, yet just possible, that the next ten to fifteen years will see breakthroughs in nanotechnology that fairly quickly lead to inexpensive large-scale hydrogen or solar power. Such unexpected advances would dial back some of the economic pressures coming from fast-rising energy demand and slow-rising oil investment and production, possibly make climate change a problem that countries could manage without major sacrifices, and reduce the geopolitical leverage of Russia and the Middle East.

If nano-enabled clean energy isn't in our near future, it is also conceivable that biotechnology advances could change what it means to grow old just as baby boomers in most countries reach the age when serious medical problems become commonplace. Scientists drawing on the recent, breakthrough successes in deciphering the human genome, for example, could come up with therapies that will repair or replace defective genes that bring on many diseases, including cancers. Perhaps a bit more possible are advances in stem cell therapies not only for conditions like Parkinson's and Huntington's diseases, spinal cord injuries, and cystic fibrosis, but also heart disease, diabetes,

Alzheimer's, strokes, burns, osteoarthritis, and rheumatoid arthritis. We cannot know now whether these kinds of medical leaps would help relieve the financing problems threatening national health systems or make them worse. If they do arrive, however, they could redefine what it means to grow old and ease the demographic burdens on many nations.

THE TERRORISM WILD CARD—WHOM TO FEAR

America's war on terrorism will help define the geopolitics of the next decade, regardless of who is president. The form that that war takes, however, will depend on how each U.S. administration defines the enemy and its threat. In the current American "us-versus-them" paradigm, the "them" has been defined broadly as Islamic fundamentalists and extremists. The future path of the war on terrorism will depend in good part on how American presidents and Congresses distinguish between Islamic fundamentalists whom the West can learn to live with and the violent Islamist extremists who will not coexist with anyone else.

Islam has deep divisions, as Iraq's current turmoil attests, that go back to a debate among Muhammad's original followers, following the prophet's death in 632 AD, over whether his successor or "caliph" should be someone from his direct bloodline or the person considered most worthy, learned, and pious.[1] When tribal leaders in Arabia declared that one of Muhammad's close companions, Abu Bakr, would be the first caliph, a majority agreed and called themselves "Sunni" after "Ahl as-Sunnah wa'l-Jamā'h," or those who follow the example of Muhammad. But a minority insisted that Muhammad's cousin and son-in-law, Ali ibn Abi Talib, should succeed, and they called themselves Shi'a or Shitties from "shi'at Ali," or helpers of Ali. For three decades, there were two competing caliphs, culminating in a series of murders and battles that formally split Islam into the two sects. The Shi'a remained a small minority ever since, comprising today 10 to 15 percent of Muslims centered in Iran and parts of Iraq, Yemen, and central and southern Asia.[2]

Sunni and Shi'a disagree on many points, but the most important for twenty-first century geopolitics is that Shi'a leaders retain political as well as religious authority, while Sunni caliphs traditionally are subordinate to Muslim kings and governments. Sunni Islam also developed numerous sects, based originally on various interpretations of Allah's admonition to help the poor. In the way that theological debates can eventually become matters of consequence for everyone else, one of the interpretations formed the basis for

Sunni Wahhabism and a fundamentalist approach to the Quran, whose
founder, Abd al Wahhab, made a fateful political and religious alliance in the
eighteenth century with an Arabian leader named Muhammad bin Saud.
Some two centuries later, when bin Saud's descendent unified Arabia in one
kingdom, Wahhabism became the official religion of Saudi Arabia.[3]

Globalization has created a sense of crisis for many Wahhabi Muslims who,
like many Christian and Jewish fundamentalists, adhere to a strict and literal
reading of their sacred text. Living by the Quran's laws has come to mean rejecting western values, in a time when globalization brings those values into
every facet of business life and most Muslim homes. What the proper response should be now divides Muslim fundamentalists in many ways, but
most importantly by spurring new "Islamist" movements with political as
well as religious goals.[4] While all fundamentalists want to restore a more
pure Islamic faith, Islamists pursue political power in order to radically reshape their societies while most other Sunni fundamentalists are willing to
coexist with whatever political system is in place.

This divide narrowed a bit in the late 1980s and early 1990s with a new
"neo-fundamentalist" movement that seeks to persuade Muslims to gradually
change their societies into strict Islamic states.[5] To westerners, neofundamentalists may sound much like Islamists. Their visions of an ideal Islamic society are similar; and they all reject democracy, because men (much less women)
have no right to usurp the laws of God. But neofundamentalists want to
achieve the ideal society by changing people's hearts and minds, while Islamists take direct aim at Muslim leaders they consider traitors, along with
all non-Muslim foreign powers.

On the fringe of the Islamist movement are the groups prepared to do battle with the West and kill anyone standing in their righteous way. These radical, violent groups emerged in the 1980s and 1990s, at the same time as the
neofundamentalists, from the catalytic war against the Soviets in Afghanistan. The battle to expel the Soviet infidels and create an Islamic state in Afghanistan brought together Islamists and jihad—in a struggle that received
extensive U.S. material support—and provided the setting and conditions to
radicalize and train thousands of young Muslim fundamentalist men. Moreover, their victory over the Soviets validated the *jihadi* approach and provided
a large corps of battle-tested followers who now could fight infidels in other
places. The United States soon became the central target, for its secular, globalist behavior and values, for establishing western military bases in the land
of Mecca for the first Gulf War—and for twelve years after—and lastly for its
support of Israel.

Over the next decade, the United States and much of Europe will face at least three distinct challenges from the Islamic world. In Saudi Arabia and other Sunni countries, millions of Islamic fundamentalists, many educated in Saudi-financed Wahhabi madrassas, will regularly express their view of the West as a direct threat to the life dictated by the Quran. Western leaders and publics can learn to live with this opprobrium, as they did for a long time when many millions of Chinese and Russians excoriated the West. Among all the Islamic fundamentalists, there also will be tens of thousands of neofundamentalists and hard-core Islamists committed as well to remaking their societies into strict Islamic states. If they succeed, it will present a larger challenge, because strict Sunni Islamist states are more likely to defy America's global security policies, to resist key aspects of globalization such as free flows of foreign investment, and to come into conflict with Shi'a Iran. Dealing with this challenge will be the realm of American and European diplomacy and statecraft, if their successful practice can be revived.

Among all the Islamists, there also are thousands of violent fanatics who will pose the third challenge and most immediate threat. The current difference between America's Republicans and Democrats in geopolitics might be framed as whether the war on terrorism should target both Islamists and jihadists—the second and third challenges—or focus on threats from the jihadists while applying conventional diplomatic and political approaches to the Islamist challenge.

Whether the next decade of America's war on terrorism should target only the violent extremists or also Islamist groups could be said to depend on what happens when Islamists create strict, Islamic states—do they become active allies of terrorists, or Islamic societies content to attend to their own affairs? Despite the rhetoric of America's neoconservatives, Islamist opposition to democracy is not tantamount to aiding terrorism, since most of America's allies in the Middle East also fiercely oppose it. In fact, of fifty countries with majority Islamic populations in the world today, only nineteen have anything like democratic government. And the West coexists happily with other implacable foes of democracy, notably China, as it did throughout the cold war. Even focusing on active support for terrorists doesn't produce easy answers beyond the extreme case of the Taliban and their aggressive protection of the terrorists who had just perpetrated a hideous attack on the United States. Iran's Shi'a Islamist government is hard to classify, supporting terrorists at some times and not at others. There's also no simple response to Hamas, a terrorist group that has assumed political power by democratic means, and Hezbollah, which combines jihadist tactics against Israel with social activism inside Lebanon.

The complexities of Middle Eastern and American politics suggest that the U.S. war on terrorism in coming years will be an ad hoc affair focused mainly on groups determined to express their implacable opposition to the West by committing mass murders. Al Qaeda is still the most influential and well known. While the United States has not managed to capture Osama bin Laden, it has destroyed the group's state support in the Taliban, decimated its original leadership, and reduced its role to providing ideological, technical, and financial support for other terror groups. In effect, Al Qaeda has returned to its origins in the Maktab al-Khadamat, a mujahadeen group in Afghanistan that did little actual fighting but raised funds and recruited and trained mujahedeen for more active fighting groups.[6]

The next several years of the war on terrorism also will likely focus on hundreds of homegrown cells and new organizations that sympathize with Al Qaeda and sometimes receive its support, especially in Europe.[7] The General Intelligence and Security Service of the Netherlands (AIVD), for example, estimates that there are twenty jihadist networks in that country alone, with links to some three to four hundred more outside.[8] One encouraging sign is evidence that many of these groups now communicate with parts of organized crime, in order to use their "supply, transport, and money-moving networks."[9] These connections may create a new vulnerability, since criminal groups have long been sources of intelligence for U.S. and European law enforcement.

The most troubling trend is the emergence out of Iraq of new terrorist groups that one day could be as effective worldwide as Al Qaeda was. Like Afghanistan in the 1980s, Iraq is now a magnet for terror recruiters for both the jihad against American forces and the struggle between Shi'a and Sunni. And just like Afghanistan during the Soviet occupation, Iraq has become an ideal place to train untested fighters under live conditions. Thousands of young Islamists are learning how to follow orders, gather intelligence and build explosive devices as they observe how American and British soldiers fight, and develop the most effective ways to kill them. These new terror groups appear to be organized mainly in semi-independent cells, which may prove to be harder to track and better suited to terror warfare than top-down organizations of the Al Qaeda type.

THE TERRORISM WILD CARD: WHAT TO FEAR

Extreme, violent terrorism has a long lineage, going back at least to the clandestine Jewish cult, the "Zealots of Judea," or *Sicarii* ("dagger men"), in the first century AD, who attacked Roman soldiers occupying Judea (and their

Jewish collaborators) until they were wiped out at Masada. The French Revolution gave terrorists their modern name, although the Reign of Terror was a form of government policy more akin to Stalinism in the late 1930s. The loose networks of violent anarchist cells in the late nineteenth century were closer to modern terrorism, managing in the years from 1878 to 1913 to assassinate a French president, two Spanish prime ministers and a Russian prime minister, Tsar Alexander II, King Umberto I of Italy, King George I of Greece, and U.S. President William McKinley.

Most modern terrorism, however, has been domestic rather than geopolitical, beginning with the Irish terrorists of the late 19th century who formed a secret army to fight an occupying power much as did the Zealots of Judea. The model of a secret domestic army spread over the twentieth century to Spain, Greece, Germany, Japan, Mexico, Jordan, Syria, Lebanon, and Palestine. What distinguishes most of the terrorism facing the West now from those groups are its geopolitical agenda and its practitioners' willingness to use extraordinary weapons of mass destruction. (Doomsday weapons are the exclusive province of geopolitical terrorists with enemies far away, since no group would explode a nuclear device or set loose a plague in a country it hopes to reclaim.)

Terrorists don't need weapons of mass destruction to have a large impact. But several decades of experience has shown that while any major terrorist act will make millions of people anxious, the vast majority have little long-term economic or political effects. For that to happen, terrorism has to be fairly pervasive across a society and protracted over time—as it has been in Colombia, Northern Ireland, the Basque region of Spain, Israel, and now Iraq. In those places, terrorism has depressed economies and stunted development. Everywhere else, where terrorism has been occasional and localized, its concrete impact has been surprisingly modest. So long as Al Qaeda and its successors are unable to use weapons more powerful than airliners or to carry out multiple attacks regularly for years, as they do in Iraq, their ambitions to seriously damage the United States or other large western countries will fail.

Even the immediate economic costs of terrorism are rarely high. Small operations—a political murder or bombing that kills a few people—have negligible economic effects. Even a huge strike is a blip in a large country. The hard data show that the 9/11 attacks did not move the U.S. economy, with consumer spending, investment, and GDP accelerating in the following quarter. And modern economies regularly absorb much greater human and financial losses from bad weather and natural disasters—the 1988 heat wave that took the lives of more than 5,000 Americans, for example, or the 1999 earthquake in Izmit, Turkey, that killed 17,000.

Terrorism uses violence to create an expectation of more to come; and where those expectations take hold, people's fears can dampen investment. These effects are greatest in small countries where no place seems safe, and in small economies that depend on foreign capital. So foreign capital fled Colombia for twenty years, and its per-capita income is still about 40 percent below Latin America's average. When religious violence raged through Belfast, Northern Ireland became the U.K.'s poorest region—and it began to recover in the 1990s when the violence abated. Israel is the other major case of a terrorism-impaired economy. The intifada has created a wartime political and economic climate; and the Bank of Israel figures it costs Israel 4 percent of GDP a year by depressing foreign investment and tourism and increasing budget deficits. In places larger than Colombia, Northern Ireland, and Israel, however, the economic effects are localized. Political violence has plagued the Basque region of Spain since the early 1970s, and while per-capita GDP has grown 10 percent more slowly in that region, the rest of Spain has grown fast enough to become the world's ninth-largest economy.

In the United States, Japan, and other large economies, acts of terrorism scare away investment from industries and local areas thought to be particularly vulnerable, and towards safer sectors and places. The World Trade Center attacks dealt a temporary blow to Manhattan's economy, but not to Boston or Chicago. Even in Manhattan, the impact was concentrated downtown, where the terrorists destroyed nearly 30 percent of class A office space. September 11 also set back the airline, hotel, and insurance industries. But investment and demand shifted to other sectors, especially once the Federal Reserve eased credit to calm the markets. And six years later, real estate in downtown Manhattan is much more valuable than it was in the summer of 2001. Similarly, when the Red Brigade's attacks spiked in Germany and Italy in the late 1970s and early 1980s, tourism suffered but not the overall economies.

Large modern economies can roll with occasional acts of destructive terror because their markets quickly redistribute capital and jobs away from the violence. It may be politically incorrect to say so but the economic impact of an attack like 9/11 most closely resembles a natural disaster, with sudden losses followed by the stimulus of rebuilding. What disrupts an economy like the United States, Japan, or Britain are not local attacks, but shocks that hit all of their markets at once, as when OPEC tripled the price of energy overnight.

In principal, the Internet could present such a target. In October 2002 and again in February 2007, for example, an unknown group or persons mounted a sophisticated "denial of service" attack on a series of Web "root servers" aimed at disabling the American military networks that use them.[10]

Moreover, Al Qaeda computers seized by American soldiers in Afghanistan contained reams of information about the digital devices that remotely control U.S. power, water, pipeline, transportation, and communications grids, and how to reprogram them.[11] U.S. officials also have concerns about a small cell of terrorists trained in computer sciences using the Internet to access the "digital distributed control systems" that throw railway switches, close circuit-breakers in power grids, and adjust valves in dams and the pipelines carrying water, oil, and natural gas.

Serious cyberterror attacks on the Internet itself or on a country's critical infrastructure systems accessed through the Internet are possible, and attacks on power, pipeline, or dam systems could produce billions of dollars in damage. Larger and longer-lasting costs and effects, however, are unlikely. The Internet's decentralized, open-access structure, which makes attacks possible, also makes it easier to sequester damaged areas of the Web and repair them. And while infrastructure attacks could produce considerable local damage, the disruptions would pass fairly quickly, again much like a natural disaster. To wreak serious damage, cyberterrorists would have to disrupt the Web or infrastructure systems for a sustained period, which for now seems nearly impossible.

The limited impact of most terrorism won't matter, however, if terrorism should become much more deadly in the next decade. Among the possibilities, terrorists unleashing nuclear or biological weapons on American or European cities, especially if it happened more than once, could shift the paradigm of terrorism to all-out war. And modern wars have very large political, social, and economic costs. A nuclear or bioterror attack on a major American or European city would be a "black swan" in the current phrase—a rare and hard-to-predict event, and one with large implications for the paths of globalization and geopolitics.

The most likely attack (all still low probability) would involve nuclear materials but not nuclear weapons, with significant but probably not nation-changing effects.[12] For example, nuclear power plants in America, Europe, and Japan are considered quite vulnerable to terrorist attacks or sabotage, but probably not to events with catastrophic effects. Numerous times, deranged people have crashed through the fences around nuclear plants and even driven their cars through the doors of buildings. These plants also can be attacked by air like the Pentagon and the World Trade Center. But the chances are vanishingly small that such kamakazi attacks could trigger the release of large amounts of deadly radiation. For that to happen, terrorists would have to breach the four-foot-thick concrete walls of the buildings containing the reactors and the seven- to twelve-inch-thick steel walls around the reactors, and then set off a large chemical explosion or fire to disperse the radiation into the atmosphere. In the 1979 Three Mile Island accident, the coolant failed and the fuel melted, but without large breaches in the containment walls and a substantial fire or explosion, very little radioactivity escaped. And experts say that even a large airliner cannot do all that.

Terrorist sabotage that replicated the human and mechanical errors that destroyed Chernobyl in 1986 would seem a more reliable way to wreak real havoc. The Chernobyl meltdown spread radiation over 50,000 square miles, rendered over 2,300 square miles indefinitely uninhabitable, and ultimately will cause the deaths of an estimated 4,000 people. The Union of Concerned Scientists has estimated that an attack that released as much radiation from the Indian Point plant in Westchester County, thirty-five miles from Manhattan, could kill 44,000 Americans.[13] But the radiation release at Chernobyl was so large and serious because the Soviets built their reactors without containment; so the hydrogen explosion and graphite fire could blow part of the radioactive core directly into the atmosphere. There's nothing like Chernobyl in Western Europe, Japan, or the United States.[14] Terrorist sabotage of a nuclear plant in France, the country with the most such plants, or another advanced country, would be a terrifying and spectacular event that would trigger much more stringent security measures. But the real, long-term economic and social impact, as with Three Mile Island, would be limited.

A terrorist group would find it easier technically to build a "dirty bomb" by attaching plutonium to conventional explosives, than to replicate Chernobyl's damage. The obstacle is securing the plutonium and moving and handling it. Even if a group can pull off all of that, the real damage, once the debris settled, might still be limited. A radiological bomb could permanently contaminate the British Parliament or the N.Y. Stock Exchange; and with enough plutonium and explosive force, several surrounding square blocks. But buildings and blocks are fungible, and members of Parliament or brokers would move elsewhere. The larger effects would come from terrifying everyone in the world, because everyone might well believe it could happen again and again. Every major city would install elaborate radiation sensors, as they have in Washington, D.C.; every government would drastically tighten security around nuclear materials—which they should have done long ago— and the military reprisals against the culprits or those suspected of it would be deadly and devastating. Yet even here, it's doubtful that the reactions would change the paths of globalization or even geopolitics.

All bets will be off, however, if terrorists manage to set off a real nuclear bomb. The most likely weapon would be an "improvised nuclear device" (IND) assembled at the target using highly enriched uranium (HEU).[15] An

IND would probably be a "gun assembly" device resembling a crude version of the Hiroshima bomb, which had a barrel six inches in diameter and six feet long capped at both ends, with standard explosives at one end, a mass of HEU next to those explosives, and a second HEU mass at the other end. Detonating the explosives propels one mass of HEU into the other, creating a large enough mass to support a fission chain reaction. It's not as simple to build and detonate as this sounds, but close enough for five Los Alamos nuclear weapons experts to conclude that terrorists with no previous experience, a few critical skills, and access to the uranium could build one and set it off.[16] The will is certainly there in Al Qaeda's case: Osama bin Ladin has told his followers that their religious duties include helping him acquire nuclear material.[17]

The terrorists' major hurdle will be securing the highly enriched uranium.[18] The most likely way is to steal it, something made possible by the spectacular failure of the world's nuclear nonproliferation regime. Since the 1950s, the non-proliferation effort has allowed any country to develop peaceful nuclear power; and today there are 435 nuclear power plants in thirty countries, all potential sources of stolen, highly enriched uranium. The most likely sources, however, are found in Russia's crumbling nuclear infrastructure, where some experts believe only 50 percent of fissile materials are safeguarded adequately. Al Qaeda, the Japanese terror cult Aum Shinrikyo, Iraq, and Iran all reportedly sought nuclear materials in Russia.[19] And the U.S. National Research Council concluded a few years ago that "large inventories of SNM (special nuclear material) are still stored at many sites (in Russia) that apparently lack inventory controls, and indigenous threats have increased."[20] The United States created the Cooperative Threat Reduction program in 1991 to secure and destroy nuclear materials and weapons in the former Soviet Union; but its funding has been so limited that the job is not expected to be done until 2018.[21]

The failures of nonproliferation also create opportunities for terrorists to steal fully made weapons. With nuclear power technologies transferable to nuclear weapons production and only inspections to prevent those transfers, nuclear weapons have spread already to Israel, India, Pakistan, and North Korea. And, in all likelihood, Iran will be a nuclear power within the next decade. Within ten to fifteen years, many experts expect the nuclear club could also include a significant number from a growing group of Japan, South Korea, Indonesia, Saudi Arabia, Egypt, South Africa, Brazil, and perhaps others.[22] Today the weapons in the arsenals of the United States, Britain, France, China, and probably India and North Korea are locked down tight. But there are serious concerns about the numerous multikiloton devices among the "special nuclear materials" not fully secured in the former Soviet Union, including many much more powerful than the Hiroshima and Nagasaki bombs.[23] Strategic and civil defense experts also worry about Al Qaeda sympathizers in the Pakistani military and intelligence services arranging access to one of their country's weapons or to the country's enriched uranium for an IND.[24]

The encouraging fact is that no terrorist group has managed yet to buy or steal a weapon or weapons-grade uranium. If that changes, there's little doubt that terrorists could deliver and detonate at least an improvised nuclear device almost anywhere in the world. The United States inspects perhaps 2 percent of the seven to eight million containers that arrive at its ports every year, and most other countries inspect even less.[25] There are plans to subject many more containers to radioactive screening; but the sensors work better with the plutonium in state-made weapons than with highly enriched uranium, which emits little radiation.[26] Every country also has hundreds or thousands of miles of borders without sensors or monitors.

A weapon also can be detonated in a port before it could be screened or inspected. The RAND Corporation has analyzed the probable effects of a ten-kiloton nuclear detonation—a small bomb, but more than an improvised device—in the Port of Long Beach, twenty miles south of Los Angeles.[27] Some 60,000 people would die instantly; 150,000 more would be exposed to radioactive water and sediment, 2 to 3 million people would have to be relocated from contaminated areas, and a total of 6 million might try to flee the Los Angeles area. The direct monetary costs would exceed $1 trillion, the equivalent of about 8 percent of U.S. GDP, or 40 to 45 percent of the annual output of Britain, France, or China. A crude nuclear device set off in lower Manhattan would produce costs at least as great.[28]

The indirect global costs if this happens in the next decade will be far greater. Every country would close its ports and most of its borders, and these closures could go on for a long time. The first casualty would be oil shipments, and then worldwide trade. The stock and bond markets would crash; and while the Federal Reserve and other central banks might be able to take a little air out of the panic, they couldn't undo the unprecedented real losses.[29] The world almost certainly would fall quickly into deep recession.

Economies would recover eventually with most of their capacities intact, but globalization and geopolitics would never be the same. Financial transfers would slow, because so much capital would flow to the areas affected directly and because governments would scrutinize all financial transfers much more strictly. Governments also would clamp down hard on trade in all technologies that conceivably could be used by terrorists, including most information technologies. No one can know the precise impact on geopolitics; but nations with significant numbers of terrorists or their sympathizers, including much of the Middle East, would be at risk of large-scale U.S.-led policing operations or attacks. The domestic aftereffects also could be seismic. A successful nuclear attack by terrorists in any advanced country could lead quickly to major changes in traditional political freedoms, including enhanced executive powers associated with world wars, covering surveillance, detention, and press censorship.

A nuclear detonation is the worst-case wild card, but certain forms of bioterrorism in the next decade also conceivably could affect the current paths of globalization, geopolitics, and even global demographics. Dr. Anthony Fauci, director of the U.S. National Institute of Allergies and Infectious Diseases, and America's leading public health physician, has said—even before 9/11—that "a bioterrorism attack against the civilian population in the United States is inevitable . . . the only question is which agent(s) will be used and under what circumstances will the attack(s) occur."[30]

Nothing in politics is inevitable, but releasing a deadly biological agent in an American or European city is a tactic with a number of features that would appeal to terrorists. It doesn't involve large equipment or radiation that can be noticed or detected easily. Escape for those carrying it out also is simple, because the target won't even know it's been attacked for some time. A biological attack also requires fewer resources than nuclear terrorism; and if it works, it's easy to replicate.[31] Perhaps most important, there are many potential sources of deadly biological agents that could be stolen or even transferred. The Center for Nonproliferation Studies at the Monterey Institute of International Studies counts eleven states with suspected or confirmed offensive biological programs, including Iran, Syria, Egypt, Algeria, Libya, Sudan, and Pakistan.[32]

Like other forms of terrorism, the bioterror attacks easiest to carry off would have limited effects. Many experts believe that the most likely target would be American and allied troops in the Middle East, or an agroterror attack on a country's food.[33] An attack on a U.S. military installation abroad would have limited economic or health costs—except for those infected, of course—but it could have geopolitical effects by triggering demands from host governments that the U.S. military withdraw.[34] Another tactic might involve infecting imported foodstuffs. In advanced countries, however, such agroterror can be detected fairly quickly, and the kinds of foodstuffs affected can be avoided. In 1996, almost 1,500 Americans in twenty states were accidentally infected with a parasite that causes severe intestinal distress, called cyclosporoiasis, from raspberries imported from Guatemala, with no social or economic aftereffects.[35]

What worries public health experts are the biological agents that the U.S.

Center for Disease Control and Prevention (CDC) classifies "high risk," because they have high mortality rates and could trigger public panics.[36] The top candidates are anthrax, botulism, and smallpox. Anthrax is worrisome because it's deadly and simple to mass-produce: The bacteria infest the soil in many parts of the world, and when a grass-grazing animal ingests it, it can be extracted from the animal's blood and then grown in quantity in any lab. Step-by-step instructions can be purchased on the Internet for $18.11,[37] and the 9/11 Commission staff found that Al Qaeda "was making advances in its ability to produce anthrax prior to September 11."[38]

But it's very difficult to infect large numbers of people with anthrax, because a person has to be directly exposed to a high density of its spores. The spores are hardy and can be transported on clothing or anything that comes in contact with an infected person, but the disease cannot be otherwise spread directly from person to person. To use anthrax as a serious weapon of bioterror, a terrorist group might manufacture several hundred pounds of it and smuggle it into the United States or Britain in carefully lined bags, target one or more major cities, and disperse the toxin by plane or automobiles. With just the right conditions, a great many people could be infected, overwhelming the medical system.[39] But that's very remote, since this kind of operation is well beyond the capacity of any terror group operating today—and since the 2001 anthrax attacks in America, the United States and other governments have stockpiled anthrax vaccine.

Like anthrax, botulism toxins are easy to secure and so deadly that they're considered the most toxic protein in nature. But terrorists would find it very difficult to infect large numbers of people with botulism as well, because it degrades quickly when exposed to the air. That's why Japan's Aum Shinrikyo cult abandoned a multimillion-dollar effort to manufacture it in quantities and turned instead to the nerve agent sarin, which it used to kill twelve people and injure thirty-eight hundred in the Tokyo subways in 1995.[40] Moreover, since July 2004, the American government has stockpiled treatments for botulism, as well as anthrax and smallpox, under "Project Bioshield."[41]

Smallpox is much more communicable than anthrax or botulism, spreading easily from person to person, especially since most people in the world either have never been inoculated against it or their inoculations have expired. If a terrorist group managed to unleash the smallpox virus in quantities, it could have devastating effects. It also would be very difficult for any terror organization to do so. Since the disease was eradicated almost thirty years ago, the only known stocks of the virus have been held at the U.S. CDC labs in Atlanta and in Russian state labs. While there are unconfirmed reports of security breaches at these sites, the World Health Organization (WHO) has concluded that a single outbreak could probably be contained. Smallpox vaccine can prevent the infection or reduce its severity, if administered up to four days after exposure. That makes it possible to "break the transmission chain and halt a smallpox outbreak within a relatively short time," so long as the country where it starts has a "strong surveillance system" sensitive to smallpox cases and a public health system that can isolate the cases and administer the vaccine to everyone who's had contact with them.[42] The vaccines available today wouldn't cover most people—the United States has about 15.4 million doses, the WHO stores another half-million doses in the Netherlands for an emergency, and other nations reportedly have 60 million more doses of varying quality and potency[43]—but are sufficient to contain an epidemic.

Like the detonation of a nuclear weapon in a major urban center, there is a bioterror worst-case scenario. It involves an agent such as smallpox released simultaneously in a number of places without "strong surveillance systems," or ones that didn't work as planned. If a smallpox epidemic started in countries or cities unprepared and then spread out of control, it could be a nation-changing event. Air and rail travel would be suspended to try to contain the outbreaks; and large shares of trade would halt as countries, states, and provinces tried to seal their borders. Public schools and public events would be suspended for fear of contagion; and businesses would close because so many of their employees would be sick or terrified of contracting the disease. The only booming sector would be health care, which would be overwhelmed quickly, and the global economy would likely fall abruptly into recession. In the worst cases, the death toll could change national demographics. Otherwise, when the crisis passed in three, six, or nine months, economic and political life would return to something like what it was before.

The two other terror-related wild cards that could reshape the trajectory of globalization and geopolitics involve extreme Islamist upheavals or takeovers in Saudi Arabia or Pakistan. Al Qaeda is still an influential force in both countries, because after the western coalition overthrew the Taliban, most of Bin Laden's followers fled across the border into Pakistan or back to Saudi Arabia. Once home, many joined the affiliated QAP, or "Al Qaeda on the Arabian Peninsula."[44] A Saudi police crackdown following the May 2003 bombings of western compounds in Riyadh weakened the QAP, but most members remain free and loyal to Bin Laden.

The real prospects of Al Qaeda-inspired terrorism in Saudi Arabia are unknown. Bin Laden's deputy, Dr. Ayman al-Zawahiri, has called regime change in Saudi Arabia a priority, urging followers in a July 2007 video to "work seriously to change these corrupt regimes and corrupters" by winning "popular sympathy for a change to Islamic Jihadism."[45] A leading young European expert on Middle Eastern terrorism, Thomas Hegghammer, sees most Saudi militants as nationalistic and therefore more inclined to attack the United States than their own regime.[46] But American experts see it differently, emphasizing the extreme austerity and intolerance of the Wahhabi Islamist ideology as a source of deep resentments against a royal family known for its opulence and corruption.

Saudi Arabia's Shi'a population, estimated at 10 percent of the country and marginalized by the official Sunni Wahhabism, are another breeding ground for anti-royalist feelings.[47] The Shi'a Saudis are concentrated in the nation's eastern region where most of the oil facilities are located and vulnerable to attack, especially the huge processing complex at Abqaiq that handles two-thirds of Saudi oil output, the major export terminals at Ras Tanura, and the Petroline pipeline from the Abqaiq and Ghawar fields to Yanbu on the Red Sea.[48] If extremists overthrow the Saud family—or perhaps worse, if terrorists take out any of the key facilities—the oil disruption could last for a year or even two. The Saudi government has recently seen this threat as growing, and announced in August 2007 a new 35,000-strong security force to protect its oil infrastructure from attack, to be trained and equipped by the U.S. defense giant Lockheed Martin.[49] The world's major economies could weather a brief interruption in Saudi oil by drawing on their strategic reserves; but one lasting a half year or longer would trigger a deep global recession with potentially serious political consequences.

The arithmetic of a cutoff in Saudi oil is sobering. Oil producers worldwide could pump about 6 million additional barrels a day by pulling out all the stops, and the Saudis account for 4 million of those 6 million barrels. If half of current Saudi production went off line—about 5 million barrels a day—that would leave the world 3 million barrels a day short, if everyone else pumped all they could. Under those conditions, one industry expert, Julian Lee, has said "it would be difficult to put an upper limit on the kind of panic reaction that you would see in the global oil markets." Conservatively, oil prices would go to $150 per barrel and beyond. The IMF estimates that a $10 per-barrel increase for a year reduces the world's GDP by 0.6 percent, with a 0.8 percent decline in America, the Euro area, and developing Asia.[50] By this calculation, $150 per-barrel oil would cut five to six percentage points off GDP in the United States and Europe, which would mean very serious recessions. Moreover, a sudden jump of $50, $70, or more per barrel would have cascading effects that could be even more disruptive. It could arrest China's modernization program, overturn governments including the Saudis themselves, and vastly increase the political leverage of Russia and Iran, the world largest producers after the Saudis.

The overthrow of the current Saudi regime by Islamist extremists some-time in the next ten to fifteen years would present other risks. Based on Iran's record, the new government could not afford to turn its back on the oil markets. But as with Iran, antiwestern and anti-American feelings would heighten, perhaps throughout the region,[51] and Saudi Arabia would go from a leading U.S. ally to potential foe.[52] While the Islamists' economic program is vague beyond their opposition to the corruption and conspicuous con-sumption of the Al Saud family, their foreign policy agenda is clearer: Im-placable opposition to the United States, Israel, and all western presence and influence in the Middle East.[53] This agenda from the world's largest oil pro-ducer would pose serious problems for America's determination to protect Israel and maintain stable oil supplies for the global economy.

An extremist takeover of the Pakistani government would have different but equally profound effects on the geopolitics of the next decade. Pakistan is already "a country that has failed five or six times," in the phrase of U.S. for-eign policy expert Stephen Cohen;[54] and extreme Islamists are very powerful there. As recently as July 2007, the "Red Mosque militants" held off nearly twelve thousand of President Musharraf's heavily armed security forces in a weeklong battle. If the militants and their legions of Islamist sympathizers came to power in Kirachi, it could mean a nuclear-armed, Taliban-like gov-ernment with strong ties to Al Qaeda. Some analysts believe they would quickly precipitate a confrontation in Kashmir, which could spur a preemp-tive strike by India and end in a regional nuclear exchange.[55] An equally alarming scenario that concerns some western military experts is the military collapse of the U.S.-led coalition campaign in Afghanistan, which could soon mean, as Lord Ashdown, the former British Liberal Democrat leader, puts it, "Pakistan goes down . . . (and) you could not then stop a widening regional war that . . . would become essentially a war between Sunni and Shi'a right across the Middle East."[56] For these reasons, the United States in coming years will go to great lengths to protect the Musharraf government or a successor much like it. China, a good friend of Pakistan, also would likely try to stop any slide to Islamist extremism, if only to prevent Pakistani support for the Afghan Islamists who in turn support Muslim separatists in China's Xinjiang province.[57]

While it's unlikely that terrorists will pull off any of these wild-card events in the next decade, if they did, the economic and political costs would be enormous, and the secondary effects could reshape the paths of globaliza-tion and geopolitics. Such high stakes should focus the war on terror in com-ing years on these threats, especially once the United States winds down its role in Iraq. The sobering consequences of such spectacular terrorist acts also may provide a political context and agenda for more extensive geopolitical co-operation between the United States, the European Union, and China, as well as Japan.

THE TECHNOLOGY WILD CARD—WHAT TO HOPE FOR

As unpredictable episodes of extreme terrorism could change the world for the worse, equally unpredicted breakthroughs in technology in the next ten to fifteen years could strengthen worldwide growth and development, and possibly help solve the energy, environmental, and health-care crises that many countries will soon face.

Technological innovations that change the way millions of people live and work are neither regular nor rare; and the features of the next important advance, when it will happen, and how it will affect people are all unknown. But there are serious reasons to believe that the overall pace of technological innovation has been quickening, increasing the likelihood that some conse-quential advance will emerge in the next ten to fifteen years. One reason is that physicists, biologists, geneticists, and chemists today, as never before, can harness the enormous power of new information technologies for their fields. Another factor is that for the first time in the history of science, tech-nological progress has gone truly global. Tens of thousands of the world's most inventive engineers, scientists, entrepreneurs, and clever people without portfolios in San Jose, Shanghai, Tel Aviv, Stockholm, and almost everywhere else, read the same analyses, participate in the same discussions, or observe them, and then work together or compete against each other, all routinely.

Globalization has another, less obvious effect on the pace of technological progress. By enabling so many societies to become more prosperous, it vastly expands the market for more powerful software, new medical treatments, more fuel-efficient automobiles, and other new technologies. A larger poten-tial market raises the potential return on developing innovations, which in-creases the resources devoted to doing that—and ultimately the number of new technologies that make it to offices, factories, and homes. So it's not sur-prising that the world's most return-driven and market-based major econ-omy, the United States, has become the world's innovation hothouse (with a few notable exceptions such as stem cell research).

America's position as sole superpower also plays a part. Over two centuries ago, Immanuel Kant observed that wars often drive progress by spurring the development and spread of new technologies. That relationship persists today in the United States, where the Pentagon's $74 billion a year R & D budget is greater than the entire military budget of any other country, except perhaps China. The main office for basic military R & D, the Defense Advanced Research Projects Agency (DARPA), famously came up with what became the Internet. Its current projects include new ways to sense the components of explosives for the war on terrorism, which could lead to medical diagnostic chips; research in supercomputing that could change the architecture of PCs; and techniques to accelerate the development of vaccines from years to months.

While we cannot predict what the next critical breakthrough will be, we can identify where much of world's research talent and financing are focused today. Many of the largest R & D bets in nanotechnology are placed on alternative energy technologies, which could have a major global impact if they ever pay off. Similarly, scientists on the frontiers of biotechnology are targeting new approaches to treat or cure cancer, heart disease, and diseases with genetic components, which eventually could help bail out national health-care systems. Much of the current energy in IT innovation seems focused on increased mobility and miniaturization, trends that could further spread Internet and communications technologies in developing countries and accelerate their growth and globalization.

NANOTECHNOLOGY AND THE WORLD'S ENERGY CRISIS

Nanotechnology is the science of almost unimaginably small "machines" and components—one-ten-millionth to one-billionth of a meter (which is a "nanometer").[58] Researchers can build nano things top-down, for example by depositing ultra-thin films on a silicon wafer and then etching away material to make a fifteen-nanometer computer chip. They also can build them bottom-up, using chemical processes to combine particular atoms or molecules.[59] One of the field's pioneers, Berkeley Professor Paul Alivisatos, calls the dawn of nanotechnology "an important moment in the history of science" because "instead of breaking pieces apart, we're putting them back together."[60] The next decade may establish whether Professor Alivisatos and other enthusiasts are right when they compare nanotechnology's potential impact to the advent of electricity and computers.

Governments worldwide clearly are paying attention, expanding their nanotechnoloy R & D budgets ninefold since 1997, from $432 million to about $4.1 billion in 2005; and companies in at least sixty countries are conducting serious nanoresearch. Much of this research focuses on building the basic components, mainly nanotubes and nanolayers, putting them together in simple nanostructures, and understanding why at the nano level of molecules and atoms, a nanostructure's basic physical characteristics often change as it "operates." They are hard to visualize, but in labs around the world today, there are nano-biodevices, nanotransistors, nanoamplifiers, nanodrugs and nanochemicals, molecular machines and molecular motors, and nanodevices that emit lasers and store energy.[61]

Without realizing it, millions of people already use products that have nanoscale components. Nanomaterials are used in magnetic recording tapes, sunscreens, automobile sensor and catalytic systems, corrosion and scratch-resistant paints, long-lasting tennis balls and lightweight tennis rackets, stain-free treatments for clothes, and dressings for burns and wounds.[62] The hard drives in most computers have nano-thin layers of magnetic molecules that increase their storage capacity. In a short time, nano-coatings also will be used in many medicines, wear-resistant floors, and diamond-hard parts for airplanes and rockets that are so light they save fuel.[63] The uses are expanding so rapidly that the nanomaterials market passed $1 billion in 2007, and industry forecasters project gains of another 40 percent in 2008 and a $35 billion market by 2020.[64] And a joint project by the Mitsubishi Research Institute, Deutsche Bank, and Lux Research in the United States recently estimated that by 2015, $1 trillion in products worldwide will contain nanocomponents, produced by two million nanotechnology workers.[65]

The nanotechnology motherlode for researchers and investors is a practical, scalable, affordable, nano-based way to produce hydrogen or solar energy, which could wean the world from fossil fuels and appreciably slow global warming. The use of clean, renewable hydrogen fuels may be ten, twenty, or thirty years away—too late at the far end to help much with climate change—but if and when it happens, it will likely depend on nano-materials. For example, the most promising research projects today for extracting the hydrogen from water—without burning fossil fuels to do it—involve either nanoscale electrodes to produce chemical energy or nanowires to produce electricity.[66] Nobel Laureate Richard Smalley predicts that in a decade or less, scientists will develop successful prototypes of fuel cells and batteries built from nanotubes, which are rolled-up "sheets" of carbon atoms, as well as solar energy converter panels covered with nanostructures. Once fully developed, they could make hydrogen a practical, clean, sustainable energy resource, and "solve the entire world's energy problem with just six solar energy plots smaller than a fraction of Arizona."[67] If that doesn't pan out, developers also are working on membranes built up from nanostructures that can remove CO_2 (and nitrogen) from natural gas as it is being produced.[68]

Nanocomponents also are central to many research projects into how to store the hydrogen to power an automobile. It would take about ten pounds of hydrogen to generate the energy of a full tank of gasoline, but that much hydrogen at room temperature would be a gas that would fill 50,000 liters, or 14,000 gallons. In order to use hydrogen as fuel, automakers have to build a car that can keep it under enormous pressure—a concept car being developed by Honda would store the hydrogen at pressures equivalent to the hydraulic pumps on an Airbus[69]—or use metal or chemical compounds that can absorb the hydrogen and then release it through a thermal, electrical, or chemical reaction to power the car. So far, the best hydrogen-absorbing material anyone has come up with are metal and chemical nanotubes.[70] Here, too, Dr. Smalley believes that nano-based storage devices for hydrogen are less than a decade away.

The other area where nanotechnology breakthroughs could have far-reaching effects are nanoscale machines. One type is a very miniature version of an existing machine. A group of scientists at the University of California at Berkeley have built a nano-conveyor belt for ferrying atoms, and the world's smallest nanomotor, including pistons, electromechanical switches, and transistors all created from nanotubes. A nanoscale "cantilever" created by the group could bend like a diving board when a particular molecule binds to it, creating a highly sensitive chemical or biological wireless sensor. And other researchers are working on capsules a few hundred atoms long that would be injected like a tiny submarine into the bloodstream to seek out and destroy diseased cells. Another type of nanoscale machine is the molecular or nanofabricator, which theoretically will be able to select and arrange individual atoms, one by one, to produce another nanostructure.[71] The model here is nature itself: Ralph Merkle, a nanoscience pioneer, notes that, "nature can grow complex molecular machines using nothing more than a plant."[72]

Most scientists see this kind of nanotechnology as a long way off—if it ever comes—but also critical to commercializing nanotechnology. The Berkeley group's lead physicist, Dr. Alex Zettl, explains, "Right now, we make most nanodevices one at a time, sometimes . . . atom by atom . . . but if you can't scale up then it becomes a curiosity instead of a viable technology."[73] One of the field's pioneers, Don Eigler at IBM, who conducted a famous nanoscale demonstration in 1989 by arranging thirty-five individual xenon atoms to form the IBM logo, is still cautious: "We can imagine a whole bunch of wild applications. The chances that we're going to nail some of the big ones on the head are, I think, fairly low."[74] For nanotechnology to affect the world's pressing energy and environmental problems in the next ten to fifteen years, a recent prediction by the research director of the Center for Responsible

Nanotechnology, Chris Phoenix, will have to be right, that nanofabricators and nanofactories will soon be produced routinely in laboratories and ready for industrial-scale production by 2013.[75]

A nanotechnology breakthrough in the energy area could have large effects in many countries. A competitively priced, scalable form of solar or hydrogen power adapted to fuel manufacturing plants and to heat and cool buildings and homes could halt or even reverse the upward spiral of energy prices, providing additional impetus for China's modernization. The United States, with the most energy-intensive economy of the advanced nations, would also benefit significantly. But an energy breakthrough would not be a quick or an easy panacea for many places, such as India or Latin America, since moving to new fuel sources would require trillions of dollars in new investments. The largest effects, especially in the first decade, could be environmental and geopolitical rather than economic. New, nano-based energy sources could contribute greatly to efforts to contain CO_2 emissions and their concentration in the atmosphere. The spread of such new energy sources also could be a body blow to the oil-producing states, including Russia and Iran, and over time reduce the geopolitical significance of the Middle East.

BIOTECHNOLOGY AND THE HEALTH-CARE CRISIS

Potentially powerful breakthroughs also could occur in biotechnology, especially if another leap of progress in manipulating genetic materials produces practical advances in medical care.

The current international public and commercial efforts to map the human genome by analyzing the order, chemical characteristics, spacing, and function of the more than 23,000 genes on human chromosomes is one of the great scientific ventures of modern times. The project was originally conceived, strangely enough, in the U.S. Department of Energy and a handful of universities in the 1980s, and formally began in 1990 as a project of the Energy Department, the U.S. National Institutes of Health, and a consortium of research institutions in France, Germany, Japan, Britain, and China, as well as the United States. Propelled in the latter 1990s by advances in software, the Human Genome Project (HGP) announced a rough first draft of the genome in 2000 and more complete and detailed versions in April 2003 and May 2005, covering about 92 percent of all gene sequences. Private companies also took on the mission of mapping the genome, including Celera Genomics, which published its own gene map at the same time as the HGP, and Human Genome Sciences, which applied the results to develop medical treatments.

Decoding the remaining 8 percent of sequences, including millions of base DNA pairs at the core and ends of each chromosome, will have to wait for new technologies; but many scientists believe that most of those sequences are "junk DNA" that may not even contain genes.

While Celera has never publicly released its DNA sequencing map to medical researchers, the HGP's analysis is stored in "GenBank" databases available to anyone on the Internet, along with scientific literature on each sequence. The interpretations of the data are still fairly basic, but private and government researchers are developing new software to analyze and mine the results for insights that could eventually help produce practical new medical treatments. The major applications so far have been new tests for genetic predispositions to breast cancer, cystic fibrosis, and liver diseases; and there also has been progress in developing tests for other cancers and Alzheimer's disease. By one estimate, 200 million people worldwide had used genetically engineered tests and pharmaceuticals by 2001, and the number today is several times larger.[76] Over the next ten years, genetic testing could become a routine medical practice for patients in all advanced countries.

Another, related frontier of biotechnology involves work on identifying the 200,000 proteins in the human body and their interactions with genes, since many diseases are thought to be caused by defective genes that incorrectly code a protein. So far, scientists have developed experimental treatments targeting about 500 proteins, or one-quarter of 1 percent of the total. Many researchers and the biotech companies that employ them believe they can map most important proteins in another five years and create gene therapies to repair or replace the defective genes that cause particular diseases in a decade's time after that.[77] Another line of genetic research involving proteins, touted by the founder of Human Genome Sciences, Dr. William Haseltine, aims to avoid most adverse drug reactions—which in the United States alone result in some 100,000 deaths each year, or more than diabetes, AIDS, pneumonia, or automobile accidents. This line of research is designed to identify and isolate particular genes that could be injected into cells and induce them to produce a desired protein that would "deliver" the medicine.[78]

Stem cell applications to treat and perhaps cure "single-cell diseases" are closer at hand.[79] Stem cells are used today in bone marrow transplants for some cancers, leukemia, and serious blood diseases. In the near future, according to the U.S. National Institutes of Health, stem cells may become "a renewable source of replacement cells and tissues" to treat Parkinson's and Alzheimer's diseases, spinal cord injury, stroke, burns, heart disease, diabetes, osteoarthritis, and rheumatoid arthritis.[80] On the frontier of research in this area, British scientists have used stem cells from the umbilical cords of newborns to grow small versions of the human liver in their laboratories. They hope to be able to use stem cells to repair damaged livers in another five years and to grow new livers for transplant a decade after that.

Similar research also is proceeding in other countries. Argentine doctors have successfully injected bone marrow stem cells into the pancreas of diabetes patients, enabling them to produce their own insulin and lower their blood glucose levels. Italian researchers have alleviated the symptoms of muscular dystrophy in golden retrievers by injecting them with stem cells. Perhaps most important, Spanish scientists have used stem cells to treat people suffering from angina and heart failure, Swiss scientists have grown heart valves in their labs from amniotic fluid, and American scientists recently used embryonic stem cells to successfully rebuild heart muscles and improve their functioning in lab rats, four days after a heart attack.[81] There are many scientific and technical obstacles, but it is possible that within the next ten to fifteen years, scientists will be able to use stem cell therapies to successfully treat a fair share of current heart disease and cancers, and grow new organs for transplants.[82]

Predictions in medical science are often off-target. As Dr. Eric Juengst at Case Western Reserve University notes, "Biology is always more complicated than we think."[83] Dr. Eric Lander, the author of the first published human genome sequence, believes it could take several more decades to develop effective drugs from the genome project.[84] Other scientists predict that the first generation of "designer," genetically enhanced babies will be born in the next decade, and new generations of genomic drugs will be developed that will "switch off" the genes that produce certain types of cancers.[85]

The surest prediction is that cultural, religious, and philosophical objections will continue to be raised in these areas and will delay some lines of research. But with the globalization of science and technological development, restrictions on research in one or even several countries cannot permanently stop the science. The Bush administration's refusal to fund work with embryonic stem cells has shifted many lines of research to stem cells from umbilical cords and to facilities in other countries, which is apparent in the recent advances in stem cell treatments coming from Europe, Latin America, and Asia. And if the prediction of a prominent geneticist at the University of London, Dr. Stephen Jones, is right that the first human cloning will occur in the next five years, it won't happen in the United States.[86] Similarly, European resistance to genetically modified food will not stop the rapid biotechnological progress expected in that area.[87] In the near future, for example, genetically enhanced strains of "golden rice" high in beta-carotene should be available in developing countries with widespread vitamin A deficiencies;[88] and according to one of the fathers of the "green revolution," Dr. Norman Borlaug, very soon, "we will have the know-how to produce the food that will be needed to feed the population of 8.3 billion people that will exist in the world in 2025.[89]

If safe and effective genetically engineered or stem cell treatments for diabetes, heart disease, and cancers come to pass in the next decade, they could help bail out strapped national health-care systems by supplanting millions of expensive surgeries and long-term treatments. If these or other breakthroughs occur in treating and even curing the more common serious conditions—again, heart disease, cancers, and diabetes—countries that adopt them could see stronger growth and productivity in just a few years, as millions of people recover and work again full time. Great advances in those areas also could help some countries deal with their demographic pressures by extending the period when older people can work productively. But these treatments also could be so costly to develop and use that they worsen the financial burdens that advanced nations already face in their health-care systems and force some countries to forgo their broad use.

INFORMATION TECHNOLOGIES AND THE OUTLOOK FOR GLOBALIZATION

While nanotechnologists and biotechnologists write regularly about their groundbreaking research horizons, information technology scientists generally seem more modest. After decades of astounding breakthroughs, many IT scientists see more powerful hardware and software coming soon, but no revolutionary changes. Greater computing power and software capacities alone, however, will mean falling real prices for both; and that should further accelerate their spread in many developing countries.

In the advanced nations, most of the development dollars are going to software. The president of Microsoft International, Jean-Philippe Courtois, sees smart software coming that can process the way a person uses it and then automatically customize itself for each user.[90] A much-followed blogger on IT trends, Robert Scoble, sees new software that will attach metadata to photographs so they can be searched as easily as text, and programs that will convert text to speech and then back again so seamlessly that a Web search on a cell phone or computer for a restaurant or hotel will produce a verbal answer.[91]

Other industry experts see another trend that Microsoft, for one, won't welcome—the spread of the open-source movement, where people get free access to software programs and their source codes, so they can revise them at will or hire someone else to make the changes they want. The president and COO of Sun Microsystems, Jonathan Schwartz, predicts a "tidal wave washing over the marketplace" as "every product at Sun at some point will be free or open-sourced,"[92] and others see the same wave coming.[93] Another potential development that would shake the firms that currently dominate software markets, according to a former executive editor of the *Harvard Business Review,* Nicholas Carr, is the emergence in the next few years of IT "utilities," where business and consumers will be able to access and use thousands of software programs online, in a shift similar to the one a hundred years ago, when manufacturers with their own electric generators began buying electricity from central utilities.[94]

The accelerating spread of broadband also is changing the Internet from a text-based medium to a video-based one. Today, Internet video is still largely entertainment; within the next few years, it will dominate mail and business communications as well.[95] This shift will likely change the way most people pay for their Internet access, and more important, increase telecommuting or teleworking.

How Internet providers charge for Web access is a sensitive political issue in the United States today, in the guise of a debate over "network neutrality," or whether Internet access providers should be able to charge Web sites different prices based on how much bandwidth their content requires. The economic pressures come from the fact that even as the numbers of new Internet subscribers slow in most advanced countries, where access rates are topping out at 70 to 80 percent, demand for bandwidth is rising rapidly with the spread of video. One minute of text browsing takes between 2 and 200 KB of bandwidth, compared to at least 4,000 KB for a minute of video, even with advanced compression technologies. Already, video accounts for 50 to 60 percent of all worldwide bandwidth traffic; and with the advent of new video applications including Internet-based high-definition TV, its share is expected to reach 80 to 90 percent by 2010.[96]

By then, bandwidth demand could hit up against the Internet's capacity to handle it, causing congestion and serious slowdowns unless Web providers and backbone companies increase their investments by a factor of three to four. Since most of the new demand comes from video-based applications, not new subscribers, the standard monthly flat fees for Internet access won't support large increases in investment. Since real Internet congestion would be intolerable for tens of millions of people and businesses, the way people pay for Internet access will change. Either all flat fees will go up, people will pay in tiers based on how much bandwidth they use, or content providers will pay in tiers based on how bandwidth-intensive their offerings are.

More important, the growing use of Internet-based video will affect the way businesses operate in many countries. In a few years, the Internet will transmit live, large-screen moving images almost as clearly as a theatrical film. That will allow millions more people than today to work from remote locations, especially their homes, and still maintain regular face-to-face contact with their bosses and colleagues. Once the technology is in place, the economics of teleworking suggest it could expand quickly. Companies could save on not only business travel, but also basic overhead for office space and utilities. Large-scale teleworking also could ease some of the conflicts between family and job for millions of two-earner families with children. AT&T, which has promoted its use for several years, figures that in 2005 teleworking saved the company $30 million in office space, reduced job turnover by half among those doing it, and increased productivity worth $150 million.[97] Experts in this area estimate that more than 12 million Americans worked remotely at least one day a month in 2006[98] and by 2008 or 2009, an additional 24 million Americans and 100 million people worldwide will telework on a regular basis.[99]

If the technology and economics play out in this way, by 2015 and 2020 the workdays of tens of millions of company men and women in America, Europe, and Japan will be very different than today. The savings from so many people working full time from home or from satellite centers, and so many others splitting their working time between home and the office, could help increase profits, investment, and growth. Perhaps as important, teleworking at significant levels could play a role in reducing greenhouse gas emissions.

However, the largest changes expected by most IT experts over the next decade involve continuing increases in the power and capacity of both hardware and software. These increases will translate into falling prices for the technologies already in place in America, Europe, and Japan, especially wireless and mobile IT, and their more rapid spread in large developing countries. The economic and social impact of wireless phone and Internet could be greater in Asia and Africa where they constitute the first modern communications system in many places, than in advanced countries where they supplant or supplement old, wired versions. And the main reason why wireless communications technologies will become part of the basic infrastructure across China, India, and other developing countries over the next decade is their low cost. Wireless systems cost so much less to build than copper landline systems that wireless service costs about one-third that of copper or fiber service.

This increasing use of wireless in developing nations already has produced large economic dividends, according to a recent McKinsey study, by boosting productivity in 2006 enough to add an estimated 2 percent to India's GDP, 5 percent to China's GDP, and 7.5 percent to that of the Philippines.[100] Cabbies in Beijing and Manila for example, waste less time driving around because they can call for directions and their regular customers can call them for service.[101] Similarly, Indian fishermen use cell service to determine which port to bring their fish into, Philippine farmers use it to compare prices for feed and equipment, and poor Latin American day laborers call around to find out which nearby villages might need their services.[102] The increasing use of mobile communications in developing countries also may ease some of the strains of modernization, enabling Chinese peasants, for example, to move to cities for jobs and still stay in touch with their families. (In the same way, prepaid international calling cards support immigration to the United States by providing an affordable way for Latin Americans and Asians to maintain contact with home.[103]) The only reason that the gains in India are less so far than in China and the Philippines is that wireless use is less widespread in India, because the people are poorer and the phones and service are more expensive there. But those costs are coming down in India, too, with Indian wireless subscriptions passing the 100 million mark in April 2006 and expected to top 200 million by 2010.[104]

Over the next decade, the same economics will almost certainly produce inexpensive wireless Internet access in these countries. Building out a wireless broadband system such as Wi-Fi costs about one-ninth as much as a landline system, since it can serve larger numbers of people with less equipment, and can be deployed and repaired more quickly and inexpensively.[105] Allow multiple wireless Internet providers to compete with themselves and with landline providers, and all access prices fall—as they have in advanced countries, where broadband prices are down and broadband use sharply up. From 2003 to 2006, U.S. households with broadband jumped from less than 21 million to more than 50 million; and industry experts predict that worldwide broadband use will go from some 250 million households in 2007 to 474 million by 2010.[106] Much as lower-income American households have gone online at the same rates as higher-income households, but five or six years later once prices have fallen, so poorer countries should be able to go online on a large scale, especially on wireless systems, over the next decade.

This trend may be reinforced by the imminent release of a laptop computer priced so low that it could become broadly available to children and adults in developing countries around the world. The MIT Media Lab launched the "One Laptop Per Child" project three years ago at the World Economic Forum in Davos, Switzerland.[107] The project's "XO" computer uses free, open-source software, and costs $176 per unit to produce.[108] It also incorporates innovations designed for conditions in developing countries, but which likely will become standard in many laptops—a screen display readable in bright sunlight, high resistance to rain and dust, a protected Wi-Fi antenna that allows very broad range, and a twelve-hour battery with power alternatives including a solar panel or a solar-powered battery charger.[109] Prototypes of the XO already are in use in schools in Brazil, Nigeria, Thailand, and Uruguay. With the machine's formal launch, governments that adopt a policy of one laptop per child will be able to buy tens of thousands of the machines at cost, including operating systems and other software in their own languages.[110] In hopes of further bringing down the price for poor countries, the XO also will be sold in the United States and Europe for $350 to $525 per machine.[111]

The trends in IT, like the prospects of breakthroughs in nanotechnology and biotechnology, could affect the paths of many countries over the next decade. The convergence of mobile telephone, mobile Internet, and inexpensive laptops could drive another burst of growth and development over the next decade—especially in places that welcome foreign companies to build the networks and provide the hardware and software, and invest in educating their children and workers to use the technologies. This development could be especially powerful in parts of Latin America that recently eased restrictions on foreign investors, in areas of Africa not torn by civil conflict, and perhaps in India. But the prospect of widespread Internet access will present a problem for China's leadership, who consistently work to restrict their people's unfettered Web access. In advanced countries, more computing power and falling software prices also should provide another boost to productivity, especially in the United States and other less-regulated economies where companies are better able to adjust their operations to take advantage of new technology. But if there are other kinds of breakthroughs coming in IT, on the order of clean, cheap energy from advances in nanotechnology or potentially revolutionary treatments for cancers and heart disease from biotechnology labs, the scientists developing them aren't talking about them yet.

THE FUTURE THAT MAJOR NATIONS CANNOT AVOID

This is a new epoch, and every business, government, and society will have to find their places in it. From time to time, singular political figures can inspire nations to follow new paths—Franklin Roosevelt, Winston Churchill,

Charles De Gaulle, Mao Zedong, Martin Luther King, and Mikhail Gorbachev—and change their societies for generations. Outside politics, creative geniuses and their teams can achieve technological breakthroughs that ultimately reorganize aspects of everyone's life. Any of this could happen tomorrow almost anywhere. However, the three seismic developments examined here are already embedded deeply in the social, economic, and political arrangements of every major country, and many of their important consequences are virtually certain.

Since no one can appreciably affect the size of generations already born, most of the direct effects from the historic aging of national populations cannot be averted. The biggest consequence for most people is that the era of governments promising and providing ever larger retirement and health-care benefits has ended. Ten or fifteen years from now, new retirees across Western Europe will receive pension checks less than they expect today, the taxes to finance those checks will be higher, and budget deficits once again will be growing larger year after year. The most serious problems, however, will hit health-care programs as the costs of procedures and other treatments for the sharply rising numbers of elderly people in every advanced country expand beyond every country's real capacity. Americans, Europeans, and Japanese will all see large increases in their health-care-related taxes, insurance premiums, deductibles, and out-of-pocket costs—and many governments and insurers will begin to chip away at broad access to the most expensive and advanced treatments.

This aging of nations will almost certainly also mean less economic progress, especially for the next generation of Europeans and Japanese, because it will begin to shrink their labor forces and reduce their national savings, which in turn will tend to lower their investment rates and productivity gains. The drag on productivity may be offset in coming years by how much more education Europeans and Japanese in their twenties and thirties have, compared to their parents—from 1970 to 2000, the number of people attending college in France, Britain, and Japan, as a share of their college-age populations, went from less than 20 percent to nearly 50 percent or more.[112] But it won't matter much unless their governments ensure that they have high-paying jobs to go to, by reforming their economies and increasing public investments. And that's not likely to happen in most places, because most Europeans and Japanese with jobs will resist reforms that might undercut their own job security, and the resources to finance more public investment would have to come out of the same pot that will fund the burgeoning boomers' retirement benefits.

The United States will feel less demographic pressure than Europe or

Japan for at least the next decade—setting aside health care—and China as well will be able to contain the impact of its own rapid aging on its economy and budget. But China's decision to let its old pension and health-care arrangements unravel may produce popular demands for change that it cannot safely resist. Moreover, America's and China's advantages are temporary. Looking ahead ten or fifteen years after 2020, both societies will face aging-driven economic and fiscal pressures comparable to what Europe and Japan have to deal with much sooner.

Across the world, globalization will change people's lives even more than demographics. It may be said that the essential drive of globalization is to break down the barriers that countries use to protect their businesses from competition and their workers from risk, and that its central value is competition at home and throughout the world. Its promise is that by breaking down barriers and embracing competition, workers and businesses will create more wealth—and the early returns suggest that on a global scale, that promise is realistic: As globalization has taken full hold in the last five to ten years, world output has grown faster than during any comparable period on record. And economies are paying large costs for trying to hold themselves back from globalization, such as Russia and parts of Latin America, or to hold on to their old arrangements, as in much of Europe and Japan.

Over the next decade, the economic imperative for most nations will continue to be: Open your economy to the ambitions of your own people and the capacities of others. The countries that will prosper in the next decade will be those that open their markets as much as practicable to the ideas and drive of their entrepreneurs, the investments and operations of foreign companies, and the goods and services produced through global networks. Following this course—or not doing so—is not a matter of right or wrong, but simply the best way the world has at this time to generate growth and wealth.

Globalization will produce tens of millions of winners who follow its rules. The most prosperous advanced countries over the next decade will be those that focus most successfully on what advanced economies do best—creating, adopting, and adapting to powerful new technologies and production processes; to new ways to finance, market, and distribute things; and to new approaches for organizing and managing businesses. Here's where the extensive regulatory protections for workers and smaller companies that much of Europe and Japan still provide will take their largest toll. Compared to them, the United States, Ireland, and a few others should see stronger growth, their companies will generally rack up higher profits and larger market shares, and their consumers will pay lower prices. And within every advanced country, the biggest gains will go to those who work in, manage, or own a piece of the enterprises that develop or commercialize innovations or that simply use them effectively.

China and other developing nations that stick with the rule book of globalization will continue to modernize and grow at a historic pace, and move several hundred million more people out of poverty over the next decade. By 2020, China may overtake the United States as the world's largest trading nation, and most of its world-class manufacturers will be native Chinese companies rather than American or Japanese transplants—so long as serious domestic unrest or its backward service sector and financial system don't derail the rapid progress.

Over the next decade, China also will become a large investor and owner of American and European companies. China's foreign direct investment (FDI) abroad grew twentyfold from 2000 to 2006, when they topped $16 billion.[113] That's still small compared to the $217 billion Americans invested abroad in 2006, and thus far most of the hundreds of billions of dollars, euros, and yen that flow into China every year come back through purchases of western government securities or investments in oil projects in the Middle East and Africa. But since 2006, FDI from China has been more diverse and upscale, including the purchase of large stakes in Bear Stearns, Barclay's, the Blackstone Group, and 3M, and the outright purchases of Britain's MG brand by the auto maker Chery and IBM's PC division by Chinese computer maker Lenovo. Over the next ten years, Chinese state enterprises or private companies will buy stakes in scores of marquee American and European banks, manufacturers, and service businesses—or buy them outright—especially in western markets, such as autos, aircraft, electronics, machinery, and software, that China plans to break into.

The next decade of globalization also will exact large costs in every country. Developing nations that remain largely closed and don't invest seriously in their infrastructure and education—most notably Russia, parts of Latin America, and much of Africa—will slip further behind. Many African countries, for example, could face waves of emigration by educated young people and middle-class professionals, so large that they further weaken their prospects and, in turn, the stability of some regimes. Even in China, globalization will not leave everyone better off. The last decade saw tens of millions of peasants driven out of agriculture, tens of millions of factory workers thrown out of work as the government closed down state-owned enterprises, and the end of broad-based pension and health-care coverage. The next ten years will produce more dislocations that will be equally wrenching.

The next decade also won't be easy for many Americans, Europeans, and Japanese. Over the last five years, the real wages of average workers fell in

France and Germany, moved little in Japan, and rose less than 1 percent a year even in the United States, where productivity gains averaged 3 percent a year. This wage squeeze is an unexpected side effect of the competition and growth that globalization promotes, and it's not going away. With thousands of companies scouring the world for the cheapest and most efficient sources of labor, materials, financing, and services, and striving to grab a piece of their rivals' business by coming up with the next new innovation, all the heightened competition makes it harder for everybody to raise their prices. (That's an important reason why overall inflation has been tame for some time and likely will remain so for the next decade.) And as the costs of energy and health care rise sharply for companies that often can't raise their own prices, they're forced instead to hold down their labor costs. There's nothing on the horizon to change any of these forces, and tens of millions of Europeans, Japanese, and Americans can expect, at best, only modest wage increases over the next decade. Even that will be out of reach for people in advanced economies whose education and training stops at high school or, worse, stops short of that.

Other aspects of globalization will exact additional costs, especially in Europe. Regardless of what any government does, the great bulk of basic manufacturing operations and jobs are not coming back to advanced countries. Nor can anyone prevent the additional losses of millions of service jobs that will soon be performed electronically and remotely. The United States, Ireland, Sweden, and other countries open to globalization and innovation will create new jobs to offset the losses. Tire and steel factories in the United States and Europe won't be hiring, but some of their American workers will become cable installers or medical machinists—while others looking for work will find it at fast-expanding firms like Google, which launched a new product nearly every week for the last three years and tripled its workforce. But if most European companies continue to largely ignore the world outside Europe and America, they will lose more of their home and foreign markets. Recent promises by France's Sarkozy to roll back some labor restrictions, like recent moves in Japan to unravel some of its barriers on foreign direct investments— if they follow through—could help in coming years. But in most respects, domestic pressures in Europe and Japan to subject their native companies to more competition will remain weak. Yet if these countries don't shift course, globalization will be unforgiving. On top of the cost squeeze of wages, these problems may well leave the real incomes of average Europeans and Japanese lower in 2020 than today, a remarkable failure for an advanced economy.

As globalization pitilessly divides workers, companies, and countries into winners and losers, the gaps between them will narrow or widen. In a process that economists call "convergence," successful developing countries led by

Korea, Singapore, Taiwan, and Portugal will close some of their income gap with advanced economies. And inside the fast-growing developing nations, large numbers of very poor people will become much better off, even if the standard economic measures of equality do not improve much. (China has become one of the world's least equal societies even as several hundred million people have moved up and out of poverty, because some 10 million high-skilled workers have made even more progress and a handful at the top have become very rich, very fast.) But the more successful developing economies also will pull further away from the laggards in Latin America, Africa, and parts of Asia.

Inequality between advanced countries also will increase as America, Sweden, Ireland, and a few others will continue to grow faster than places like Germany, France, and Japan. Inequality also will intensify inside most advanced countries, especially the United States. High returns on capital are making the rich much richer, and the salaries of those who develop or work with the panoply of twenty-first-century innovations are increasing faster than the salaries of everyone else—while Americans, Europeans, and Japanese who have to compete with remote workers in developing countries see their paychecks shrink, and the salaries of most other working people are squeezed by the intense competition that globalization promotes.

Globalization will present advanced countries with some formidable political challenges. Many domestic political battles of the next decade, especially in America, Britain, and Germany, will focus on the spiraling prices for energy and health care, driven in part by globalization. However, the largest political challenge will entail maintaining popular support for open economies even as many people's incomes stagnate. The hard truth is that the only course in this period that won't lead to national decline is to recognize the way the world is changing and do what can be done—keep the economy open while, for example, providing broad training in new technologies, supporting new business formation to create more jobs, and easing pressures on wages by slowing the rising costs of health care and energy for businesses. Societies that work with the forces driving globalization—the United States, Ireland, and Sweden come to mind, and Korea and China among developing nations—should find the next ten to fifteen years a period of real prosperity.

The core elements of today's geopolitics also will be with us at least through the year 2020. America's position as a superpower with no near peer will not change, whatever the outcome of the U.S.-led campaigns in Iraq and Afghanistan; and for another generation or longer, no other nation or plausible alliance of nations will be able to create a real balance of power with the United States. The current global mission of the United States to preserve security across the world is very unlikely to change, either, however American voters ultimately judge the Bush administration's policies and whomever they next elect president.

America's global power in this period is not built on its policy finesse or execution, but on how its global mission serves the needs of most other nations and its disproportionately large capacity to carry out that mission. Whatever the leaders of other nations sometimes say for domestic political reasons, almost every country welcomes America's tacit pledge to prevent regional powers from overwhelming their neighbors—as in Kuwait and Bosnia—and safeguard the sea lanes and air routes that carry the world's trade and oil supplies. Governments throughout the world accept that unless someone assumes those responsibilities, their own security could be at risk—and for the foreseeable future, no nation but the United States can or will. So, as Iran or later Syria move further down the road to nuclear weapons—or if radical Islamists should seize power in Pakistan—France, Germany, China, and even Russia will accept and privately welcome American-led campaigns to stop it.

Other developments unknown today will affect the course of geopolitics over the next decade, and when it happens only the United States will be in a position to lead the world's response. Europe and Japan have the resources to create a serious geopolitical role for themselves, but decades ago they put aside the inclination and commitment to do so. The power to balance or challenge the United States on a policy like the Iraq war, as the Soviet Union could do during the cold war, requires the capacity to back up and carry out that challenge. Given Europe's and Japan's looming domestic demands and economic problems adjusting to globalization, there is no prospect that either will commit those resources to geopolitics. Russia under Putin has the inclination to challenge the United States and does so regularly in words, but it is much too weak militarily and too small economically to back up its bombast. Looking ahead, Russia's disastrous demographics and its pervasive economic distortions and corruption will effectively preclude it from rebuilding the resources required to constrain the world's only superpower. A RAND Corporation study entitled *Assessing Russia's Decline* concludes that the once-superpower has become "a weak state" in "military, social, and political decline" and with little "ability to resolve its economic troubles."[114] China is the only country with the means and the will to eventually make itself heard and heeded in geopolitics. But China will be in no great hurry to do so in the next ten to fifteen years, since its leaders are convinced that their political authority depends on rapid economic progress, which in turn depends vitally on good relations with the United States.

China's global influence will be felt in at least one important way in coming years. Its spectacular economic success and the prospect that someday it will be able to challenge the United States will make it a political and economic model for other countries, in stark and direct competition with the American model. The international appeal of the U.S. approach—combining open markets, individual political freedom, pluralism, and the rule of law—rests as much on its record of political and social stability as on its current economic and military dominance. Most of the reconstituted nations of central and eastern Europe, much of Asia, and a good part of Latin America have adopted the basic American model. China offers authoritarianism combined with western technology and business methods, and the promise that it can generate faster growth and development than the compromises and delays of democracy. The approach and the promise will appeal almost certainly to some Asian, African, and Latin American governments over the next decade. There is a third model as well in the world today: the new Islamist approach that uses authoritarian politics in the service of religion, combined with hostility to western values and sometimes economic ties with western nations. This model in its strict form, as in Afghanistan under the Taliban, has limited international appeal and reach; but combined with western investment and trade, the Islamist model is already a prominent approach in the Middle East and could spread to parts of Africa and Asia over the next decade.

For all of America's disproportionate economic and military power, its ability to wield that power to shape worldwide developments will continue to be limited by domestic opinion, especially should the United States once again try to go to war largely alone. As Zbigniew Brzezinski has recently observed of the United States in the Middle East, "While (America's) power is incomparably greater than that of any state . . . it cannot, for domestic reasons, mobilize on a sufficient scale to impose its will by force."[115] Whatever the next U.S. president believes concerning foreign affairs, she or he will have to rebuild domestic support for a sole superpower's inescapable mission of preserving global stability. Al Qaeda and its imitators could make that task politically easier by carrying out another high-profile attack on the order of 9/11. The essential element, however, will be the next administration's virtually certain return to the traditional U.S. approach of addressing threats by building and leading expansive international coalitions.

Over the next decade, America will use broad collaborative approaches to deal with, for example, growing pressures to reduce the risks of climate change and the next episode of global financial turmoil that threatens trade and investment flows. Dealing with future incidences of domestic instability in oil states or attacks on their regimes also will involve both American leadership and the active participation by key Muslim countries, France and Germany, and perhaps China and Russia. Moreover, even while the United States takes the lead role, all of the world's major nations will be actively involved in trying to slow the continued spread of nuclear capabilities, avoid clashes between nuclear-armed developing countries, and prevent terrorists from securing nuclear materials or weapons. The next decade also may demonstrate the limits of such coalitions—and perhaps even the efficacy of a lone American response—if political turmoil in the Russian Federation or domestic unrest in China should threaten to destabilize those governments.

Ancient Greeks taught their children that the union of Zeus and Themis, the goddess of necessity, produced three daughters known as the Moirai or Fates, who determined the span and course of people's lives. Most of us now see the world differently, confident that human decisions along with some random events determine the conditions that individuals and societies confront and the paths they take. But nature always has a powerful hand in all human matters, and the three historic shifts of our time have the practical force of nature in their own spheres. Yes, political leaders with particularly good or bad ideas will trigger unexpected turns in our geopolitics, businesses and governments will come up with new ways to address the stresses of globalization and capitalize on its opportunities, and even the inexorable aging of most societies will produce unanticipated social, economic, and political responses. Yet these historic developments have staked their claims on our lives, and in this time, at least, they will not be denied.

NOTES

I. THE GLOBAL BLUEPRINT

1. *Population and the American Future: The Report of the Commission on Population Growth and the American Future,* The Center for Research on Population and Security, http://www.population-security.org/rockefeller/001_population_growth_and_the_american_future.htm#TOC.
2. Interview, January 5, 2006.
3. Bureau of Labor Statistics (BLS), Data on Mass Layoffs, http://www.bls.gov/home.htm.
4. Alexander Hamilton, James Madison, John Jay, *The Federalist Papers,* Paper Four (New York: Signet Classics, 2003).
5. Julie DaVanzo, Olga Oliker, and Clifford A. Grammich, *Too Few Good Men: The Security Implications of Russian Demographics* (Santa Monica, CA: RAND Corp, 2003).
6. Ibid.
7. Interview, September 12, 2006.
8. Interview, December 1, 2005.

2. THE DEMOGRAPHIC EARTHQUAKE

1. DeVanzo, Oliker, and Grammich, *Too Few Good Men.*
2. Ibid.
3. Diane Farrell, Sacha Ghai, and Tim Shavers, "The demographic deficit: How aging will reduce global wealth," *McKinsey Quarterly Online,* March 2005, http://www.mckinseyquarterly.com/.
4. Quoted in Robert Stone England, *Aging China: The Demographic Challenge to China's Economic Prospects* (Westport, CT: Praeger/Greenwood, 2005).
5. Interview, January 25, 2006.

3. THE PRIMACY OF GLOBALIZATION

1. International Labour Organization, Key Indicators of the Labour Market (KILM), chapter 6, "Wage and Labour Cost Indicators," figure 15c, "Real manufacturing wage indices, selected Asian economies, 1995 to latest available years," http://www.ilo.org/public/english/employment/strat/kilm/download/kilm15.pdf.
2. World Bank, Human Development Indicators, 2005.
3. "China Boosts Road Construction," *People's Daily,* http://www.english.people.com

.cn/english/200012/20/eng20001220_58274.html; and "Road Financing in China," http://www.unescap.org/ttdw/common/TIS/AH/files/egm06/financing_china_2nd.pdf.

4. Telecom Regulatory Authority of India, cited in Candace Lombardi, "Cell phone subscriptions surge in India," CNET, http://news.com/Cell+phone+subscriptions+surge+in+India/2110-1037_3-6059482.html.

5. World Bank, Human Development Indicators, 2005.

6. Interview, August 2006.

7. Jonathan Watts, *Guardian*, March 15, 2006.

8. *Fortune,* "Global 500," http://money.cnn.com/magazines/fortune/global500.

9. "Studying McDonald's abroad: overseas branches merge regional preferences, corporate directives," http://www.findarticles.com/p/articles/mi_m3190/is_15_39/ai_n13649042

10. Ibid.

11. William W. Lewis, *The Power of Productivity: Wealth, Poverty and the Threat to Global Stability* (Chicago: University of Chicago Press, 2004).

12. Ibid.

13. Ibid.

14. Lowell L. Bryan and Michelle Zanini, "Strategy in an era of global giants," *McKinsey Quarterly,* no. 4 (2005).

15. International Telecommunications Union, http://www.itu.int/ITU-D/ict/statistics.

16. "China Boosts Road Construction," *People's Daily.*

17. David Wessel, *Wall Street Journal,* April 2, 2004, cited in Diana Farrell, *The Emerging Global Labor Market* (McKinsey Global Institute, June 2005).

18. Northeast Human Resources Association, http://www.nehra.com/articlesresources/article.cfm?id=1106.

19. Shabana Hussein, "And now a domestic outsourcing boom," Expressindia.com, http://www.expressindia.com/fullstory.php?newsid=39601.

20. Farrell, *The Emerging Global Labor Market.*

21. Bureau of Economic Analysis, "Summary Statistics for Multinational Companies," http://www.bea.gov/bea/newsrel/mncnewsrelease.html.

22. *The Impact of Offshore IT Software and Services Outsourcing on the U.S. Economy and the IT Industry* (Lexington, MA: Global Insight, March 2004).

23. Ibid.

24. Farrell, op. cit.

25. Ibid.

26. http://www.privatehealth.co.uk/.

27. Kristin Gerencher, "Vital Signs: Going abroad for medical and dental care," Market Watch, May 5, 2006, http://www.marketwatch.com/News/Story/Story.aspx?guid=%7B0 C1BE8DD-24FE-4AFE-AB03-B728D9A59990%7D&siteid=bigcharts&dist=.

28. Catherine L. Mann, "Offshore Outsourcing and the Globalization of US Services: Why Now, How Important, and What Policy Implications," in *The United States and the World Economy,* ed. C. Fred Bergsten (Washington, DC: Institute for International Economics, January 2005).

29. Ibid.

30. Bureau of Labor Statistics, "Hourly Compensation Costs for Production Workers in

Manufacturing, 33 Countries or Areas, 22 Manufacturing Industries, 1992–2005," http://www.bls.gov/fls/flshcindnaics.htm.

31. Ibid.

32. "Andy Stern Introduction Booklet," http://www.seiu.org/docUploads/Andy%20Stern%20Introduction%20Booklet%2001312006%2Epdf.

33. Alexander Jung, "Taming the Globalization Monster," *Spiegel* Online, http://www.spiegel.de/international/0,1518,392276,00.html.

34. International Monetary Fund (IMF), *World Economic Outlook* (2006), Statistical Appendix, tables 1, 20, 43; and Mangal Goswami, Jack Ree, and Ina Kota, "Global Capital Flows: Defying Gravity," *Finance and Development* 44, no. 1, March 2007, http://www.imf.org/external/pubs/ft/fandd/2007/03/picture.htm.

35. McKinsey Global Institute, "$118 Trillion and Counting: Taking Account of Global Capital Markets."

36. Senate Committee on Banking, Housing, and Urban Affairs, *Federal Reserve's First Monetary Report for 2005,* 109th Cong., 1st sess., 2005, http://frwebgate.access.gpo.gov/cgi-bin/getdoc.cgi?dbname=109_senate_hearings&docid=f:21981.wais.

37. International Monetary Fund, *World Economic Outlook* (April 2006), chapter 1, "Global Prospects and Policy Issues."

38. Thomas Helbling, Florence Jaumotte, and Martin Sommer, "How Has Globalization Affected Inflation?" International Monetary Fund, *World Economic Outlook* (April 2006).

39. DeAnne Julius, "Back to the Future of Low Global Inflation," http://www.bankofengland.co.uk/publicatons/speeches/1999/speech57.pdf.

40. Helbling, Jaumotte, and Sommer, "How Has Globalization Affected Inflation?"

41. *The Economist Housing Survey,* June 16, 2005.

42. Chiaku Chukwuogor-Ndu, "Stock Market Returns Analysis, Day of the Week Effects, Volatility of Returns: Evidence from European Financial Markets 1997–2004," *International Research Journal of Finance and Economics,* no. 1 (2006).

43. Bureau of Economic Analysis.

44. National Intelligence Council, *Mapping the Global Future: Report of the National Intelligence Council's 2020 Project,* December 2004, http://www.dni.gov/nic/NIC_globaltrend2020.html.

45. Ibid.

4. THE TWO POLES OF GLOBALIZATION: CHINA AND THE UNITED STATES

1. World Bank Group, EdStats, http://www1.worldbank.org/education/edstats/.

2. World Bank, World Development Indicators.

3. United Nations, *World Investment Report 2005,* http://www.unctad.org/en/docs/wir2005annexes_en.pdf.

4. World Bank, World Development Indicators.

5. World Bank data.

6. Christopher Koch, "Making It in China," *CIO,* http://www.cio.com/archive/101505/china.html.

7. Ibid.

8. Interview, July 19, 2006.

9. Lai Nai-keung, "It's time to take seriously a US-led global recession," *China Daily,* http://www.chinadaily.com.cn/english/doc/2005-10/06/content_482807.htm.

10. Ibid.

11. Jinglian Wu, *Understanding and Interpreting Chinese Economic Reform,* trans. Wang Jianmao (Mason, OH: Thomson/South-Western, 2005).

12. "Protests in China, the Cauldron Boils," *Economist,* September 29, 2005.

13. Charles Pigott, *China in the World Economy: The Domestic Policy Challenges* (Paris: Organisation for Economic Co-operation and Development, 2002).

14. Elizabeth C. Economy, *The River Runs Black* (Ithaca, NY: Cornell University Press, 2005).

15. Andrew Batson, "China Warns Pollution Will Grow with the Economy," DowJones Newswire, October 25, 2005.

16. Emmanuel Pitsilis, Jonathan Woetzel, and Jeffrey Wong, "Checking China's Vital Signs" (McKinsey Global Institute, December 2004), http://www.mckinseyquarterly.com/Economic_Studies/Country_Reports/Checking_Chinas_vital_signs_1483.

17. *China Daily,* January 12, 2006.

18. Diana Farrell, Ulrich Gersch, and Elizabeth Stephenson, "The Value of China's Emerging Middle Class" (McKinsey Global Institute, June 2006), http://www.mckinseyquarterly.com/Retail_Consumer_Goods/The_value_of_Chinas_emerging_middle_class_1798.

19. Dorothy Guerrero, "China: Beyond the Growth Figures," *Globalist,* February 21, 2006, http://www.theglobalist.com/DBWeb/StoryId.aspx?StoryId=5095.

20. World Bank, World Development Indicators, 2006.

21. Ibid.

22. Ibid.

23. U.S. Department of Commerce, International Trade Administration data, 2006.

24. World Trade Organization data, 2006.

25. United Nations Conference on Trade and Development (UNCTD), 2006.

26. Ibid.

27. As America's population continues to age, those profits will help support the retirement of tens of millions of elderly people—and thereby also help support the overall domestic demand that helps create the jobs whose wages are rising only very slowly. And Europeans also miss out on this benefit, since the elderly there depend much more heavily on state pensions and hold few stocks, directly or indirectly. There, the only way to boost demand by the elderly will be to increase their pension checks, which would only reduce the incomes of everyone who would to pay higher taxes.

28. See, for example, "The Economic Impact of Wal-Mart," Business Planning Solutions Global Insight Advisory Services Division (Boston, MA: Global Insight, November 2, 2005).

29. The OECD takes all the jobs statistics from around the world and applies standard definitions. *OECD Employment Outlook 2005* (Paris: Organisation for Economic Cooperation and Development, 2005).

30. International Telecommunications Union, 2006.

31. R. Sean Randolph, "The Innovation Edge: Meeting the Global Competitive Challenge," in *The Innovation Edge: Meeting the Global Competitive Challenge* (San Francisco: Bay Area Economic Forum, September 2006).

32. National Science Board, Science and Engineering Indicators, 2006.

33. PricewaterhouseCoopers and the National Venture Capital Association, Money Tree Report, 2006.

34. Clyde Prestowitz, "America's Technology Future at Risk: Broadband and Investment Strategies to Refire Innovation," Economic Strategy Institute, 2006.

35. National Science Board, "Science and Engineering Indicators," 2006.

36. Randolph, "The Innovation Edge."

37. Raffaella Sadun and John Van Reenen, "Intellectual property, technology and productivity: It ain't what you do it's the way you do I.T." (Discussion Paper No. 002, EDS Innovation Research Programme, October 2005).

38. World Bank Group, EdStats, http://www1.worldbank.org/education/edstats/.

39. Nivedita Das Kundu, "Resurgence of the Russian Economy," *IDSA Strategic Comments*, http://www.idsa.in/publications/stratcomments/NiveditaKundu100706.htm.

40. Mehmet Ögütçu, "Attracting Foreign Direct Investment for Russia's Modernization: Battling Against the Odds," http://www.oecd.org/dataoecd/44/45/1942539.pdf.

41. See "Russian Federation: 2005 Article IV Consultation" at http://www.imf.org/external/pubs/ft/scr/2005/cr05377.pdf; see also the IMF's World Economic Outlook database at http://www.imf.org/external/pubs/ft/weo/2006/01/data/index.htm.

42. Bank of Finland, "Russia: Growth Prospects and Policy Debates," http://www.bof.fi/.

43. See http://mosnews.com/money/2006/08/08/gdpforecast.shtml; and "Economic Ministry upgrades GDP forecast," http://top.rbc.ru/english/index.shtml?/news/english/2006/08/08/08123623_bod.shtml.

44. William W. Lewis, *The Power of Productivity: Wealth, Poverty and the Threat to Global Stability* (Chicago: University of Chicago Press, 2004).

45. Ibid.

46. E. Andreev, S. Scherbov, and F. J. Willekens, "The Population of Russia: Fewer and Older," Demographic Report 22, Groningen Demographic Reports, http://www.rug.nl/prc/publications/demographicreports/abstract22.

47. Nicholas Eberstadt, "Growing Old the Hard Way: China, Russia, India," *Policy Review*, no. 136 (April–May 2006), http://www.hoover.org/publications/policyreview/2912391.html.

48. United Nations, Human Development Indicators, 2004.

49. Norbert Walter, "Why I Worry About Russia," http://www.dbresearch.com/servlet/reweb2.ReWEB?rwkey=u6929389.

50. Lewis, *The Power of Productivity.*

51. Ibid.

52. United Nations, *World Investment Report 2005,* http://www.unctad.org/en/docs/wir2005annexes_en.pdf.

53. "Domestic Constraints on International Participation," in T. N. Srinivasan and Suresh D. Tendylkarm, *Reintegrating India with the World Economy* (Washington, DC: Peter G. Peterson Institute, March 2003).

54. "India's Economy," *Economist* (September 22, 2005).

55. http://www.buyusa.gov/china/en/power.html.

56. http://www.cslforum.org/china/htm; and http://www.cslforum.org/india.htm.

57. Diana Farrell and Aneta Marcheva Key, "India's lagging financial system," *McKinsey*

Quarterly, no. 2 (2005), http://www.mckinseyquarterly.com/Indias_lagging_financial _system_1600.

58. Lewis, *The Power of Productivity.*

59. Jagdish Bhagwati and Arvind Panagariya, "Defensive plays simply won't work," http:// www.economictimes.indiatimes.com/.

60. Rajat Gupta, "India's economic agenda: An Interview with Manmohan Singh," *McKinsey Quarterly* (September 2005), http://www.mckinseyquarterly.com/Indias_economic _agenda_An_interview_with_Manmohan_Singh_1674.

61. India Brand Equity Foundation, http://www.ibef.org/industry/autocomponents.aspx.

62. Shashank Luthra, Ramesh Mangaleswaran, and Asutosh Padhi, "When to make India a manufacturing base," *McKinsey Quarterly* (September 2005), http://www.mckinseyquarterly .com/Automotive/When_to_make_India_a_manufacturing_base_1650.

63. Diana Farrell, Noshir Kaka, and Sascha Sturze, "Ensuring India's offshoring future," *McKinsey Quarterly* (September 2005), http://www.mckinseyquarterly.com/Operations/ Outsourcing/Ensuring_Indias_offshoring_future_1660.

64. Prabuddha Ganguli, "The Pharmaceutical Industry in India," http://www.touchbriefings .com/pdf/17/pt031_p_ganguli.pdf.

65. Cited, for example, in Raj Jayadev, "To Silicon Valley Indian Entrepreneurs," AsianWeek .com, http://www.asianweek.com/2001_12_07/opinion_voices.html.

66. Ibid.

67. Lawrence Klein and T. Palanivel, *Economic Reforms and Growth Prospects in India* (Canberra: Australian National University, Publishers, 2000).

68. Dani Rodrik and Arvind Subramanian, "Why India Can Grow at 7 Percent a Year or More: Projections and Reflections" (IMF Working Paper, International Monetary Fund, 2004), http://www.imf.org/external/pubs/ft/wp/2004/wp04118.pdf.

69. V. T. Bharadwaj, Gautam Swaroop and Ireena Vittal, "Winning the Indian consumer," *McKinsey Quarterly* (September 2005).

70. Jagdish Shethm, "Making India Globally Competitive," *Vakalpa*, 2004.

71. P. N. Mari Bhat, "Demographic Scenario, 2025," http://www.planningcommission.nic .in/reports/sereport/ser/vision2025/demogra.pdf.

72. Jean Druze and Amartya Sen, "India's Economic Development and Social Opportunity," http://www.questia.com/.

73. Morgan Stanley, October 31, 2005.

5. THE NEW ECONOMICS OF DECLINE FOR EUROPE AND JAPAN

1. Martin Baily and Diana Farrell, "A road map for European economic reform," *McKinsey Quarterly* (September 2005), http://www.mckinseyquarterly.com/A_road_map_for _European_economic_reform_1679.

2. Bureau of Labor Statistics, *Comparative Civilian Labor Force Statistics, Ten Countries, 1960–2006* (2007), 18, table 4, "Civilian Labor Force Participation Rates Approximating U.S. Concepts by Sex, 1960–2006," http://www.bls.gov/fls/lfcompendium .pdf.

3. Organisation for Economic Co-operation and Development, "Annual Houses Worked," in *OECD in Figures,* 2004.

4. Organisation for Economic Co-operation and Development, OECD Statistics, "OECD Estimates of Labour Productivity Levels," http://stats.oecd.org/WBOS/Default.aspx ?DatasetCode=PDYGTH.

5. Bureau of Labor Statistics, *International Comparisons of Manufacturing Productivity and Unit Labor Cost Trends, Supplementary Tables,* table 1.1, "Output per hour in manufacturing, 16 countries or areas, 1950–2006," http://www.bls.gov/fls/prodsupptabletoc .htm.

6. For example, see "Reaching Higher Productivity Growth in France and Germany" (McKinsey Global Institute, October 2002).

7. United Nations, Comtrade database.

8. U.S. National Science Foundation, Science and Engineering Indicators.

9. "How France Can Win from Offshoring" (McKinsey Global Institute, August 2005).

10. The best recent treatment of these issues comes from the OECD: Randall Jones and Tae-sik Yoon, "Strengthening the Integration of Japan in the World Economy to Benefit More from Globalisation" (Economics Department Working Paper no. 526, Organisation for Economic Co-operation and Development, November 29, 2006).

11. Christine Tierney and Ed Garsten, "Toyota, GM locked in fight for worldwide supremacy," *Detroit News,* February 13, 2005, http://www.detnews.com/2005/ specialreport/0502/13/A01-87977.htm.

12. The results are apparent in the number of U.S. patents issued to American or foreign companies and citizens, especially as America is an indispensable market for everyone else. The 950,000 U.S. patents issued to U.S. firms and citizens since 1990 are, again, nearly three times the number issued to Japanese filers and eight to twenty times the number granted to German, French, or British companies and citizens.

13. Andres Fuentes, Eckhard Wurzel, Margaret Morgan, "Improving the Capacity to Innovate in Germany" (Economics Department Working Paper no. 407, Organisation for Economic Co-operation and Development, October 22, 2004).

14. World Bank Group, EdStats, http://www1.worldbank.org/education/edstats/.

15. Stephen Machin and Sandra McNally, "Tertiary Education Systems and Labour Markets," Organisation for Economic Co-operation and Development, January 2007.

16. Jim Hull, *More than a horse race: A guide to international tests of student achievement,* Center for Public Education, http://www.centerforpubliceducation.org/site/c.kjJXJ5MPIwE/b .2422943/k.3608/More_than_a_horse_race_A_guide_to_international_tests_of_student _achievement.htm.

17. "How France Can Win from Offshoring" (McKinsey Global Institute, August 2005).

18. "Japan and Globalization," *Globalist,* August 5, 2005, http://www.theglobalist.com/ StoryId.aspx?StoryId=4559.

19. Louis Hayes, *Introduction to Japanese Politics,* 4th ed. (Armonk, NY: M. E. Sharpe, 2005).

20. Organisation for Economic Co-operation and Development, *Economic Survey of Japan 2005,* http://www.oecd.org/document/61/0,3343,en_33873108_33873539_34274621 _1_1_1_1,00.html.

21. Ryosei Kokubun, "China and Japan in the Age of Globalization," *Japan Review of International Affairs* 17, no. 1 (Spring 2003).

22. World Bank, World Development Indicators, 2006.

23. Paul Tansey, *Productivity: Ireland's Economic Imperative* (prepared for Microsoft Ireland, 2005).

24. Gerry Pyke, FAS Ireland (presentation to Workforce Innovations Conference, Philadelphia, PA, 2005); FAS (Irish National Training and Employment Authority), "Quarterly Labour Market Commentary, First Quarter, 2005."

25. "The luck of the Irish," *Economist* (October 14, 2004).

26. Ibid.

27. Timothy Barnicle, "Ireland Case Study, National Center on Education and the Economy" (manuscript, November 2005).

28. Sean Dorgan, "How Ireland Became the Celtic Tiger," Heritage Foundation, Backgrounder #1945, June 23, 2006.

29. Ibid.

30. For the Irish case, see David Bloom and David Canning, "Contraception and the Celtic Tiger," *Economic and Social Review* 34, no. 3 (Winter 2003).

31. Ibid.; see also David E. Bloom, David Canning, and Jaypee Sevilla, "Economic Growth and the Demographic Transition" (NBER Working Paper no. 8685, National Bureau of Economic Research, December 2001).

32. United Nations, Human Development Indicators, 2006.

33. United Nations Conference on Trade and Development, *World Investment Report* (2005); and Organisation for Economic Co-operation and Development data.

34. International Labour Organization, Laborstat, 2005.

35. Organisation for Economic Co-operation and Development, 2005.

36. FAS (Irish National Training and Employment Authority), "Quarterly Labour Market Commentary, 1995–2005."

37. Ibid.

38. Tansey, *Productivity: Ireland's Economic Imperative*.

39. Hun Joo Park, "Between Development and State: Recasting Korean Dirigisme" (working paper, KDI School of Public Policy and Management, 2003).

40. Bruce Cummings, *Korea's Place in the Sun: A Modern History* (New York: W. W. Norton, 1997).

41. Charles Harvie and Hyun-Hoon Lee, "Export-Led Industrialization and Growth—Korea's Economic Miracle, 1962–89" (Economic Working Paper Series 03-01, University of Wollongong, 2003).

42. World Bank, World Development Indicators, 2006.

43. Cummings, *Korea's Place in the Sun*.

44. Ibid.

45. Ibid.

46. Ibid.

47. Ibid.

48. World Bank, World Development Indicators, 2006.

49. Ibid.

50. Fortune Global 500, 2007.

51. World Bank Group, EdStats, 2006, http://www1.worldbank.org/education/edstats/.

52. Ibid.

53. Organisation for Economic Co-operation and Development, *Economic Survey of Korea 2005*, http://www.oecd.org/document/18/0,2340,en_2649_201185_35428626_1_1_1_1,00.html.

54. Economic equality is measured by the GINI Index, with a higher value signifying greater inequality. Korea's GINI rating is 31.6, substantially lower than Thailand (42), Singapore (42.5), and China (44.7), and lower than advanced countries such as France (32.7), Italy, the U.K. (36), and the U.S. (40.8). World Bank, World Development Indicators, 2006; see also http://www.earthtrends.wri.org/text/economics-business/variable-353.html.

55. Harvie and Lee, "Export-Led Industrialization and Growth." In addition, 97 percent of all commodities were produced in sectors where three or fewer companies accounted for at least 60 percent of the market.

56. World Bank, World Development Indicators, 2006. The top eight: The United States, Japan, Germany, Finland, Sweden, Switzerland, Israel, and Iceland.

57. Andrew Eungi Kim and Innwon Park, "Changing Trends of Work in South Korea," *Asian Survey* 46, no. 3 (May/June 2006).

58. Yeonho Lee, "Participatory Democracy and Chaebol Regulation in Korea," *Asian Survey* 45, no. 2 (March/April 2005).

59. Organisation for Economic Co-operation and Development, *Economic Survey of Korea 2005*.

60. Organisation for Economic Co-operation and Development, Productivity Statistics, 2005.

61. Kim and Park, "Changing Trends of Work in South Korea."

62. World Bank, World Development Indicators, 2006.

63. United Nations, Population Division, World Population Prospects, 2006 Revisions, December 27, 2006, http://www.esa.un.org/unpp/.

64. Organisation for Economic Co-operation and Development, *Economic Survey of Korea 2005*.

65. The gap in opportunities here is greatest for the country's college-educated women: Only 55 percent of them are in the workforce today, compared to 82 percent in other OECD countries.

66. On average, OECD countries commit about 21 percent of R & D to service sectors, and the average gap between manufacturing and service sector productivity is just 7 percent.

67. U.S. Bureau of Labor Statistics, International Comparisons of Hourly Compensation Costs for Production Workers in Manufacturing, Supplementary Tables, 2006.

68. Korea's service sectors account for just 13 percent of the nation's R & D. Organisation for Economic Co-operation and Development, ICT Indicators, 2006.

69. World Bank, World Development Indicators, 2006; and Organisation for Economic Co-operation and Development, Broadband Statistics, 2006. In 2005, Korea trailed only Denmark, the Netherlands, and Iceland in broadband use.

6. THE NEW GEOPOLITICS OF THE SOLE SUPERPOWER: THE PLAYERS

1. Stephen M. Walt, *Taming American Power, The Global Response to U.S. Primacy* (New York: W. W. Norton, 2005).

2. The Venusberg Group, *Beyond 2010: European Grand Strategy in a Global Age* (3rd Venusberg Report, July 2007).

3. Thomas B. M. Barnett, *The Pentagon's New Map, War and Peace in the Twenty-first Century* (New York: Berkley Books, 2005).

4. Leonard Weiss, "Atoms for Peace," *Bulletin of the Atomic Scientists* 59, no. 6 (November 1, 2003).

5. Scott Sagan, "Why Do States Build Nuclear Weapons? Three Models in Search of a Bomb," *International Security* 21, No. 3 (Winter 1996/1997).

6. Samuel R. Berger, "A Foreign Policy for the Global Age," *Foreign Affairs* 79, no. 6 (November–December 2000).

7. U.S. Department of Defense, *Quadrennial Defense Review Report,* February 6, 2006.

8. Ibid.

9. Peter Spiegel, "Review ordered into vulnerability of U.S. satellites," *Los Angeles Times*, April 22, 2007.

10. Demetri Sevastopulo, "Chinese military hacked into Pentagon," *Financial Times,* September 3, 2007.

11. Ibid.

12. National Security and the Threat of Climate Change, CNA Corporation, 2006.

13. Leon Aron, "Where is Russia Headed?" http://www.aei.org/publicaitons/filer.all.pubID.22473/pub_detail.

14. Peter Lavelle, "U.S. and Russia—What's Next?" http://www.spacedaily.com/news/russia-04b.html; and Janusz Bugajski, "American vs. Russian imperialism," http://www.examiner.com/a-173697~Janusz_Bugajski__American_vs__Russian_imperialism.html.

15. World Bank, World Development Indicators, 2007.

16. Murray Feshbach, Woodrow Wilson International Center for Scholars.

17. Nicholas Eberstadt, "Growing Old the Hard Way: China, Russia, India," *Policy Review*, no. 136 (April–May 2006), http://www.hoover.org/publications/policyreview/2912391.html.

18. E. Andreev, S. Scherbov, and F. J. Willekens, "The Population of Russia: Fewer and Older," Demographic Report 22, Groningen Demographic Reports, http://www.rug.nl/prc/publications/demographicreports/abstract22.

19. "Protests in China, the Cauldron Boils," *Economist*, September 29, 2005.

20. Shaoguang Wang, "The Problem of State Weakness," *Journal of Democracy* 14, no. 1 (January 2003).

21. "Protests in China, the Cauldron Boils."

22. Jeffrey Bader, "China's Emergence and Its Implications for the United States" (presentation to Brookings Council, February 14, 2006).

23. Ibid.

24. David Shambaugh, "Modernizing China's Military" (Chinese White Paper on National Security, 2002), http://www.globalsecurity.org/.

25. Demetri Sevastopulo and Mure Dickie, "U.S. fears over China's long-range missiles," *Financial Times,* May 25, 2007.

26. Eric McVadon, quoted in the *International Herald Tribune*, April 8, 2005.

27. Sevastopulo and Dickie, "U.S. fears over China's long-range missiles."

28. Bader, "China's Emergence and Its Implications for the United States."

29. Quoted in *Globalist,* August 5, 2005.

30. Mid-term Defense Program (FY2005–FY2009).

31. Mike M. Mochizuki, "Japan: Between Alliance and Autonomy," in *Strategic Asia 2004–2005: Confronting Terrorism in the Pursuit of Power,* ed. Ashley Tellis and Michael Wills (Seattle: National Bureau of Asian Research, 2004).

32. Matake Kamiya, "Nuclear Japan: Oxymoron or Coming Soon?" *Washington Quarterly* 26, no. 1 (Winter 2002–2003).

33. Nakanishi Terumasa, "Nuclear Weapons for Japan," in *Japan Echo* 30 (October 2003).

34. Andrew J. Nathan and Bruce Gilley, *China's New Rulers: The Secret Files* (New York: New York Review Books, 2003).

35. CNN, Tuesday, June 5, 2007.

36. "What to Do in Iraq: A Roundtable," *Foreign Affairs* 85, no. 4 (July–August 2006).

7. THE COMING CRISES IN HEALTH CARE, ENERGY, AND THE GLOBAL ENVIRONMENT

1. Ivo Bozon, Warren Campbell and Mats Lindstrand, "Global trends in energy," *McKinsey Quarterly*, no. 1 (February 2007).

2. World Bank, World Development Indicators, 2007.

3. Commonwealth Fund, "First Report and Recommendations of the Commonwealth Fund's International Working Group on Quality Indicators. A Report to Health Ministers of Australia, Canada, New Zealand, the United Kingdom and the United States," June 2004.

4. World Bank, World Development Indicators, 2007.

5. Christian Hagist and Laurence Kotlikoff, "Health Care Spending: What the Future Will Look Like" (report 286, National Center for Policy Analysis, June 2006).

6. World Health Organization, database, http://www.who.int/whosis/database/core/core _select.cfm.

7. Ibid.

8. Ibid.

9. Hagist and Kotlikoff, "Health Care Spending: What the Future Will Look Like."

10. Ibid.

11. United Nations, Population Division, World Population Prospects, 2004 Revisions, http://esa.un.org/unpp/.

12. "Age Distribution of Diagnosis and Death," SEER Cancer Statistics Review, 1975– 2004, National Cancer Institute, http://www.seer.cancer.gov/csr/1975_2004/.

13. National Health Expenditures, Office of the Actuary, Centers for Medicare and Medicaid Services, January 2007, http://www.cms.hhs.gov/NationalHealthExpenData/.

14. Ibid.

15. "How Changes in Medical Technology Affect Health Care Costs," Henry J. Kaiser Family Foundation, March 2007, http://www.kff.org/insurance/snapshot/chcm030807oth .cfm.

16. David M. Cutler and Mark McClellan, "Is Technological Change in Medicine Worth It?" *Health Affairs* 20, no. 5 (September–October 2001).

17. Richard A. Rettig, "Medical Innovation Duels Cost Containment," *Health Affairs* 13, no. 3 (Summer 1994).

18. Joseph P. Newhouse, "Medical Care Costs: How Much Welfare Loss?" *Journal of Economic Perspectives* 6, no. 3 (Summer 1992).

19. Barry Feder, "Heart Therapy Strains Efforts to Limit Costs," *New York Times*, July 7, 2007.

20. Elizabeth Docteur and Valérie Paris, "Pharmaceutical Pricing and Reimbursement Policies in Canada" (Health Working Papers no. 24, Organisation for Economic Co-operation and Development, February 15, 2007).

21. Peter Neumann and Milton Weinstein, "The Diffusion of New Technology: Costs and Benefits to Health Care," in *The Changing Economics of Medical Technology*, ed. Annetine C. Gelijns and Ethan A. Halm (Washington, DC: National Academy Press, 1991).

22. "How Changes in Medical Technology Affect Health Care Costs."

23. Victor Rodwin, "The Health Care System Under French National Health Insurance: Lessons for Health Reform in the United States," *American Journal of Public Health* 93, no. 1 (January 2003); and "Ambulatory Care Use/Physician Visits," National Center for Health Statistics, U.S. Centers for Disease Control, http://www.cdc.gov/nchs/fastats/docvisit.htm.

24. Thomas Buchmueller and Agnes Couffinhal, "Private Health Insurance in France" (Health Working Papers no. 12, Organisation for Economic Co-operation and Development, March 2004).

25. David Green and Benedict Irvine, "Health Care in France and Germany: Lessons for the UK," at http://www.civitas.org.uk/pdf/cs17.pdf.

26. "Medical Malpractice Policy: France," http://www.electoral-math.com/archive/200504/20050420.html. French and German doctors, however, also don't have to pay for the medical education or malpractice insurance. In America, most new doctors carry educational debts averaging $100,000 to $140,000, and U.S. doctors pay malpractice premiums that in 2002 ranged from $3,800 a year for an internist in rural Minnesota to more than $200,000 for an ob/gyn in Miami, Florida. Paul Jolly, "Medical School Tuition and Young Physician Indebtedness," Association of American Medical Colleges, March 2004, http://www.sls.downstate.edu/financial_aid/docs/MedicalSchoolDebt.pdf; and "Medical Malpractice: Implications of Rising Premiums on Access to Health Care," General Accounting Office, GAO-03-836, August 2003, http://www.gao.gov/new.items/d03836.pdf.

27. "Trouble for French medicine," *Economist*, http://www.economist.com/displaystory.cfm?story_id=2670654.

28. Didem Bernard, "Out of Pocket Expenditures on Health Care Among the Non-elderly Population, 2004" (Medical Expenses Panel Survey, Agency for Healthcare Research and Quality, Department of Health and Human Services, January 2007).

29. "Highlights on health, France 2004," World Health Organization, http://www.euro.who.int/eprise/main/who/progs/chhfra/system/20050131_1.

30. "Health Insurance Cost," National Coalition on Health Care, http://www.nchc.org/facts/cost.shtml.

31. Ibid.

32. Paul Dutton, "Health Care in France and the United States: Learning from Each Other," Center on the United States and France, The Brookings Institution, http://www .brookings.edu./fp/cusf/analysis/dutton.pdf.

33. World Health Organization, Statistical Information System Online.

34. Ibid.

35. Ibid.

36. Hideki Nomura and Takeo Nakayama, "The Japanese healthcare system," *BMJ* 331, no. 7518 (September 24, 2005), http://www.bmj.com/cgi/content/full/331/7518/648 ?etoc&eaf.

37. James R.Bean, "National Healthcare Spending in the U.S. and Japan: National Economic Policy and Implications for Neurosurgery," *Neurologia medico-chirurgica* 45, no. 1 (2005), http://www.jstage.jst.go.jp/article/nmc/45/1/45_18/_article.

38. Ibid.

39. Eddy van Doorslaer, Cristina Masseria, and Xander Koolman, "Inequalities in access to medical care by income in developed countries," *Canadian Medical Association Journal* 174, no. 2 (January 17, 2006), http://www.pubmedcentral.nih.gov/articlerender.fcgi ?artid=1329455.

40. Derek Wanless, "Securing Good Health for the Whole Population," February 25, 2004, http://www.hm-treasury.gov.uk/consultations_and_legislation/wanless/consult _wanless04_final.cfm.

41. "Health care systems in eight countries: trends and challenges," European Observatory on Health Care Systems, April 2002, http://www.hm-treasury.gov.uk/media/8/3/ observatory_report.pdf.

42. Ibid.

43. King's Fund, "An Independent Audit of the NHS Under Labour (1997–2005)," http:// www.kingsfund.org.uk/publications/kings_fund_publications/an_independent.html.

44. "Core Health Indicators," Statistical Information System Online, World Health Organization.

45. Chris Philip and Aaron Smith, "The Quality of UK Healthcare: Comparison to Europe," The Bow Group, http://www.bowgroup.org/harriercollectionitems/HealthStandards.pdf.

46. *The Economics of Health Care,* appendix v, "NHS cost by age group," table, "Estimated HCHS per capita expenditure by age group, England, 2002/03," Office of Health Economics, http://www.ohe.org/page/knowledge/schools/appendix/nhs_cost.cfm.

47. United Nations, Population Division, World Population Prospects, http://www.esa.un .org/unpp/.

48. "Health care systems in eight countries: trends and challenges."

49. United Nations, Population Division, World Population Prospects, http://www.esa.un .org/unpp/.

50. Tadahiko Tokita, "The Prospects for Reform of the Japanese Healthcare System," *PharmacoEconomics* 20, suppl. 3 (2002).

51. World Health Organization, Global InfoBase Online.

52. David Cutler, "An International Look at the Medical Care Financing Problem," in *Health Care Issues in the United States and Japan,* ed. David Wise and Naohiro Yashiro (Chicago: University of Chicago Press, 2006).

53. David Wise, "Introduction," in *Health Care Issues in the United States and Japan*.

54. Ibid.

55. Tokita, "The Prospects for Reform of the Japanese Healthcare System."

56. Elizabeth Docteur and Howard Oxley, "Health Care Systems: Lessons from the Reform Experience" (Health Working Papers no. 9, Organisation for Economic Co-operation and Development, December 5, 2003); and Tadashi Fukui and Yashushi Iwamoto, "Policy Options for Financing the Future: Health and Long-Term Care Costs in Japan," (NBER Working Paper no. 12427, National Bureau of Economic Research, August 2006).

57. Jean Woo, T. Kwok, F. K. H. Sze, and H. J. Yuan, "Aging in China: Health and Social Consequences and Responses," *International Journal of Epidemiology* 31, no. 4 (2002).

58. Michael Moreton, M.D, "Healthcare in China," http://www.medhunters.com/articles/healthcareInChina.html.

59. Ibid.

60. World Bank, Briefing Note no. 33233.

61. Howard French, "Wealth Grows, But Health Care Withers in China," *New York Times*, January 14, 2006.

62. "700 People Die from TB in China Every Day," *People's Daily*, http://english.people.com.cn/english/200106/05/eng20010605_71790.html

63. "China launches battle against hepatitis," BBC News, June 1, 2002, http://news.bbc.co.uk/2/low/asia-pacific/2020586.stm.

64. World Bank, Briefing Note no. 22325.

65. World Bank, Briefing Note no. 33234.

66. World Bank, Briefing Note no. 33232.

67. World Health Organization database; and "Assessing Government Health Expenditure in China," China National Health Economics Institute, October 2005.

68. United Nations, Population Division, World Population Prospects, 2004 Revisions and World Urbanization Prospects, http://www.esa.un.org/unpp/.

69. WHO Global Infobase Online.

70. World Bank, Briefing Note no. 33234.

71. "Crowd Protests Health Care in China—Joseph Kahn," *China Digital Times*, http://chinadigitaltimes.net/2006/11/crowd_protests_health_care_in_china_joseph_kahn.php.

72. J. Watts, "Protests in China over suspicions of a pay-or-die policy," *Lancet* 369, no. 9556 (January 13–19, 2007).

73. *International Energy Outlook 2007*, Energy Information Administration, U.S. Department of Energy.

74. Ibid.

75. "Industrial Energy Intensity by Industry," Natural Resources Department, Canada, http://www.oee.nrcan.gc.ca/corporate/statistics/neud/dpa/tableshandbook2/agg_00_6_e_1.cfm?attr=0.

76. "Total Energy Intensity," European Environment Agency, http://themes.eea.europa.eu/IMS/ISpecs/ISpecification20041007132134/IAssessment1144242696225/view_content

77. Ibid.

78. Ivo Bozon, Subbu Narayanswamy and Vipul Tuli, "Securing Asia's Energy Future," *McKinsey Quarterly Online*, April 2005, http://www.mckinseyquarterly.com.

79. Jacqueline Lang Weaver, "The Traditional Petroleum-based Economy's Eventful Future: Of Peak Oil, Big Oil, Chinese Oil, Flags and Open Doors," Public Law and Legal Theory Series 2006-W-06, University of Houston. Princeton geologist Kenneth Deffeyes set off the debate by predicting that world crude oil production would peak in 2005. It didn't. See "Is the World Running Out of Oil?" *Wall Street Journal*, October 8, 2005.

80. "Is the World Running Out of Oil?" *Wall Street Journal*, October 8, 2005.

81. *International Energy Outlook* 2007, appendix D, "High World Oil Price Projections," Energy Information Administration, U.S. Department of Energy.

82. Amory Lovins, *Winning the Oil Endgame* (Snowmass, CO: Rocky Mountain Institute, 2005).

83. Three McKinsey and Company energy experts wrote recently, "Today's tight supply and hefty prices could conceivably spark an energy boom and bust like the one that shook the world economy in the 1980s." See Bozon, Narayanswamy, and Tuli, "Securing Asia's Energy Future."

84. Lovins, *Winning the Oil Endgame.*

85. IEA Statements, International Energy Agency, http://www.iea.org/journalists/infocus .asp.

86. "Crude Oil and Total Petroleum Imports Top 15 Countries," Energy Information Administration, May 2007, http://www.eia.doe.gov/pub/oil_gas/petroleum/data_publications/ company_level_imports/current/import.html.

87. "U.S. Trade in Goods and Services Annual Revision for 2006" and "International Investment Position of the United States at Year End, 1976–2006," Bureau of Economic Analysis, http://www.bea.gov/.

88. Daniel Yergin, "Ensuring Energy Security" *Foreign Affairs* 85, no. 2 (March–April 2006).

89. Flynt Leverett and Pierre Noel, "The New Axis of Oil," *National Interest* (June 1, 2006).

90. Flynt Leverett and Jeffrey Bader, "Managing China-U.S. Energy Competition in the Middle East," *Washington Quarterly* 29, no. 2 (Spring 2006).

91. Ibid.

92. Erica Downs, "The Chinese Energy Security Debate," *China Quarterly* no. 177 (2004).

93. "Transportation Indicators for Selected Countries," Federal Highway Administration, http://www.fhwa.dot.gov/ohim/hso1/in3.htm.

94. *International Energy Outlook* 2006, Energy Information Administration; and *Quantifying Energy: BP Statistical Review of World Energy June 2006,* British Petroleum, http://www .bp.com/liveassets/bp_internet/globalbp/globalbp_uk_english/reports_and_publications/ statistical_energy_review_2006/STAGING/local_assets/downloads/pdf/statistical_review_ of_world_energy_full_report_2006.pdf.

95. Ivo Bozon, Warren Campbell, and Thomas Vahlenkamp, "Europe and Russia: Charting an energy alliance," *McKinsey Quarterly*, no. 4 (2006), http://www.mckinseyquarterly.com/ Energy_Resources_Materials/Europe_and_Russia_Charting_an_energy_alliance_1852.

96. *An Inconvenient Truth,* DVD, directed by David Guggenheim (2006; Hollywood: Paramount Pictures, 2006).

97. Kevin Baumert, "The challenge of climate protection: Balancing energy and environment," in *Energy and Security: Towards a New Foreign Policy Strategy*, ed. Jan Kalicki and David Goldwyn (Baltimore: The Johns Hopkins Press, 2005).

98. Melanie Kenderdane and Ernest Moniz, "Technology Development and Energy Security" in *Energy and Security: Towards a New Foreign Policy Strategy*.

99. Per-Anders Enkvist, Tomas Nauclér, and Jerker Rosander, "A cost curve for greenhouse gas reduction," *McKinsey Quarterly*, no. 1 (February 2007), http://www .mckinseyquarterly.com/A_cost_curve_for_greenhouse_gas_reduction_1911.

100. Ibid.

101. Andrew Batson, "China warns pollution will grow with economy," DowJones Newswire, October 25, 2005.

8. HISTORY'S WILD CARDS: CATASTROPHIC TERRORISM AND TECHNOLOGICAL BREAKTHROUGHS

1. Christopher Blanchard, "Islam: Sunnis and Shiites" (Washington, DC: Library of Congress, Congressional Research Service, December 11, 2006).

2. Ibid.

3. Christopher Blanchard, "The Islamic Traditions of Wahhabism and Salafiyya" (Washington, DC: Library of Congress, Congressional Research Service, January 17, 2007).

4. Guilain Denoeux, "The Forgotten Swamp: Navigating Political Islam," *Middle East Policy* 9, no. 2 (2002).

5. Ibid.

6. Kenneth Katzman, "Al Qaeda: Profile and Threat Assessment," (Washington, DC: Library of Congress, Congressional Research Service, February 10, 2005).

7. Spencer Ackerman, "Religious Protection: Why American Muslims Haven't Turned to Terrorism," *New Republic* 233, no. 4743 (December 12, 2005).

8. "Violent Jihad in Netherlands: Current Trends in the Islamic Terrorist Threat," Center for Strategic and International Studies, 2006.

9. Raphael Perl, "Trends in Terrorism: 2006" (Washington, DC: Library of Congress, Congressional Research Service, July 21, 2006; updated March 12, 2007).

10. Ellen Messmer, "U.S. cyber counterattack: Bomb 'em one way or the other." *Network World*, February 8, 2007, http://www.networkworld.com/news/2007/020807-rsa-cyber -attacks.html.

11. Barton Gellman, "Cyber-Attacks by Al Qaeda Feared," *Washington Post*, June 27, 2002, http://www.washingtonpost.com/ac2/wp-dyn/A50765-2002Jun26.

12. William Potter, Charles Ferguson, and Leonard Spector, "The Four Faces of Nuclear Terror and the Need for a Prioritized Response," *Foreign Affairs* 83, no. 3 (May–June 2004).

13. Eben Kaplan, "Safeguarding Nuclear Energy," Council on Foreign Relations, April 14, 2006, http://www.cfr.org/publication/10449/safeguarding_nuclear_energy.html.

14. Carl Behrens and Mark Holt, "Nuclear Power Plants: Vulnerability to Terrorist Attack" (Washington, DC: Library of Congress, Congressional Research Service, February 4, 2005).

15. Potter, Ferguson, and Spector, "The Four Faces of Nuclear Terror."

16. Jonathan Medalia, "Terrorist Nuclear Attacks on Seaports: Threat and Response" (Washington, DC: Library of Congress, Congressional Research Service, January 24, 2005), http://www.fas.org/irp/crs/RS21293.pdf.

17. William J. Perry, "Preparing for the Next Attack," *Foreign Affairs* 80, no. 6 (November–December 2001).

18. Medalia, "Terrorist Nuclear Attacks on Seaports."

19. Daniel Lindley, "The Dark Side of Playing Out the War Scenario," *Chicago Tribune,* October 27, 2002, http://www.nd.edu/~dlindley/handouts/ChicagoTribuneLindleyonIraqWar.htm.

20. *Making the Nation Safer: The Role of Science and Technology in Countering Terrorism,* Committee on Science and Technology for Countering Terrorism, National Research Council (Washington, DC: The National Academies Press, 2002), 40. http://www.nap.edu/catalog/10415.html.

21. Sam Nunn, "The Race Between Cooperation and Catastrophe," *Vital Speeches of the Day* 71, no. 12 (April 2, 2005), 369; and Lindley, "The Dark Side of Playing Out the War Scenario."

22. Nunn, "The Race Between Cooperation and Catastrophe."

23. Cham Dallas, Address to U.N. General Assembly, U.S. Federal News Service, March 27, 2006.

24. Medalia, "Terrorist Nuclear Attacks on Seaports: Threat and Response."

25. Graham Allison, "How to Stop Nuclear Terror," Foreign Affairs 83, no. 1 (January–February 2004).

26. Medalia, "Terrorist Nuclear Attacks on Seaports: Threat and Response."

27. Charles Meade and Roger Molander, "Considering the Effects of a Catastrophic Terrorist Attack" (Santa Monica, CA: RAND Corp., 2006), http://www.rand.org/pubs/technical_reports/2006/RAND_TR391.pdf.

28. Matthew C. Weinzierl, "The Cost of Living," *National Interest* (April 1, 2004), 118–122.

29. Meade and Molander, "Considering the Effects of a Catastrophic Terrorist Attack."

30. Anthony Fauci, *The Diane Rehm Show,* WAMU 88.5 FM, October 31, 2001, http://www.wamu.org/programs/dr/01/10/31.php.

31. Marc L. Ostfield, "Bioterrorism as a Foreign Policy Issue," *SAIS Review* 24, no. 1 (January 2004).

32. "Chemical and Biological Weapons: Possession and Programs Past and Present," James Martin Center for Nonproliferation Studies, http://cns.miis.edu/research/cbw/possess.htm.

33. Jim A. Davis, "The looming biological warfare storm: Misconceptions and probable scenarios," *Air & Space Power Journal* (April 1, 2003), 57–67.

34. Ibid.

35. Mark G. Polyak, "The Threat of Agroterrorism: Economics of Bioterrorism," *Georgetown Journal of International Affairs* 5, no. 2 (July 2004).

36. Robert J. Leggiadro, "Bioterrorism: A Clinical Reality," *Pediatric Annals* 36, no. 6 (June 2007).

37. Davis, "The looming biological warfare storm."

38. Robert Lenzner and Nathan Vardi, "Catching the Bad Bug," *Forbes* (August 16, 2004).

39. Davis, "The looming biological warfare storm."

40. *Public health response to biological and chemical weapons: WHO guidance,* World Health Organization (2004), http://www.who.int/csr/delibepidemics/biochemguide/en/index.html.

41. Breanne Wagner, "Germ Warfare: Agencies scramble to create vaccine market," National

Defense Industrial Association, June 1, 2005, http://www.thefreelibrary.com/Germ
+warfare:+agencies+scramble+to+create+vaccine+market.-a0165192805.

42. F. Fenner, D. A. Henderson, I. Arita, Z. Jezek, and I. D. Ladnyi, "Smallpox and its Erad-
ication," World Health Organization, 1988.

43. Laurie Garrett, "The Nightmare of Bioterrorism," *Foreign Affairs* 80, no. 1
(January–February 2001).

44. QAP is a Sunni terrorist organization, located primarily in Saudi Arabia. It was named
for Al-Qaeda and claims to be subordinate to that group and Osama bin Laden. Like Al-
Qaeda, it opposes the Saud monarchy. The group is suspected in a number of mass murders in Saudi Arabia, the kidnap and murder of Paul Johnson in Riyadh, and bombings
in Doha, Qatar, in March 2005.

45. "Bin Laden deputy calls for regime change in Saudi Arabia, Egypt," *Al Bawaba*, May 5,
2007, http://www.albawaba.com/en/countries/Egypt/214789.

46. Thomas Hegghammer, "Terrorist Recruitment and Radicalization in Saudi Arabia,"
Middle East Policy Council 13, no. 4 (Winter 2006), http://www.mil.no/multimedia/
archive/00087/Terrorist_Recruitmen_87543a.pdf.

47. John C. K. Daly, "Saudi 'Black Gold': Will Terrorism Deny The West Its Fix?" *Terrorism
Monitor* (December 4, 2003), http://www.jamestown.org/terrorism/news/uploads/ter
_001_007.pdf.

48. Ibid.

49. Andrew England, "Saudis set up force to guard oil plants," *Financial Times*, August 27,
2007.

50. Vincent Cable, "The economic consequences of war," *Observer*, February 2, 2003, http://
www.observer.guardian.co.uk/iraq/story/0,,886596,00.html.

51. Ibid.

52. Daniel L. Byman, "The Implications of Leadership Change in the Arab World," *Political
Science Quarterly* 120, no. 1 (April 2005).

53. Ibid.

54. James Steinberg, Robert Litan, Stephen Cohen, James Lindsay, "America's New War
Against Terrorism," The Brookings Institution, September 14, 2001, http://www
.brookings.edu/events/2001/0914terrorism.aspx.

55. Phil Gordon, "If Pakistan prospers, al-Qaeda won't," *Financial Times,* July 25, 2007.

56. Talha Khubaib, "Afghan failure may lead to regime change in Pakistan: UK generals,"
Daily Times (Pakistan), July 16, 2007, http://www.dailytimes.com.pk/default.asp?page=
2007%5C07%5C16%5Cstory_16-7-2007_pg1_8.

57. "Pakistan, China to set up free trade area," *China Daily*, December 26, 2005, http://
www.chinadaily.com.cn/english/doc/2004-12/26/content_403376.htm.

58. "What Is Nanotechnology?" National Nanotechnology Initiative, http://www.nano.gov/
html/facts/whatIsNano.html.

59. Michael Deal, "Nanotechnology," Stanford University Nanofabrication Facility, http://
snf.stanford.edu/Education/Overview.html.

60. "Unlocking the smallest secrets," *San Francisco Chronicle*, February 1, 2004.

61. M. C. Roco, "International Perspective on Government Nanotechnology Funding in
2005," *Journal of Nanoparticle Research* 7, no. 6 (2005), http://www.nsf.gov/.

62. "Applications/Products," National Nanotechnology Initiative, http://www.nano.gov/html/facts/appsprod.html.

63. Gary Thayer, Fred Roach, and Lori Dauelsberg, "Estimated Energy Savings and Financial Impacts of Nanomaterials by Design on Selected Applications in the Chemical Industry." Los Alamos National Laboratory, March 2006; "Nanomaterials to 2008" (Study #1887, Freedonia Group, January 1, 2005); and "Nanotechnology: Designs for the Future," *ACM: Ubiquity* (2002), http://www.acm.org/ubiquity/interviews/r_merkle_1.html.

64. Thayer, et. al., "Estimated Energy Savings"; and Freedonia Group, "Nanotechnology: Designs for the Future."

65. Roco, "International Perspective on Government Nanotechnology Funding in 2005."

66. Michael Berger, "Nanotechnology Could Clean up the Hydrogen Car's Dirty Little Secret," *Nanowerk*, July 19, 2007.

67. Tom Bearden, "Interview: Nobel Prize Winner Dr. Richard Smalley," *Online NewsHour*, PBS, October 20, 2003, http://www.pbs.org/newshour/science/hydrogen/smalley.html.

68. Thayer, et. al., "Estimated Energy Savings."

69. Ibid.

70. Berger, "Nanotechnology Could Clean up the Hydrogen Car's Dirty Little Secret."

71. George Whitesides, "The Once and Future Nanomachine," *Scientific American* (September 2001); and "Interview: Uzi Landman," *Georgia Tech Alumni Magazine* Online, Spring 2005, http://gtalumni.org/Publications/magazine/spr05/campaign.html.

72. "Nanotechnology: Designs for the Future."

73. David Pescovitz, "Nanoscience's Master Mechanic," *ScienceMatters@Berkeley* 2, no. 11 (May 2005).

74. "Unlocking the smallest secrets."

75. Rocky Rawstern, "All About Bootstrapping," *Nanotechnology Now* (November 2003), http://www.nanotech-now.com/CRN-interview-11082003.htm.

76. "Interviews—Martina McGloughlin," *Harvest of Fear*, PBS, 2001, http://www.pbs.org/wgbh/harvest/interviews/mcgloughlin.html.

77. "ISB's Ruedi Aebersold on the Challenges of Proteomics," *Science Watch* 15, no. 3 (May–June 2004), http://www.sciencewatch.com/may-june2004/sw_may-june2004_page3.htm.

78. Gregory Fahy, "Report—William Haseltine," *Life Extension Magazine* (July 2002), http://www.lef.org/magazine/mag2002/jul2002_report_haseltine_01.html.

79. Ibid.; and Susan Greenfield, "The scientist who came in from the cold," *Public Lecture* (August 18, 2004), http://www.thinkers.sa.gov.au/images/Greenfield_PubLect_Transcript.pdf.

80. "Stem Cell Basics," Stem Cell Information, National Institutes of Health, http://stemcells.nih.gov/info/basics/basics6.asp.

81. Clive Cookson, "Stem-cell advance offers hope to people with heart failure," *Financial Times*, August 27, 2007.

82. Fahy, "Report—William Haseltine"; and "Interview with a Mad Scientist." *Daily Kos*, January 6, 2006, http://www.dailykos.com/storyonly/2006/1/6/95138/89017.

83. Mark Compton, "Enhancement Genetics: Let the Games Begin," *Geneforum* (February 10, 2005), http://www.geneforum.org/node/55.

84. "Eric S. Lander's '14-Year Digression' with the Human Genome," *Science Watch* (March–April 2002), http://www.sciencewatch.com/march-april2002/sw_march-april2002_page3.htm.

85. Compton, "Enhancement Genetics."

86. Sean O'Hagen, "End of sperm report," *Observer,* September 15, 2002.

87. "Interviews—Martina McGloughlin."

88. Ibid.

89. Ronald Bailey, "Billions Served," *Reason* Online, April 2000, http://www.reason.com/news/show/27665.html.

90. *European Business Forum* no. 22 (Autumn 2005).

91. Andy Beal, "Microsoft's Robert Scoble Discusses Search Engine Technology," Search Engine Guide, http://www.searchengineguide.com/beal/2004/0204_abl.html.

92. Interview at OCSON, August 2005.

93. Federico Biancuzzi, "Freedom, Innovation, and Convenience: The RMS Interview," Linux DevCenter.com, December 22, 2004, http://www.linuxdevcenter.com/pub/a/linux/2004/12/22/rms_interview.html?page+2.

94. "The End of Corporate IT," *Computerworld* 39, no. 19 (May 9, 2005).

95. Patrick Mannion, "Ethernet's Inventor Sounds Off," *EE Times* (November 11, 2005), http://www.eetimes.com/showArticle.jhtml;jsessionid=2T0UGPW0YO1SCQSNDLO SKH0CJUNN2JVN?articleID=173601905.

96. Robert J. Shapiro, "The Internet's Capacity to Handle Fast-Rising Demand for Bandwidth," September 2007, http://www.usiia.org/pubs/Demand.pdf.

97. Ann Bednarz, "Telework Thrives at AT&T," NetworkWorld.com, December 19, 2005.

98. "Telework Trending Upward, Survey Says," WorldatWork, February 8, 2007, http://www.workingfromanywhere.org/news/pr020707.html.

99. Caroline Jones, "Teleworking: The Quiet Revolution," Gartner Dataquest, September 14, 2005.

100. "Wireless Unbound: The Surprising Economic Value and Untapped Potential of the Mobile Phone," McKinsey & Company, December 2006, http://www.mckinsey.com/clientservice/telecommunications/WirelsUnbnd.pdf.

101. "The Wireless Internet Opportunity for Developing Countries," Wireless Internet Institute, November 2003, http://www.firstmilesolutions.com/documents/The_WiFi_Opportunity.pdf.

102. Richard Koman, "The Next Revolution: Smart Mobs," OpenP2P, March 13, 2003, http://www.openp2p.com/pub/a/p2p/2003/03/13/howard.html.

103. Robert J. Shapiro, "Maintaining Contact: The Provision of International Long Distance Service for Low-Income Immigrants in the United States," December 2006, http://www.sonecon.com/.

104. "Penetration of Mobile Telephony in India & Value Added Services in Indian Mobile Telephony Market," Zinnov Research and Consulting, October 2006, http://www.zinnov.com/presentation/Mobile_VAS.pdf.

105. "Penetration of Mobile Telephony in India & Value added services in Indian Mobile Telephony market."

106. Robert J. Shapiro, "The Internet's Capacity to Handle Fast-Rising Demand for Bandwidth."

107. OLPC Web site, MIT Media Laboratory, http://laptop.media.mit.edu/.

108. One Laptop per Child Web site, http://laptop.org/en/laptop/software/.

109. "One Laptop per Child Announces Final Beta Version of Its Revolutionary XO Laptop," Business Wire, July 23, 2007.

110. Hiawatha Bray, "Let the Games Begin," *Boston Globe,* June 9, 2007.

111. Jim Finkle, "Non-profit may launch $350 laptop by Christmas," Reuters, July 23, 2007, http://www.reuters.com/article/companyNewsAndPR/idUSN2336963020070723.

112. World Bank Group, EdStats, http://www1.worldbank.org/education/edstats/.

113. United Nations Conference on Trade and Development, *World Investment Report* 2007, http://www.unctad.org/Templates/webflyer.asp?docid=9001&intItemID=4361&lang=1&mode=highlights.

114. Olga Oliker and Tanya Charlick-Paley, *Assessing Russia's Decline: Trends and Implications for the United States and the U.S. Air Force* (Santa Monica, CA: RAND Corp., 2002).

115. Zbigniew Brzezinski, *Second Chance, Three Presidents and the Crisis of American Superpower* (New York: Basic Books, 2007), 152–153.

INDEX